Yearbook of Astronomy 2027

Front Cover: Arp107 is a pair of interacting galaxies located around 450 million light years from Earth in the constellation of Leo Minor. This *James Webb Space Telescope* image, taken by the Mid-Infrared Instrument (MIRI), shows that there are millions of young stars forming. These are highlighted in blue. The older stars around the central black hole in the spiral in the right are drawn into a spiral with lanes of dust by the gravitational attraction of the black hole. (NASA, ESA, CSA, STScI)

YEARBOOK OF ASTRONOMY 2027

EDITED BY

Brian Jones

WHITE OWL
AN IMPRINT OF PEN & SWORD BOOKS LTD.
YORKSHIRE – PHILADELPHIA

First published in Great Britain in 2026 by
WHITE OWL
An imprint of
Pen & Sword Books Ltd
Yorkshire – Philadelphia

ISBN 978 1 03618 683 8

A CIP catalogue record for this book is available from the British Library.

Typeset by Mac Style
Printed and bound in India by Replika Press Pvt. Ltd.

MIX
Paper | Supporting responsible forestry
FSC™ C016779

The Publisher's authorised representative in the EU for product safety is Authorised Rep Compliance Ltd., Ground Floor, 71 Lower Baggot Street, Dublin D02 P593, Ireland.
www.arccompliance.com

For a complete list of Pen & Sword titles please contact

PEN & SWORD BOOKS LIMITED
47 Church Street, Barnsley, South Yorkshire, S70 2AS, England
E-mail: enquiries@pen-and-sword.co.uk
Website: www.pen-and-sword.co.uk
or
PEN AND SWORD BOOKS
1950 Lawrence Road, Havertown, PA 19083, USA
E-mail: uspen-and-sword@casematepublishers.com
Website: www.penandswordbooks.com

Contents

Editor's Foreword

The *Yearbook of Astronomy 2027* is the latest edition of what has long been an indispensable publication, the annual appearance of which has been eagerly anticipated by astronomers, both amateur and professional, for well over half a century. As ever, the *Yearbook* is aimed at both the armchair astronomer and the active backyard observer. Within its pages you will find a rich blend of information, star charts and guides to the night sky coupled with an interesting mixture of articles which collectively embrace a wide range of topics, ranging from the history of astronomy to the latest results of astronomical research; space exploration to observational astronomy; and our own celestial neighbourhood out to the farthest reaches of space.

The *Monthly Star Charts* have been compiled by David Harper and show the night sky as seen throughout the year. Two sets of twelve charts have been provided, one set for observers in the Northern Hemisphere and one for those in the Southern Hemisphere. Between them, each pair of charts depicts the entire sky as two semi-circular half-sky views, one looking north and the other looking south.

The *Monthly Star Charts* are followed by *The Solar System in 2027* which commences with details of the observing conditions for the Moon in *Lunar Phenomena and Eclipses*. This section has been mainly compiled by Lynne Marie Stockman and includes not only the phases and eclipses, but also features apsides, node crossings, and lunar occultations. This is followed by a trio of newly-formatted articles penned by Lynne Marie Stockman and Brian Jones, these being *Meteor Showers*, *Comets*, and *Minor Planets*, all three titles being fairly self-explanatory describing as they do the occurrence and visibility of examples of these three classes of object during and throughout the year. *The Solar System in 2027* concludes with *The Planets*, which covers all the major planetary members of our Solar System (apparition charts for which have been compiled by David Harper); further details appear in the article *Using the Yearbook of Astronomy as an Observing Guide*. The finder charts for the superior planets follow the descriptions of the planets, with horizon apparition charts for Mercury, Venus, and Mars appearing throughout the *Monthly Sky Notes*.

The *Monthly Sky Notes* have been compiled by Lynne Marie Stockman and give details of the positions and visibility of the planets for each month throughout the year. At the beginning of each of the monthly notes is a list of the significant solar system events occurring during that particular month. Each section of the *Monthly Sky Notes* is accompanied by a short article, the range of which includes items on a variety of astronomy- and space-related topics including an interesting item by Lynne, this being the fourth instalment of her fascinating *Gone But Not Forgotten* series, in which she examines Corona Meridiana. Instantly recognisable from the southern hemisphere and even from orbit, this misshapen semi-circle has been interpreted by many cultures in many different ways. Visit the celestial southern crown. Or is it a turtle? Boomerang? Llama corral?

In his article *Recent Advances in Astronomy*, regular contributor Rod Hine discusses several topics, including the potential for the Vera Rubin Observatory to revolutionise the supply of data for future research; the discovery of a new solar system in the very act of birth; and the latest revelations about Fast Radio Bursts. He also tells the story of the enigmatic supernova that simply disappeared – could this be more common than we thought possible?

This is followed by *Recent Advances in Solar System Exploration* in which Peter Rea updates us on the progress of a number of planetary missions. The *Parker Solar Probe* is coming to the end of its primary mission with a series of very close flybys of the Sun. A little further away from the Sun the joint European and Japanese *BepiColombo* mission is closing in on Mercury for a slightly-delayed entry into orbit. The Moon has become an object of interest again and in preparation for the upcoming *Artemis* missions a number of spacecraft from the USA and other countries are being sent to explore our nearest neighbour in space with a focus on the south polar region. Four spacecraft are en route to a retinue of asteroids, one of which is a metallic asteroid never before explored. Mars continues to be a focus of attention for the USA and other countries, with many long lived spacecraft still operating around the red planet. Two missions, one from the USA and one from ESA have been sent back to Jupiter to explore the icy moons, some of which may contain a sub-surface water ocean.

In 1877, when Mars had an exceptionally close, once in 15 years opposition, its two tiny moons were discovered, and observations were made which began the controversy of its non-existent 'canals'. The same year saw the death of Urbain le Verrier, the man who 'discovered a planet with the point of his pen', and the birth of Henry Norris Russell, 'the father of modern astrophysics'. In *Anniversaries in 2027*, Neil Haggath commemorates these events, as well as the death of Sir Isaac Newton 300 years ago and *Voyager*, the most successful planetary exploration mission in history.

In his article *Jocelyn Bell Discovers Neutron Stars by Their Radio Pulsations* regular contributor David M. Harland reviews the events leading up to the fortuitous discovery in 1967 by Cambridge University research student Jocelyn Bell of pulsating radio stars (pulsars), and how they became the first identified cases of neutron stars.

The article *The Vera Rubin Observatory* by Scotland's Astronomer Royal, Prof. Catherine Heymans, introduces a groundbreaking new telescope in Chile that promises to revolutionise our cosmic view of the solar system, the Milky Way, and the mysterious dark side of our Universe.

In his article *Alcor and Mizar: More than Meets the Eye* Jonathan Powell examines the observational history of one the most documented and researched optical double star pairings. Alcor and Mizar's secrets are slowly unravelled from the first naked eye observations to the advent of the telescope, then by progressive photographic means to spectroscopic technology. In time, the duo have subsequently yielded many of their secrets and in doing so have enhanced our understanding not only of double stars, but the mechanics of stellar birth and evolution.

Lynne Marie Stockman studies chemically-peculiar stars from *Across the Spectrum* in part six of her series of articles *The Astronomers' Stars*. Spectroscopy can reveal many secrets of a star, and in this article, Stockman considers unusual objects such as Campbell's Hydrogen Star, Popper's Extreme Helium Star, some of the oddball stars uncovered by master spectroscopist W.P. Bidelman, and a few objects infested with rare elements from the far reaches of the periodic table.

No-one would have believed that tiny Mars could have a significant impact on Earth's climate, but in his article *Mars Attacks! How the Red Planet Affects Earth's Climate*, David Harper looks at recent evidence from deep ocean core samples which shows that the red planet does indeed affect our planet's climate on timescales of millions of years.

In the article *What's the Matter?: It's Dark Down Boulby Mine* John McCue describes his visits to the Boulby Potash Mine which is located near his home. For over twenty years researchers working at the mine have been searching for the faint electrical signal that would reveal a dark matter particle. The role of Vera Rubin – after whom the world's most advanced telescope has been named – in proposing the existence of dark matter is also described.

The next time you're upset about an astronomical event being clouded out, ask yourself whether your luck is really so bad. In his article *The Unluckiest Astronomer Who Ever Lived*, Neil Haggath tells us the sad story of the French aristocrat Guillaume Joseph Hyacinthe Jean-Baptiste Le Gentil de la Galaisière, the man who was perhaps the unluckiest astronomer who ever lived – and undoubtedly the one with the silliest name!

In the third part of her series of articles *A History of Observatory Designs*, regular contributor Katrin Raynor tells us how the birth of the modern observatory has ushered in a fusion of science, architecture, and technology designed to serve both research and imagination. The Dominion Observatory, once Canada's timekeeper, laid groundwork for national geodesy; Palomar Observatory's Hale Telescope remains vital, now enhanced with cutting-edge spectrographs; and the *Hubble Space Telescope* continues unveiling cosmic wonders from low Earth orbit. Meanwhile, the *James Webb Space Telescope* peers ever deeper into space and time, capturing stellar births and ancient galaxies with unmatched infrared precision and ushering in a new era of discovery. We can safely conclude that the modern observatories and telescopes – spanning Earth and space – have truly revolutionised astronomy.

There are almost 6,000 confirmed exoplanets, or planets beyond our solar system. In his article *The Flight Log of Exoplanet Explorer 9: The First Landing on an Exoplanet*, Peter Rea uses a mixture of current science fact with fictional details from his own imagination, to describe the imaginary landing of the uncrewed *Exoplanet Explorer 9* spacecraft on the exoplanet Primus, orbiting around the star Gliese 581 located at a distance of 20 light years from our solar system.

In his article *Minor Planets Genesis: The Celestial Police to John Russell Hind*, regular contributor Neil Norman tells the story of the search for the supposed 'missing planet' hypothesized to exist between Mars and Jupiter. Although the best astronomers in the world tried to find it, their efforts proved fruitless, until the discovery was made, somewhat by chance, on the opening day of the nineteenth century. The article then describes how, over the following 70 years or so, not just a single planet, but dozens of the objects known as minor planets were made.

The final section of the book starts off with *Some Interesting Variable Stars* by Tracie Heywood which contains useful information on variables as well as predictions for timings of minimum brightness of the famous eclipsing binary Algol for 2027. The revised design and layout of the section *Some Interesting Telescopic Objects* is based on how this information was presented in the very early editions of the *Yearbook of Astronomy*. Rather than being comprised of lists of objects, as in recent editions, the items previously headed *Some Interesting Double Stars* and *Some Interesting Nebulae, Star Clusters and Galaxies* have been 'combined' into a feature which, although detailing fewer celestial targets, goes into much more detail for each one covered. Presenting a selection of objects for you to seek out in the night sky, the text contains much more information, including notes and comments made by astronomers of the past. Future editions of the *Yearbook of Astronomy* will include additional objects, expanding the range for the backyard astronomer to seek out and identify.

Next we have a selection of *Astronomical Organizations*, which lists organizations and associations across the world through which you can further pursue your interest and participation in astronomy (if there are any that we have omitted please let us know) and *Our Contributors*, which contains brief background details of the numerous writers who have contributed to this edition of the *Yearbook*.

New topics and themes are occasionally introduced into the *Yearbook of Astronomy*, allowing it to keep pace with the increasing range of skills, techniques and observing methods now open to amateur astronomers, this in addition to articles relating to our rapidly-expanding knowledge of the Universe in which we live. There is always an interesting mix, some articles written at a level which will appeal to the casual reader and some of what may be loosely described as at a more academic level. The intention is to fully maintain and continually increase the usefulness and relevance of the *Yearbook of Astronomy* to the interests of the readership who are, without doubt, the most important aspect of the *Yearbook* and the reason it exists in the first place. With this in mind, suggestions from readers for further improvements and additions to the *Yearbook* content are welcomed. All thoughts and comments can be sent via the *Yearbook of Astronomy* website at **yearbookofastronomy.com** After all, the book is written for you …

As ever, grateful thanks are extended to those individuals who have contributed a great deal of time and effort to the *Yearbook of Astronomy 2027*, including David Harper who, as well as compiling all the planetary apparition charts featured in the book, has also provided the *Monthly Star Charts*, which were generated specifically for what has been described as the new generation of the *Yearbook of Astronomy*. Equally important are the efforts of Lynne Marie Stockman who has compiled the *Monthly Sky Notes*, and assembled much of the information contained in *The Solar System in 2027*. Their combined efforts have produced what can justifiably be described as the backbone of the *Yearbook of Astronomy*.

Also worthy of mention is Mat Blurton, who has done an excellent job typesetting the *Yearbook*, and David M. Harland who has provided valuable assistance with several images used in this edition. My thanks are also due to Jonathan Wright, Charlotte Mitchell, Olivia Camozzi, Lori Jones, Janet Brookes, Paul Wilkinson, Charlie Simpson and Rosie Crofts of Pen & Sword Books Ltd for their efforts in producing and promoting the *Yearbook of Astronomy 2027*, the latest edition of this much-loved and iconic publication.

Brian Jones – Editor
Bradford, West Riding of Yorkshire
February 2026

Preface

The information given in this edition of the *Yearbook of Astronomy* is in narrative form. The positions of the planets given in the *Monthly Sky Notes* often refer to the constellations in which they lie at the time. These can be found on the star charts which collectively show the whole sky via two charts depicting the northern and southern circumpolar stars and forty-eight charts depicting the main stars and constellations for each month of the year. The northern and southern circumpolar charts show the stars that are within 45° of the two celestial poles, while the monthly charts depict the stars and constellations that are visible throughout the year from Europe and North America or from Australia and New Zealand. The monthly charts overlap the circumpolar charts. Wherever you are on the Earth, you will be able to locate and identify the stars depicted on the appropriate areas of the chart(s).

There are numerous star atlases available that offer more detailed information, such as *Sky & Telescope's POCKET SKY ATLAS* and *Norton's STAR ATLAS and Reference Handbook* to name but a couple. In addition, more precise information relating to planetary positions and so on can be found in a number of publications, a good example of which is *The Handbook of the British Astronomical Association*, as well as many of the popular astronomy magazines such as the British monthly periodicals *Sky at Night* and *Astronomy Now* and the American monthly magazines *Astronomy* and *Sky & Telescope*.

About Time

Before the late eighteenth century, the biggest problem affecting mariners sailing the seas was finding their position. Latitude was easily determined by observing the altitude of the pole star above the northern horizon. Longitude, however, was far more difficult to measure. The inability of mariners to determine their longitude often led to them getting lost, and on many occasions shipwrecked. To address this problem King Charles II established the Royal Observatory at Greenwich in 1675 and from here, Astronomers Royal began the process of measuring and cataloguing the stars as they passed due south across the Greenwich meridian.

Now mariners only needed an accurate timepiece (the chronometer invented by Yorkshire-born clockmaker John Harrison) to display GMT (Greenwich Mean Time). Working out the local standard time onboard ship and subtracting this from GMT gave the ship's longitude (west or east) from the Greenwich meridian. Therefore mariners always knew where they were at sea and the longitude problem was solved.

Astronomers use a time scale called Universal Time (UT). This is equivalent to Greenwich Mean Time and is defined by the rotation of the Earth. The *Yearbook of Astronomy* gives all times in UT rather than in the local time for a particular city or country. Times are expressed using the 24-hour clock, with the day beginning at midnight, denoted by 00:00. Universal Time (UT) is related to local mean time by the formula:

Local Mean Time = UT - west longitude

In practice, small differences in longitude are ignored and the observer will use local clock time which will be the appropriate Standard (or Zone) Time. As the formula indicates, places in west longitude will have a Standard Time slow on UT, while those in east longitude will have a Standard Time fast on UT. As examples we have:

Standard Time in

New Zealand	UT + 12 hours
Victoria, NSW	UT + 10 hours
Japan	UT + 9 hours
Western Australia	UT + 8 hours
India	UT + 5 hours 30 minutes
Pakistan	UT + 5 hours
Kenya	UT + 3 hours
South Africa	UT + 2 hours
British Isles	UT
Newfoundland Standard Time	UT – 3 hours 30 minutes
Atlantic Standard Time	UT – 4 hours
Eastern Standard Time	UT – 5 hours
Central Standard Time	UT – 6 hours
Mountain Standard Time	UT – 7 hours
Pacific Standard Time	UT – 8 hours
Alaska Standard Time	UT – 9 hours
Hawaii-Aleutian Standard Time	UT – 10 hours

During the periods when Summer Time (also called Daylight Saving Time) is in use, one hour must be added to Standard Time to obtain the appropriate Summer/Daylight Saving Time. For example, Pacific Daylight Time is UT – 7 hours.

Using the Yearbook of Astronomy as an Observing Guide

Notes on the Monthly Star Charts

The star charts on the following pages show the night sky throughout the year. There are two sets of charts, one for use by observers in the Northern Hemisphere and one for those in the Southern Hemisphere. The first set is drawn for latitude 52°N and can be used by observers in Europe, Canada and most of the United States. The second set is drawn for latitude 35°S and show the stars as seen from Australia and New Zealand. Twelve pairs of charts are provided for each of these latitudes.

Each pair of charts shows the entire sky as two semi-circular half-sky views, one looking north and the other looking south. A given pair of charts can be used at different times of year. For example, chart 1 shows the night sky at midnight on 21 December, but also at 2am on 21 January, 4am on 21 February and so forth. The accompanying table will enable you to select the correct chart for a given month and time of night. The caption next to each chart also lists the dates and times of night for which it is valid.

The charts are intended to help you find the more prominent constellations and other objects of interest mentioned in the monthly observing notes. To avoid the charts becoming too crowded, only stars of magnitude 4.5 or brighter are shown. This corresponds to stars that are bright enough to be seen from any dark suburban garden on a night when the Moon is not too close to full phase.

Each constellation is depicted by joining selected stars with lines to form a pattern. There is no official standard for these patterns, so you may occasionally find different patterns used in other popular astronomy books for some of the constellations.

Any map projection from a sphere onto a flat page will by necessity contain some distortions. This is true of star charts as well as maps of the Earth. The distortion on the half-sky charts is greatest near the semi-circular boundary of each chart, where it may appear to stretch constellation patterns out of shape.

The charts also show selected deep-sky objects such as galaxies, nebulae and star clusters. Many of these objects are too faint to be seen with the naked eye, and you will need binoculars or a telescope to observe them. Please refer to the table of deep-sky objects for more information.

Planetary Apparition Diagrams

The diagrams of the apparitions of Mercury and Venus show the position of the respective planet in the sky at the moment of sunrise or sunset throughout the entire apparition. Two sets of positions are plotted on each chart: for latitude 52° North (blue line) and for latitude 35° South (red line). A thin dotted line denotes the portion of the apparition which falls outside the year covered by this edition of the *Yearbook*. A white dot indicates the position of Venus on the first day of each month, or of Mercury on the first, eleventh and 21st of the month. The day of greatest elongation (GE) is also marked by a white dot. Note that the dots do NOT indicate the magnitude of the planet.

The main finder chart for Mars shows the path of the red planet from 1 January through to 1 July. The ecliptic is denoted by the dotted line running across the chart from lower left to upper right. The position of Mars is indicated on the 1st of each month as well as at opposition and at stationary points in Right Ascension. Note that the dots do NOT indicate the magnitude of Mars. In addition, there is a Mars evening apparition diagram (appearing in the *Monthly Sky Notes* for September) which shows the position of the planet at the moment of sunset from 1 September 2027 through to 1 February 2028.

The finder charts for Jupiter, Saturn, Uranus and Neptune show the paths of the planets throughout the year. The position of each planet is indicated at opposition and at stationary points, as well as the start and end of the year and on the 1st of each month (1st of April, July and October only for Uranus and Neptune) where these dates do not fall too close to an event that is already marked. Stars are shown to magnitude 5.5 on the charts for Jupiter and Saturn. On the Uranus chart, stars are shown to magnitude 8; on the Neptune chart, the limiting magnitude is 10. In both cases, this is approximately two magnitudes fainter than the planet itself. Right Ascension and Declination scales are shown for the epoch J2000 to allow comparison with modern star charts. Note that the sizes of the dots denoting the planets do NOT indicate their magnitudes.

Selecting the Correct Charts

The table below shows which of the charts to use for particular dates and times throughout the year and will help you to select the correct pair of half-sky charts for any combination of month and time of night.

The Earth takes 23 hours 56 minutes (and 4 seconds) to rotate once around its axis with respect to the fixed stars. Because this is around four minutes shorter than a full 24 hours, the stars appear to rise and set about 4 minutes earlier on each successive day, or around an hour earlier each fortnight. Therefore, as well as showing the stars at 10pm (22h in 24-hour notation) on 21 January, chart 1 also depicts the sky at 9pm (21h) on 6 February, 8pm (20h) on 21 February and 7pm (19h) on 6 March.

The times listed do not include summer time (daylight saving time), so if summer time is in force you must subtract one hour to obtain standard time (GMT if you are in the United Kingdom) before referring to the chart. For example, to find the correct chart for mid-September in the northern hemisphere at 3am summer time, first of all subtract one hour to obtain 2am (2h) standard time. Then you can consult the table, where you will find that you should use chart 11.

The table does not indicate sunrise, sunset or twilight. In northern temperate latitudes, the sky is still light at 18h and 6h from April to September, and still light at 20h and 4h from May to August. In Australia and New Zealand, the sky is still light at 18h and 6h from October to March, and in twilight (with only bright stars visible) at 20h and 04h from November to January.

Local Time	**18h**	**20h**	**22h**	**0h**	**2h**	**4h**	**6h**
January	11	12	1	2	3	4	5
February	12	1	2	3	4	5	6
March	1	2	3	4	5	6	7
April	2	3	4	5	6	7	8
May	3	4	5	6	7	8	9
June	4	5	6	7	8	9	10
July	5	6	7	8	9	10	11
August	6	7	8	9	10	11	12
September	7	8	9	10	11	12	1
October	8	9	10	11	12	1	2
November	9	10	11	12	1	2	3
December	10	11	12	1	2	3	4

Legend to the Star Charts

STARS

Symbol	Magnitude
•	0 or brighter
•	1
•	2
•	3
•	4
·	5
•	Double star
⊙	Variable star

DEEP-SKY OBJECTS

Symbol	Type of object
⁘	Open star cluster
◌	Globular star cluster
□	Nebula
⊡	Cluster with nebula
o	Planetary nebula
⬯	Galaxy
	Magellanic Clouds

Star Names

There are over 200 stars with proper names, most of which are of Roman, Greek or Arabic origin although only a couple of dozen or so of these names are used regularly. Examples include Arcturus in Boötes, Castor and Pollux in Gemini and Rigel in Orion.

A system whereby Greek letters were assigned to stars was introduced by the German astronomer and celestial cartographer Johann Bayer in his star atlas *Uranometria*, published in 1603. Bayer's system is applied to the brighter stars within any particular constellation, which are given a letter from the Greek alphabet followed by the genitive case of the constellation in which the star is located. This genitive case is simply the Latin form meaning 'of' the constellation. Examples are the stars Alpha Boötis and Beta Centauri which translate literally as 'Alpha of Boötes' and 'Beta of the Centaur'.

As a general rule, the brightest star in a constellation is labelled Alpha (α), the second brightest Beta (β), and the third brightest Gamma (γ) and so on, although there are some constellations where the system falls down. An example is Gemini where the principal star (Pollux) is designated Beta Geminorum, the second brightest (Castor) being known as Alpha Geminorum.

There are only 24 letters in the Greek alphabet, the consequence of which was that the fainter naked eye stars needed an alternative system of classification. The system in popular use is that devised by the first Astronomer Royal John Flamsteed in which the stars in each constellation are listed numerically in order from west to east. Although many of the brighter stars within any particular constellation will have both Greek letters and Flamsteed numbers, the latter are generally used only when a star does not have a Greek letter.

The Greek Alphabet

α	Alpha	ι	Iota	ρ	Rho
β	Beta	κ	Kappa	σ	Sigma
γ	Gamma	λ	Lambda	τ	Tau
δ	Delta	μ	Mu	υ	Upsilon
ε	Epsilon	ν	Nu	φ	Phi
ζ	Zeta	ξ	Xi	χ	Chi
η	Eta	ο	Omicron	ψ	Psi
θ	Theta	π	Pi	ω	Omega

The Names of the Constellations

On clear, dark, moonless nights, the sky seems to teem with stars although in reality you can never see more than a couple of thousand or so at any one time when looking with the unaided eye. Each and every one of these stars belongs to a particular constellation, although the constellations that we see in the sky, and which grace the pages of star atlases, are nothing more than chance alignments. The stars that make up the constellations are often situated at vastly differing distances from us and only appear close to each other, and form the patterns that we see, because they lie in more or less the same direction as each other as seen from Earth.

A large number of the constellations are named after mythological characters, and were given their names thousands of years ago. However, those star groups lying close to the south celestial pole were discovered by Europeans only during the last few centuries, many of these by explorers and astronomers who mapped the stars during their journeys to lands under southern skies. This resulted in many of the newer constellations having modern-sounding names, such as Octans (the Octant) and Microscopium (the Microscope), both of which were devised by the French astronomer Nicolas Louis De La Caille during the early 1750s.

Over the centuries, many different suggestions for new constellations have been put forward by astronomers who, for one reason or another, felt the need to add new groupings to star charts and to fill gaps between the traditional constellations. Astronomers drew up their own charts of the sky, incorporating their new groups into them. A number of these new constellations had cumbersome names, notable examples including Officina Typographica (the Printing Shop) introduced by the German astronomer Johann Bode in 1801; Sceptrum Brandenburgicum (the Sceptre of Brandenburg) introduced by the German astronomer Gottfried Kirch in 1688; Taurus Poniatovii (Poniatowski's Bull) introduced by the Polish-Lithuanian astronomer Martin Odlanicky Poczobut in 1777; and Quadrans Muralis (the Mural Quadrant) devised by the French astronomer Joseph-Jerôme de Lalande in 1795. Although these have long since been rejected, the latter has been immortalised by the annual Quadrantid meteor shower, the radiant of which lies in an area of sky formerly occupied by Quadrans Muralis.

During the 1920s the International Astronomical Union (IAU) systemised matters by adopting an official list of 88 accepted constellations, each with official spellings and abbreviations. Precise boundaries for each constellation were then drawn up so that every point in the sky belonged to a particular constellation.

The abbreviations devised by the IAU each have three letters which in the majority of cases are the first three letters of the constellation name, such as AND for Andromeda, EQU for Equuleus, HER for Hercules, ORI for Orion and so on. This trend is not strictly adhered to in cases where confusion may arise. This happens with the two constellations Leo (abbreviated LEO) and Leo Minor (abbreviated LMI). Similarly, because Triangulum (TRI) may be mistaken for Triangulum Australe, the latter is abbreviated TRA. Other instances occur with Sagitta (SGE) and Sagittarius (SGR) and with Canis Major (CMA) and Canis Minor (CMI) where the first two letters from the second names of the constellations are used. This is also the case with Corona Australis (CRA) and Corona Borealis (CRB) where the first letter of the second name of each constellation is incorporated. Finally, mention must be made of Crater (CRT) which has been abbreviated in such a way as to avoid confusion with the aforementioned CRA (Corona Australis).

The table shown on the following pages contains the name of each of the 88 constellations together with the translation and abbreviation of the constellation name. The constellations depicted on the monthly star charts are identified with their abbreviations rather than the full constellation names.

The Constellations

Andromeda	Andromeda	AND
Antlia	The Air Pump	ANT
Apus	The Bird of Paradise	APS
Aquarius	The Water Carrier	AQR
Aquila	The Eagle	AQL
Ara	The Altar	ARA
Aries	The Ram	ARI
Auriga	The Charioteer	AUR
Boötes	The Herdsman	BOO
Caelum	The Graving Tool	CAE
Camelopardalis	The Giraffe	CAM
Cancer	The Crab	CNC
Canes Venatici	The Hunting Dogs	CVN
Canis Major	The Great Dog	CMA
Canis Minor	The Little Dog	CMI
Capricornus	The Goat	CAP
Carina	The Keel	CAR
Cassiopeia	Cassiopeia	CAS
Centaurus	The Centaur	CEN
Cepheus	Cepheus	CEP
Cetus	The Whale	CET
Chamaeleon	The Chameleon	CHA
Circinus	The Pair of Compasses	CIR
Columba	The Dove	COL
Coma Berenices	Berenice's Hair	COM
Corona Australis	The Southern Crown	CRA
Corona Borealis	The Northern Crown	CRB
Corvus	The Crow	CRV
Crater	The Cup	CRT
Crux	The Cross	CRU
Cygnus	The Swan	CYG

Delphinus	The Dolphin	DEL
Dorado	The Goldfish	DOR
Draco	The Dragon	DRA
Equuleus	The Foal	EQU
Eridanus	The River	ERI
Fornax	The Furnace	FOR
Gemini	The Twins	GEM
Grus	The Crane	GRU
Hercules	Hercules	HER
Horologium	The Pendulum Clock	HOR
Hydra	The Water Snake	HYA
Hydrus	The Lesser Water Snake	HYI
Indus	The Indian	IND
Lacerta	The Lizard	LAC
Leo	The Lion	LEO
Leo Minor	The Lesser Lion	LMI
Lepus	The Hare	LEP
Libra	The Scales	LIB
Lupus	The Wolf	LUP
Lynx	The Lynx	LYN
Lyra	The Lyre	LYR
Mensa	The Table Mountain	MEN
Microscopium	The Microscope	MIC
Monoceros	The Unicorn	MON
Musca	The Fly	MUS
Norma	The Level	NOR
Octans	The Octant	OCT
Ophiuchus	The Serpent Bearer	OPH
Orion	Orion	ORI
Pavo	The Peacock	PAV
Pegasus	Pegasus	PEG
Perseus	Perseus	PER

Phoenix	The Phoenix	PHE
Pictor	The Painter's Easel	PIC
Pisces	The Fish	PSC
Piscis Austrinus	The Southern Fish	PSA
Puppis	The Stern	PUP
Pyxis	The Mariner's Compass	PYX
Reticulum	The Net	RET
Sagitta	The Arrow	SGE
Sagittarius	The Archer	SGR
Scorpius	The Scorpion	SCO
Sculptor	The Sculptor	SCL
Scutum	The Shield	SCT
Serpens Caput and Cauda	The Serpent	SER

Sextans	The Sextant	SEX
Taurus	The Bull	TAU
Telescopium	The Telescope	TEL
Triangulum	The Triangle	TRI
Triangulum Australe	The Southern Triangle	TRA
Tucana	The Toucan	TUC
Ursa Major	The Great Bear	UMA
Ursa Minor	The Little Bear	UMI
Vela	The Sail	VEL
Virgo	The Virgin	VIR
Volans	The Flying Fish	VOL
Vulpecula	The Fox	VUL

The Monthly Star Charts

Northern Hemisphere Star Charts

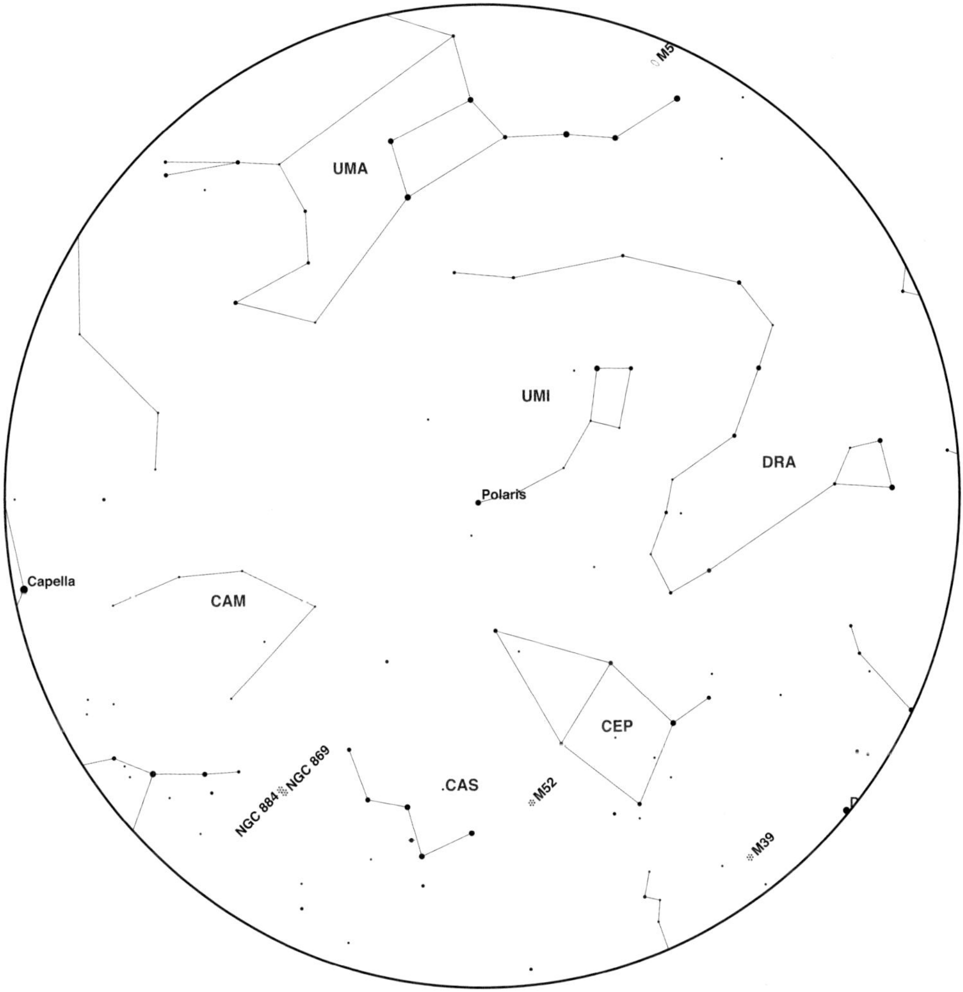

This chart shows stars lying at declinations between +45 and +90 degrees. These constellations are circumpolar for observers in Europe and North America.

1N

October 21 at 4h
November 21 at 2h
December 21 at 0h
January 21 at 22h
February 21 at 20h

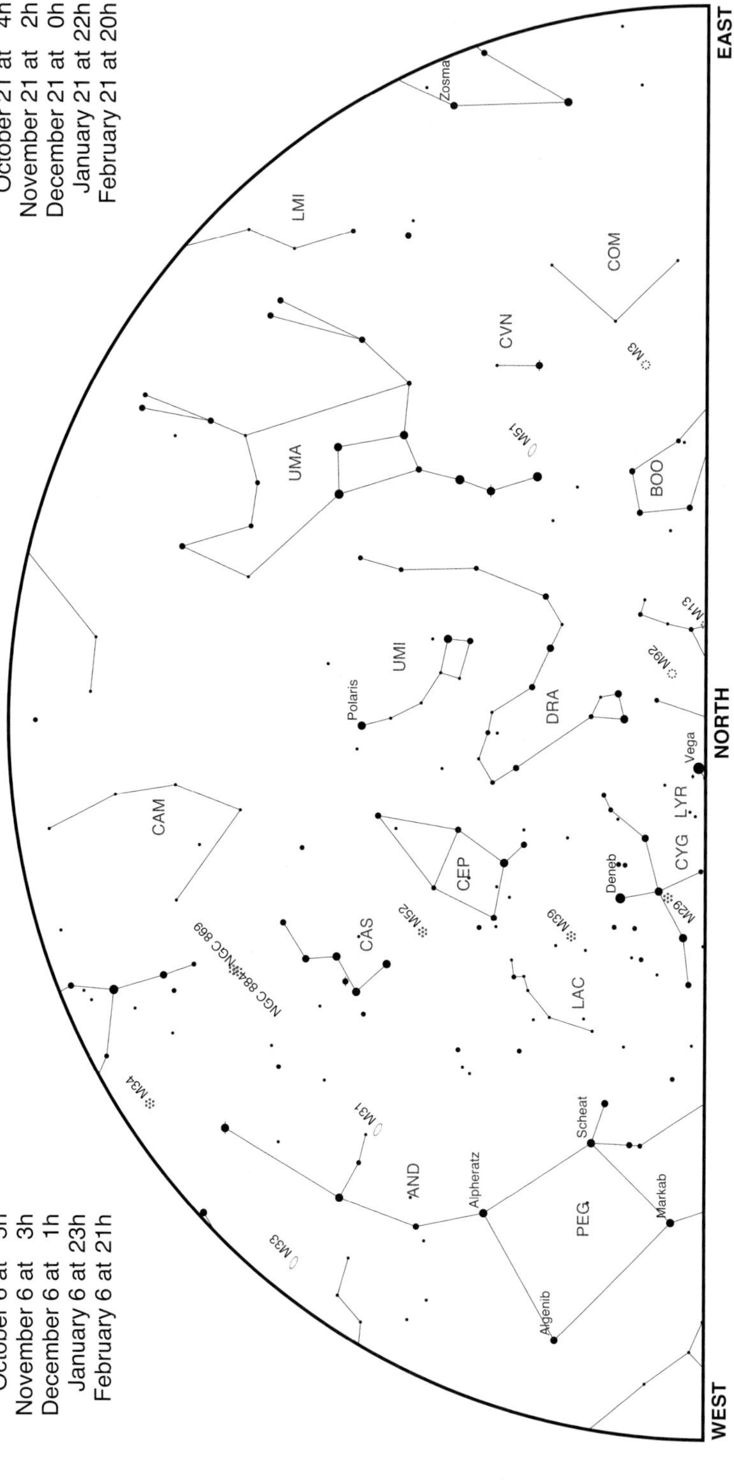

October 6 at 5h
November 6 at 3h
December 6 at 1h
January 6 at 23h
February 6 at 21h

October 6 at 5h
November 6 at 3h
December 6 at 1h
January 6 at 23h
February 6 at 21h

October 21 at 4h
November 21 at 2h
December 21 at 0h
January 21 at 22h
February 21 at 20h

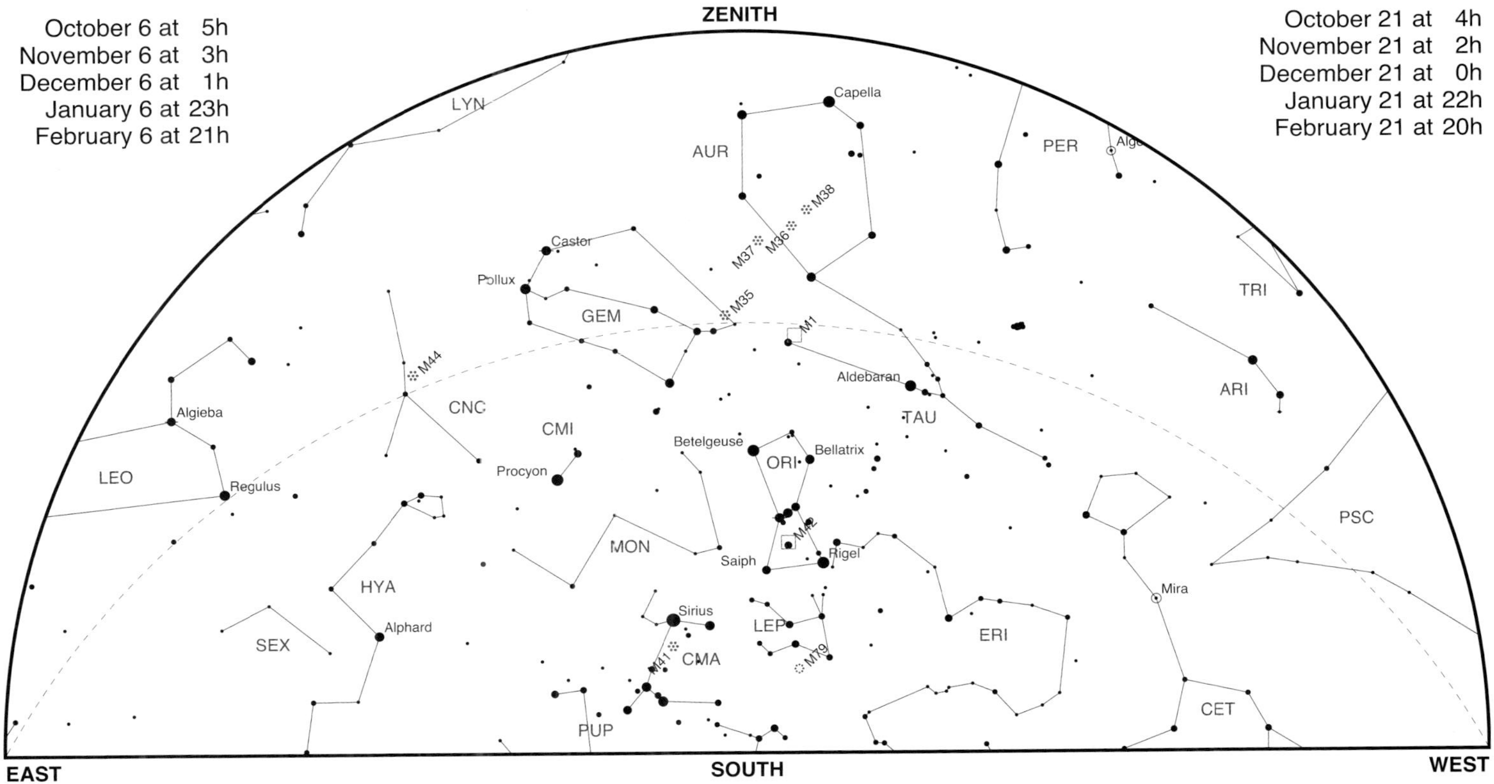

1S

2N

November 6 at 5h
December 6 at 3h
January 6 at 1h
February 6 at 23h
March 6 at 21h

November 21 at 4h
December 21 at 2h
January 21 at 0h
February 21 at 22h
March 21 at 20h

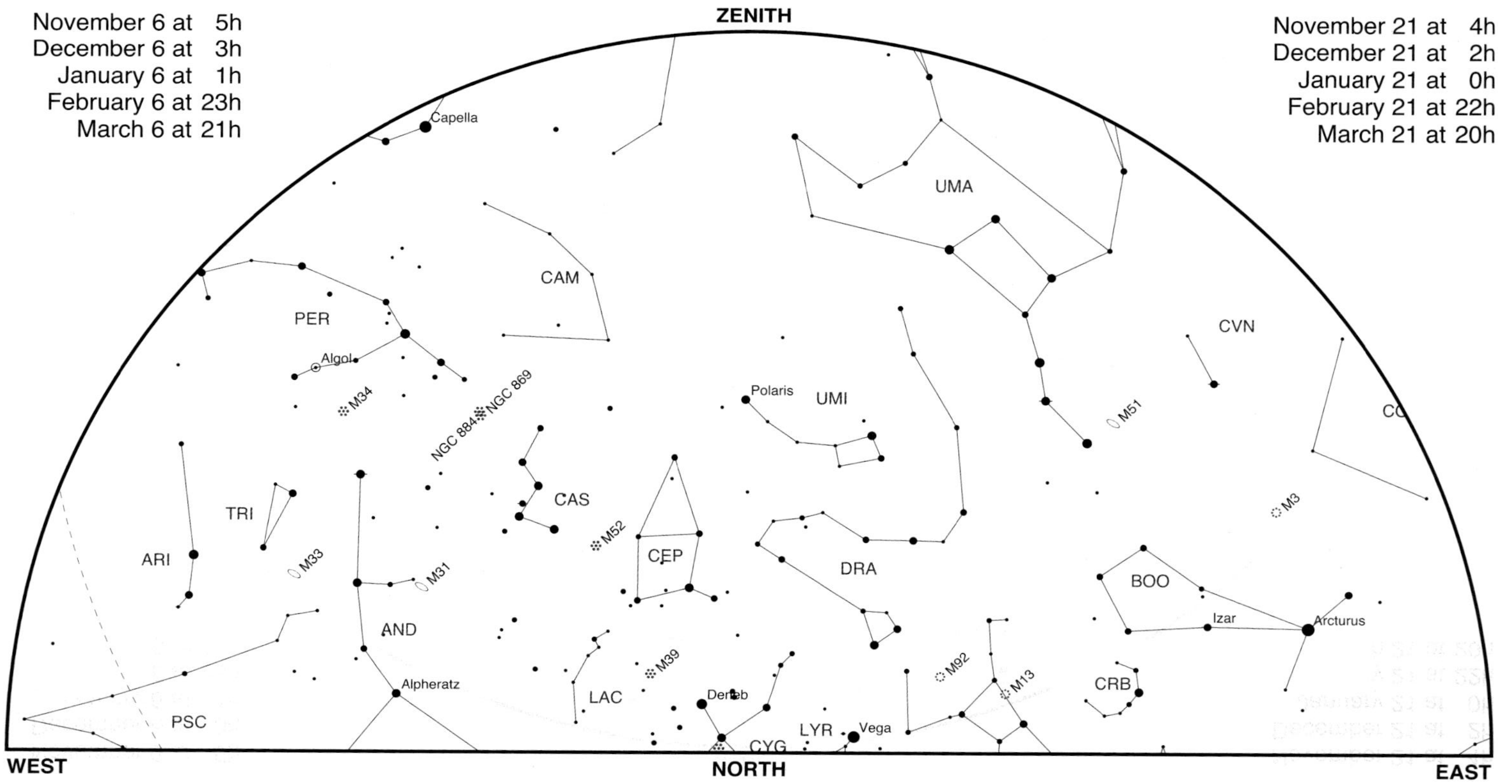

November 6 at 5h
December 6 at 3h
January 6 at 1h
February 6 at 23h
March 6 at 21h

November 21 at 4h
December 21 at 2h
January 21 at 0h
February 21 at 22h
March 21 at 20h

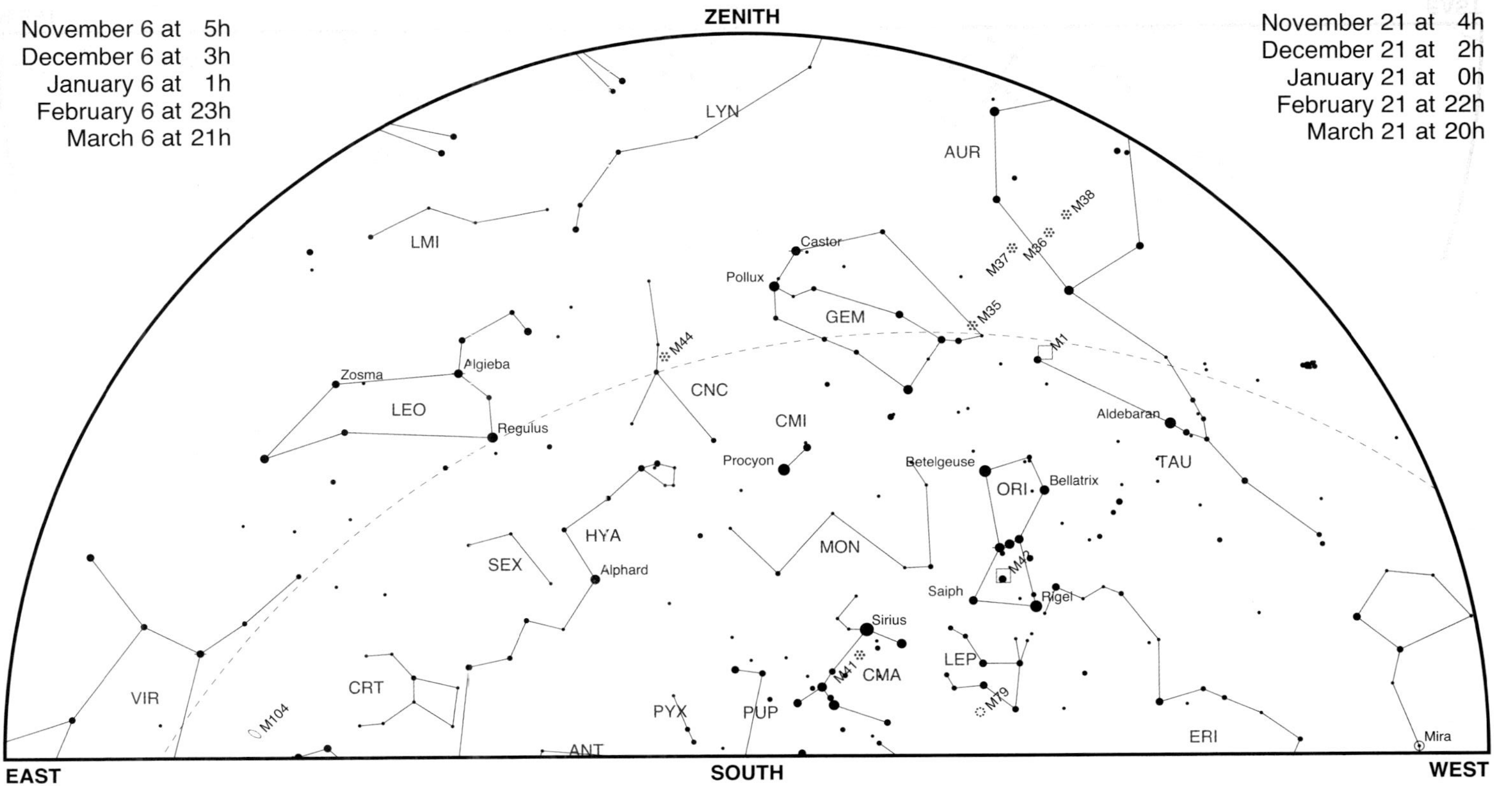

2S

3N

December 6 at 5h
January 6 at 3h
February 6 at 1h
March 6 at 23h
April 6 at 21h

December 21 at 4h
January 21 at 2h
February 21 at 0h
March 21 at 22h
April 21 at 20h

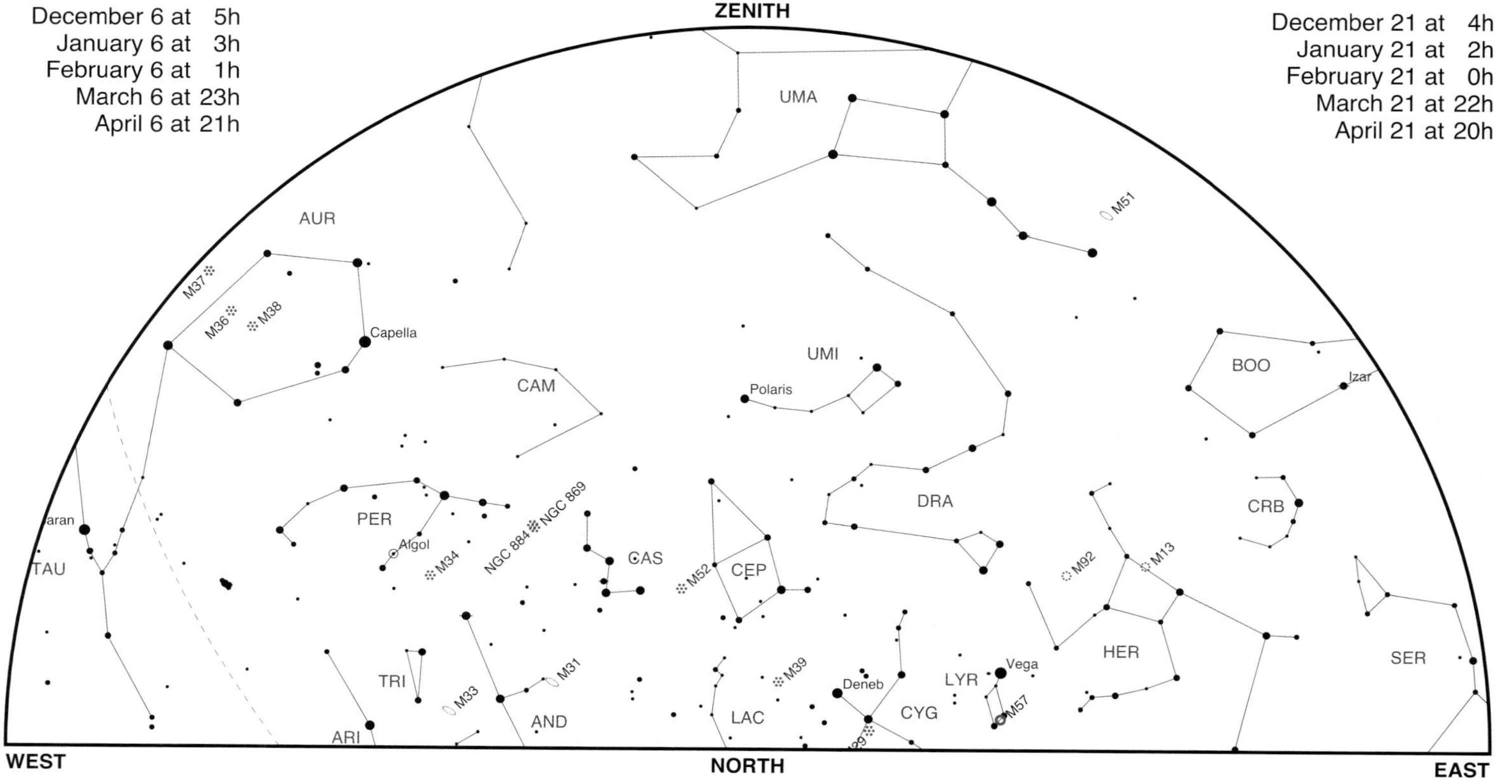

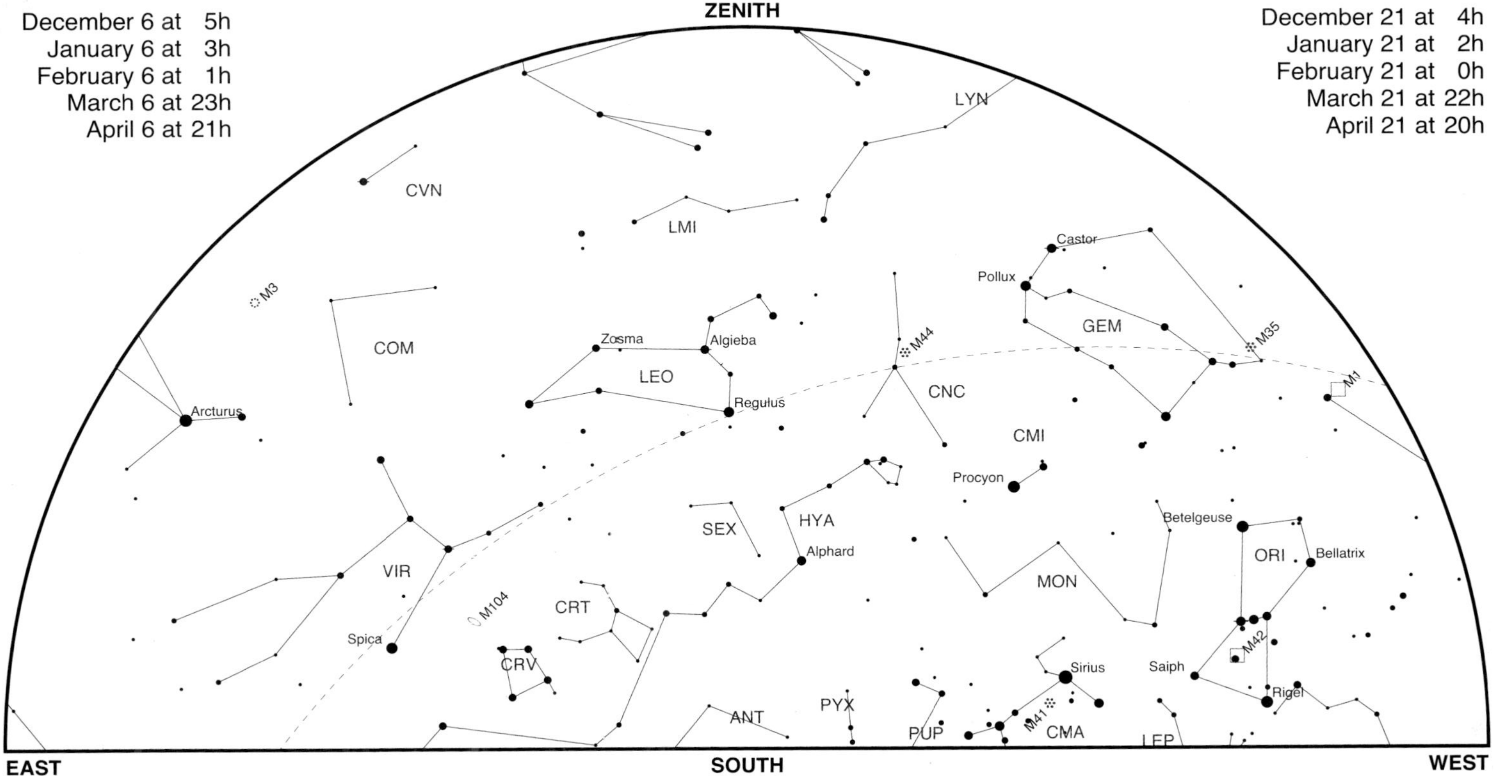
December 6 at 5h
January 6 at 3h
February 6 at 1h
March 6 at 23h
April 6 at 21h
ZENITH
December 21 at 4h
January 21 at 2h
February 21 at 0h
March 21 at 22h
April 21 at 20h
LYN
CVN
LMI
Castor
Pollux
GEM
M3
COM
Zosma
Algieba
M44
M35
LEO
CNC
M1
Arcturus
Regulus
CMI
Procyon
SEX
HYA
Betelgeuse
VIR
Alphard
ORI
Bellatrix
MON
M104
CRT
Spica
M42
Saiph
Sirius
CRV
Rigel
PYX
ANT
PUP
M41
CMA
LEP
EAST
SOUTH
WEST

3S

4N

January 6 at 5h
February 6 at 3h
March 6 at 1h
April 6 at 23h
May 6 at 21h

January 21 at 4h
February 21 at 2h
March 21 at 0h
April 21 at 22h
May 21 at 20h

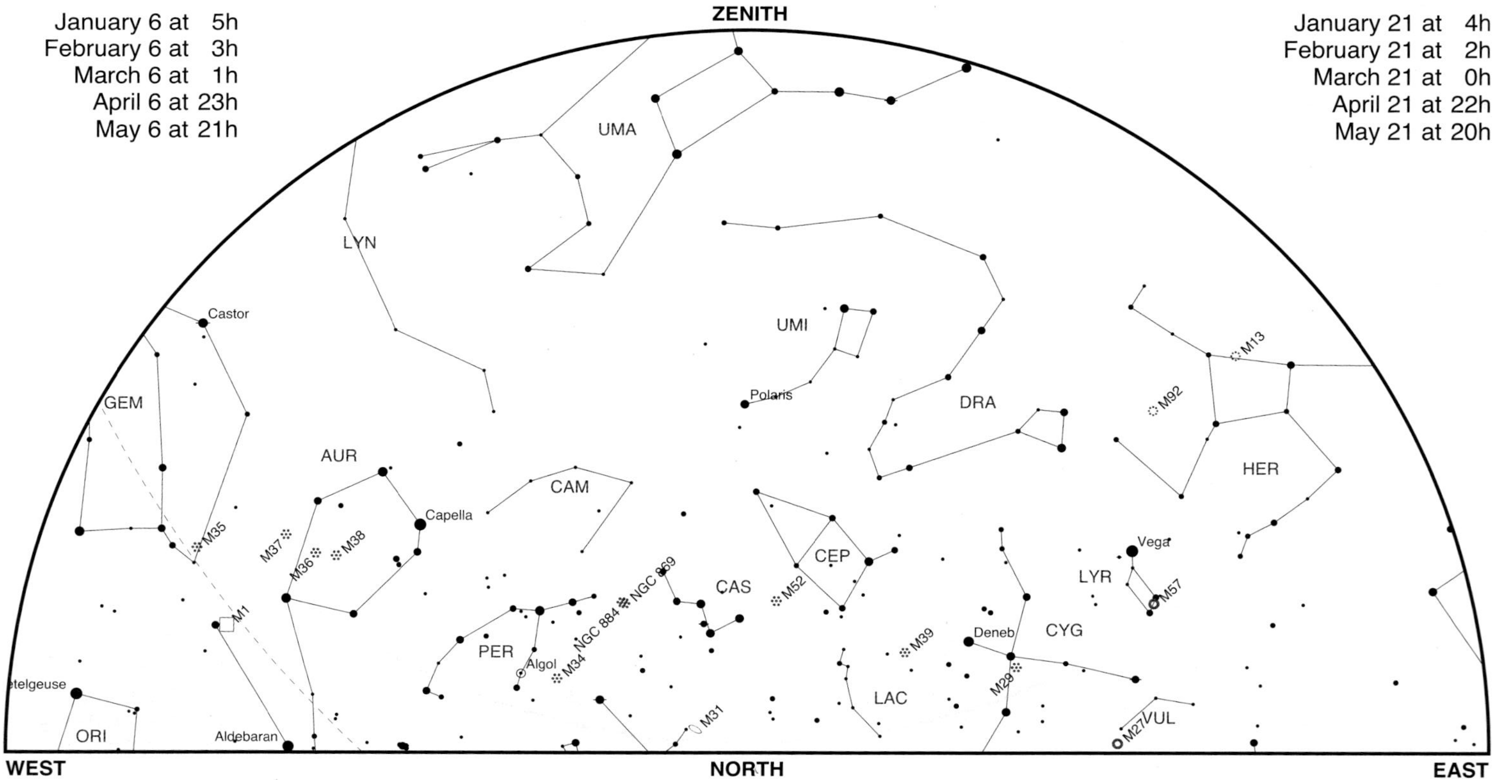

January 6 at 5h
February 6 at 3h
March 6 at 1h
April 6 at 23h
May 6 at 21h

January 21 at 4h
February 21 at 2h
March 21 at 0h
April 21 at 22h
May 21 at 20h

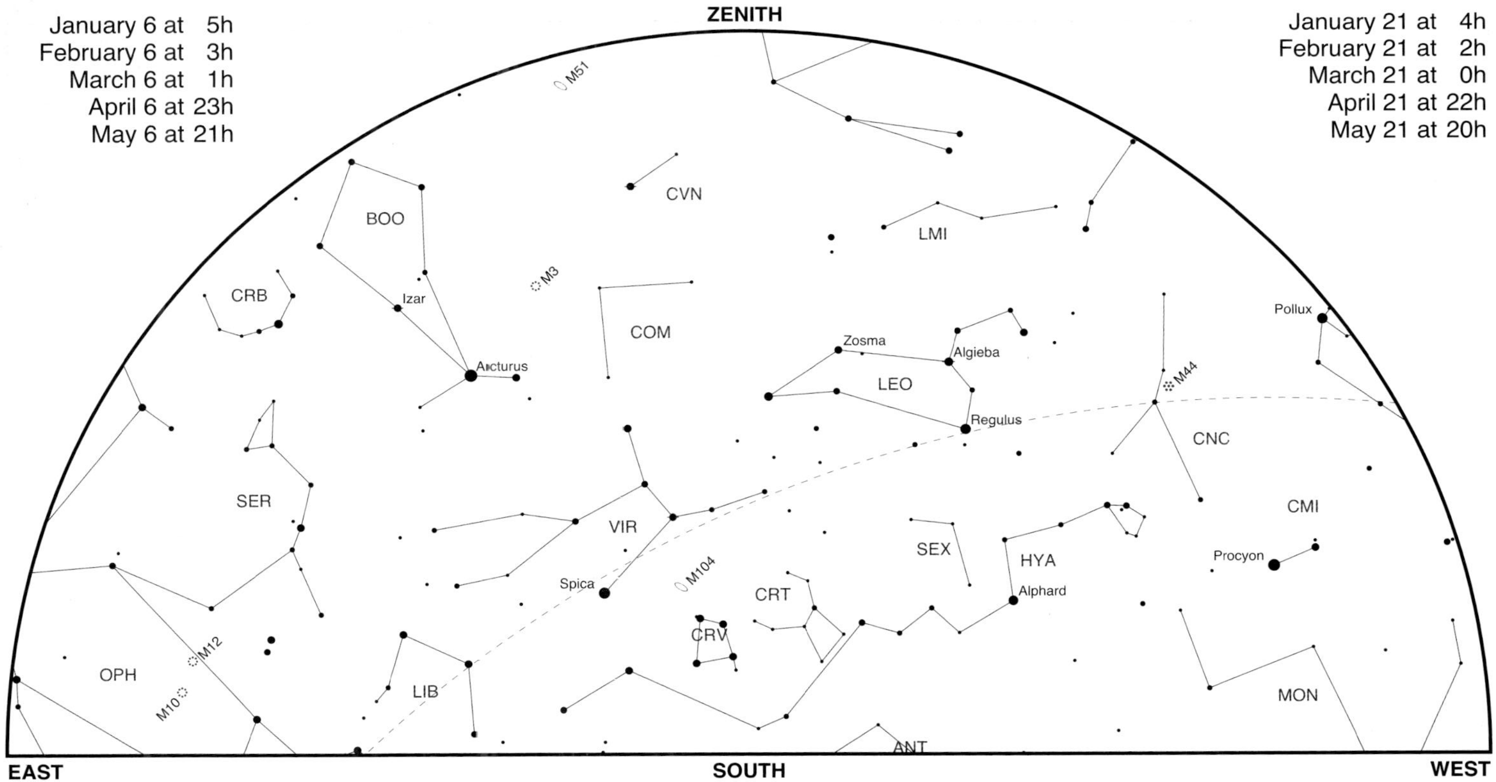

4S

5N

January 6 at 7h
February 6 at 5h
March 6 at 3h
April 6 at 1h
May 6 at 23h

January 21 at 6h
February 21 at 4h
March 21 at 2h
April 21 at 0h
May 21 at 22h

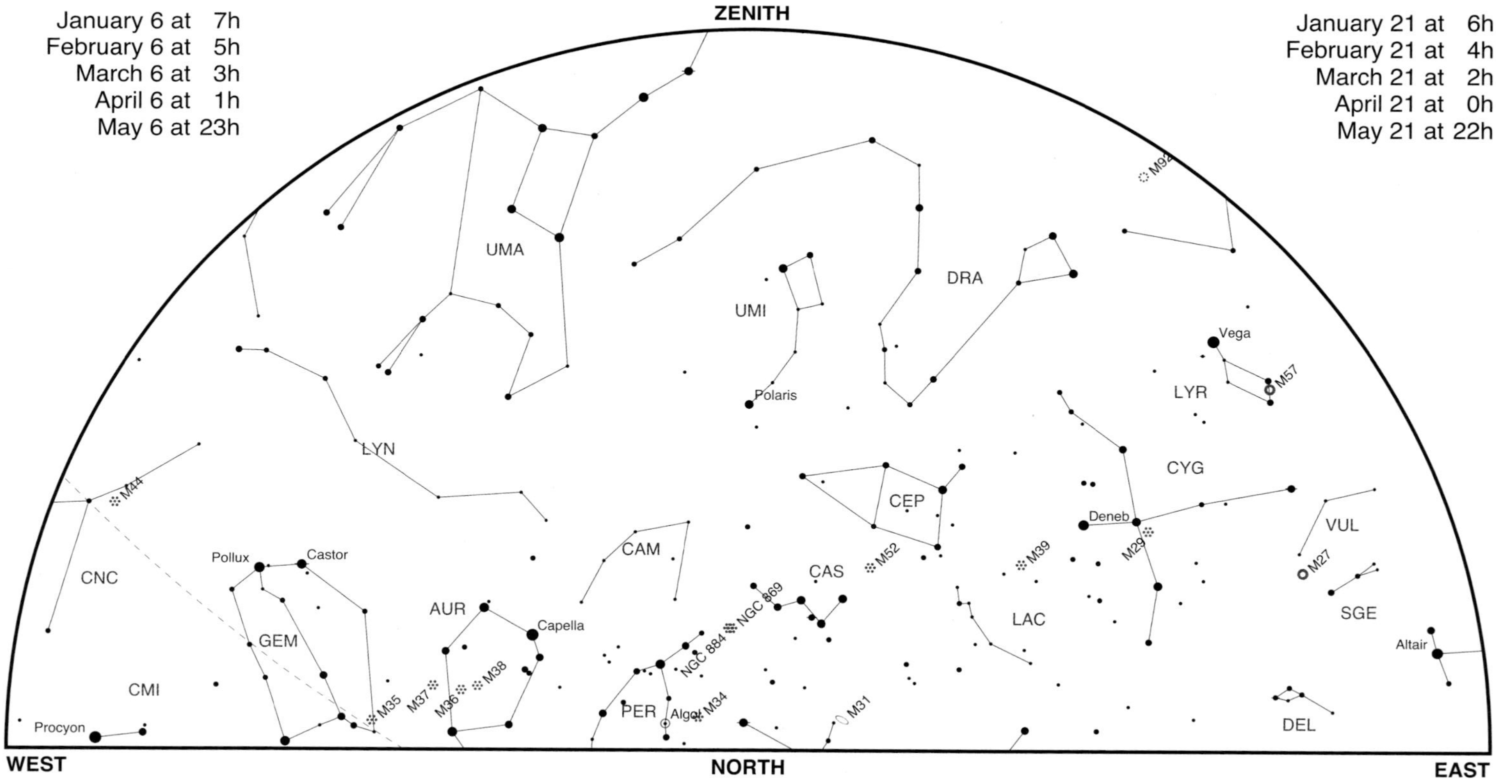

January 6 at 7h
February 6 at 5h
March 6 at 3h
April 6 at 1h
May 6 at 23h

January 21 at 6h
February 21 at 4h
March 21 at 2h
April 21 at 0h
May 21 at 22h

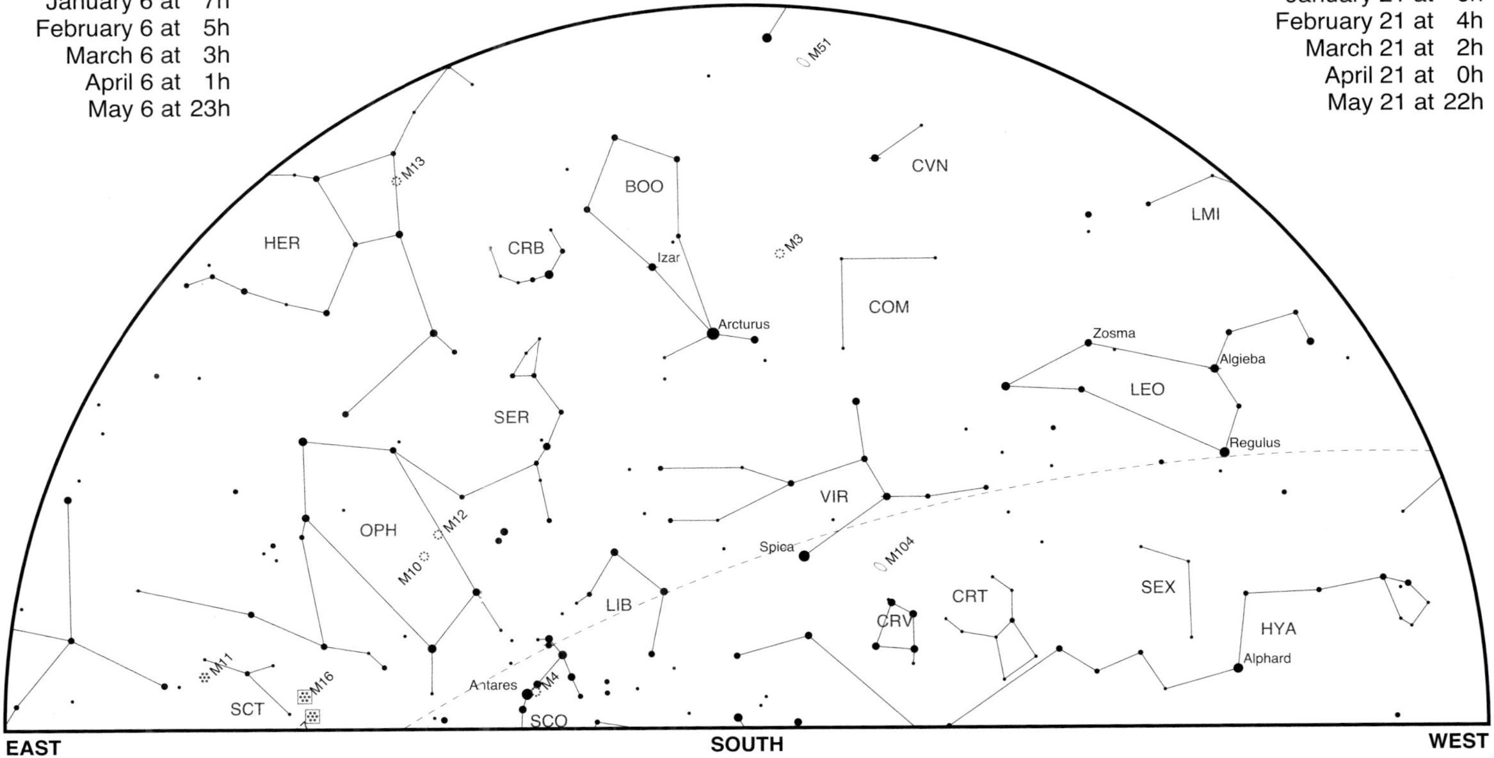

5S

6N

March 6 at 5h
April 6 at 3h
May 6 at 1h
June 6 at 23h
July 6 at 21h

March 21 at 4h
April 21 at 2h
May 21 at 0h
June 21 at 22h
July 21 at 20h

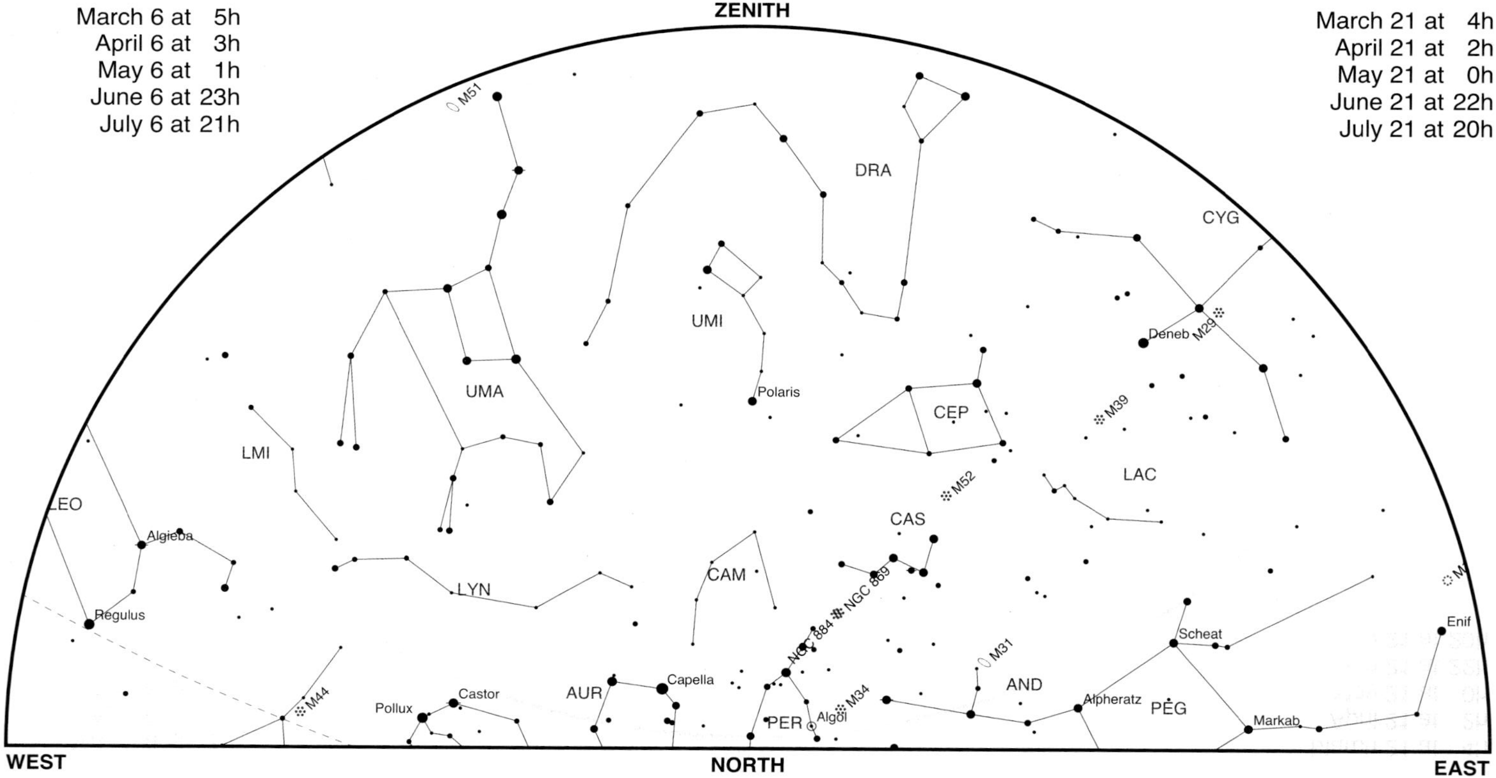

March 6 at 5h
April 6 at 3h
May 6 at 1h
June 6 at 23h
July 6 at 21h

March 21 at 4h
April 21 at 2h
May 21 at 0h
June 21 at 22h
July 21 at 20h

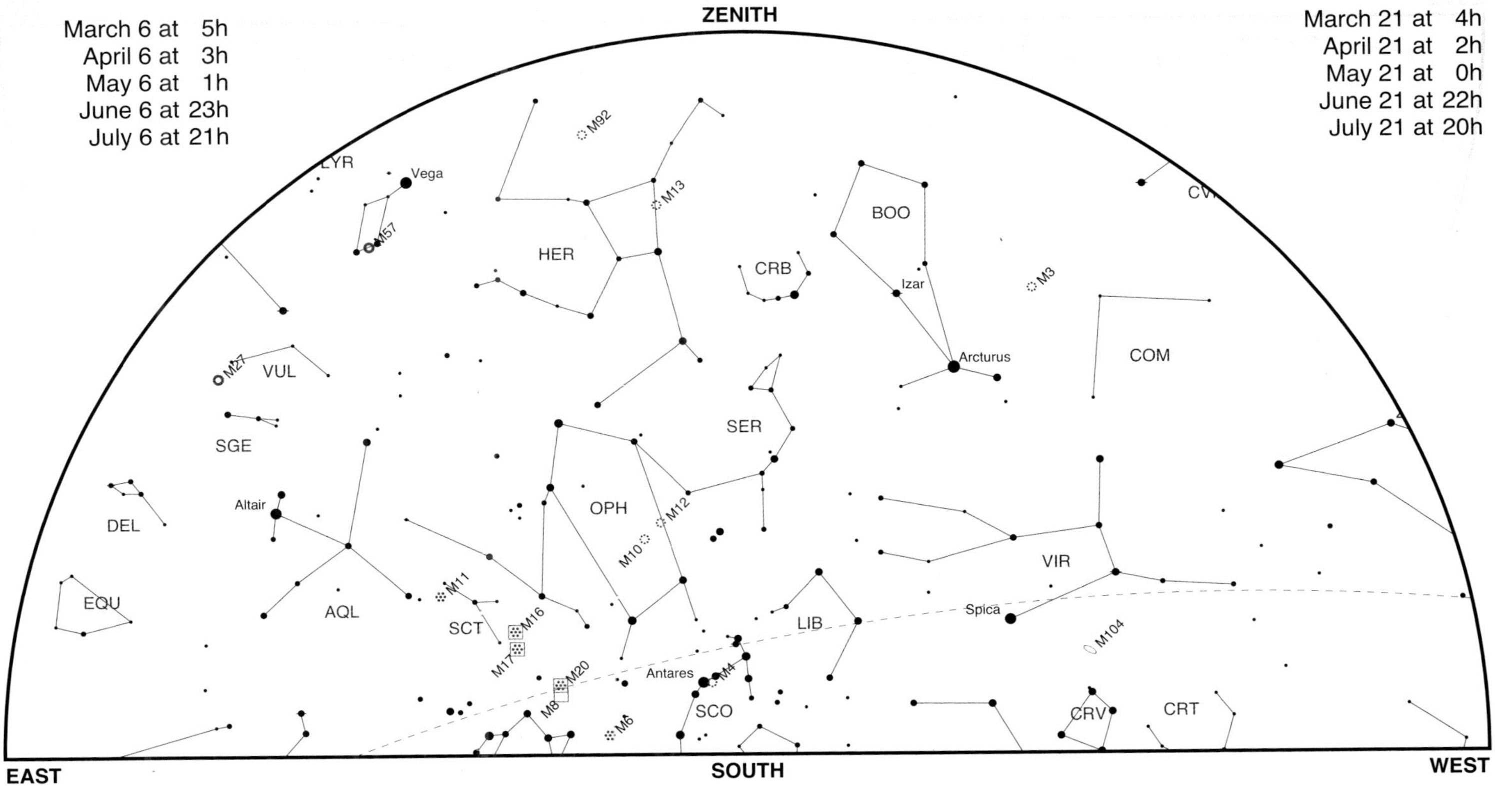

6S

7N

May 6 at 3h
June 6 at 1h
July 6 at 23h
August 6 at 21h
September 6 at 19h

May 21 at 2h
June 21 at 0h
July 21 at 22h
August 21 at 20h
September 21 at 18h

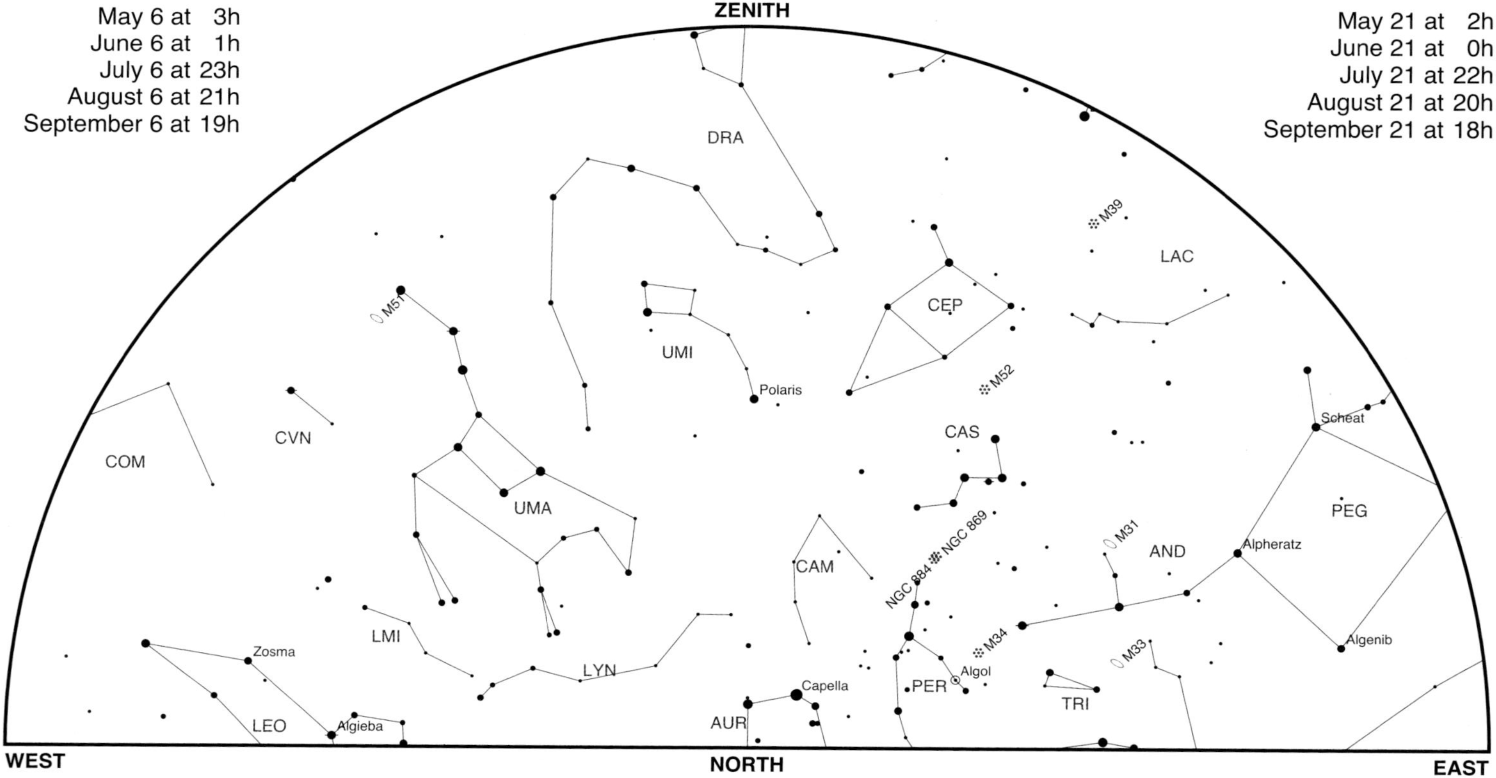

May 6 at 3h
June 6 at 1h
July 6 at 23h
August 6 at 21h
September 6 at 19h

May 21 at 2h
June 21 at 0h
July 21 at 22h
August 21 at 20h
September 21 at 18h

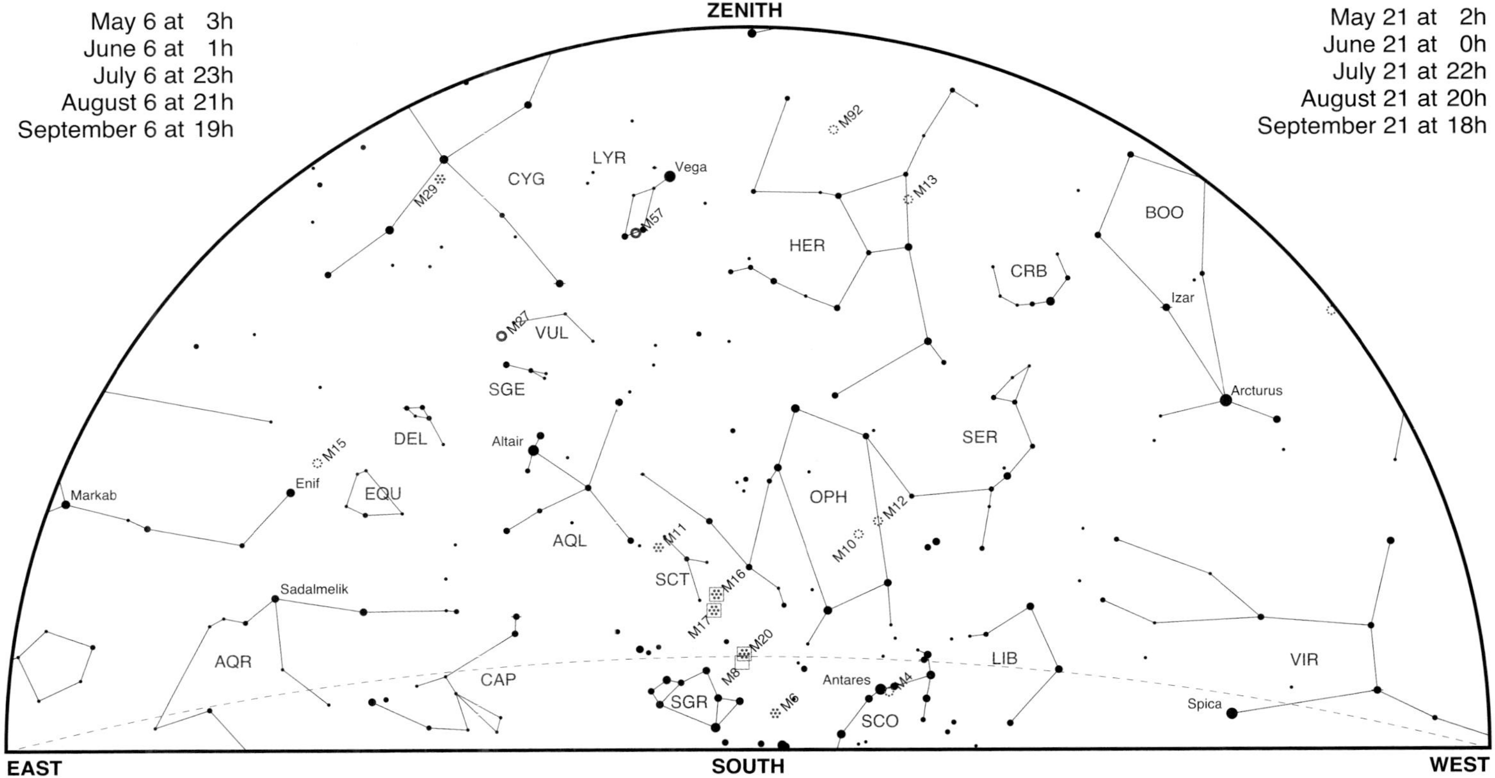

7S

8N

July 6 at 1h
August 6 at 23h
September 6 at 21h
October 6 at 19h
November 6 at 17h

July 21 at 0h
August 21 at 22h
September 21 at 20h
October 21 at 18h
November 21 at 16h

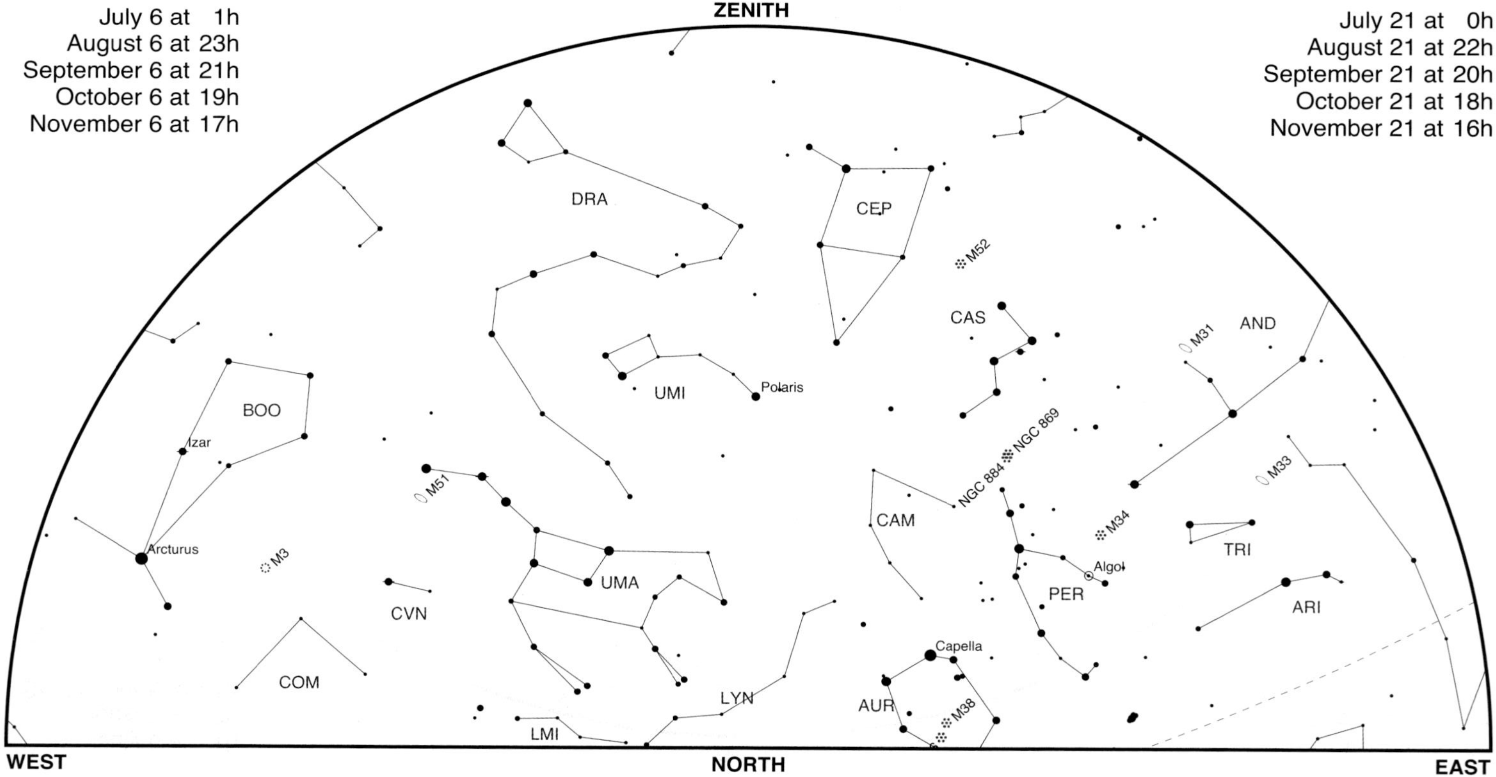

July 6 at 1h
August 6 at 23h
September 6 at 21h
October 6 at 19h
November 6 at 17h

July 21 at 0h
August 21 at 22h
September 21 at 20h
October 21 at 18h
November 21 at 16h

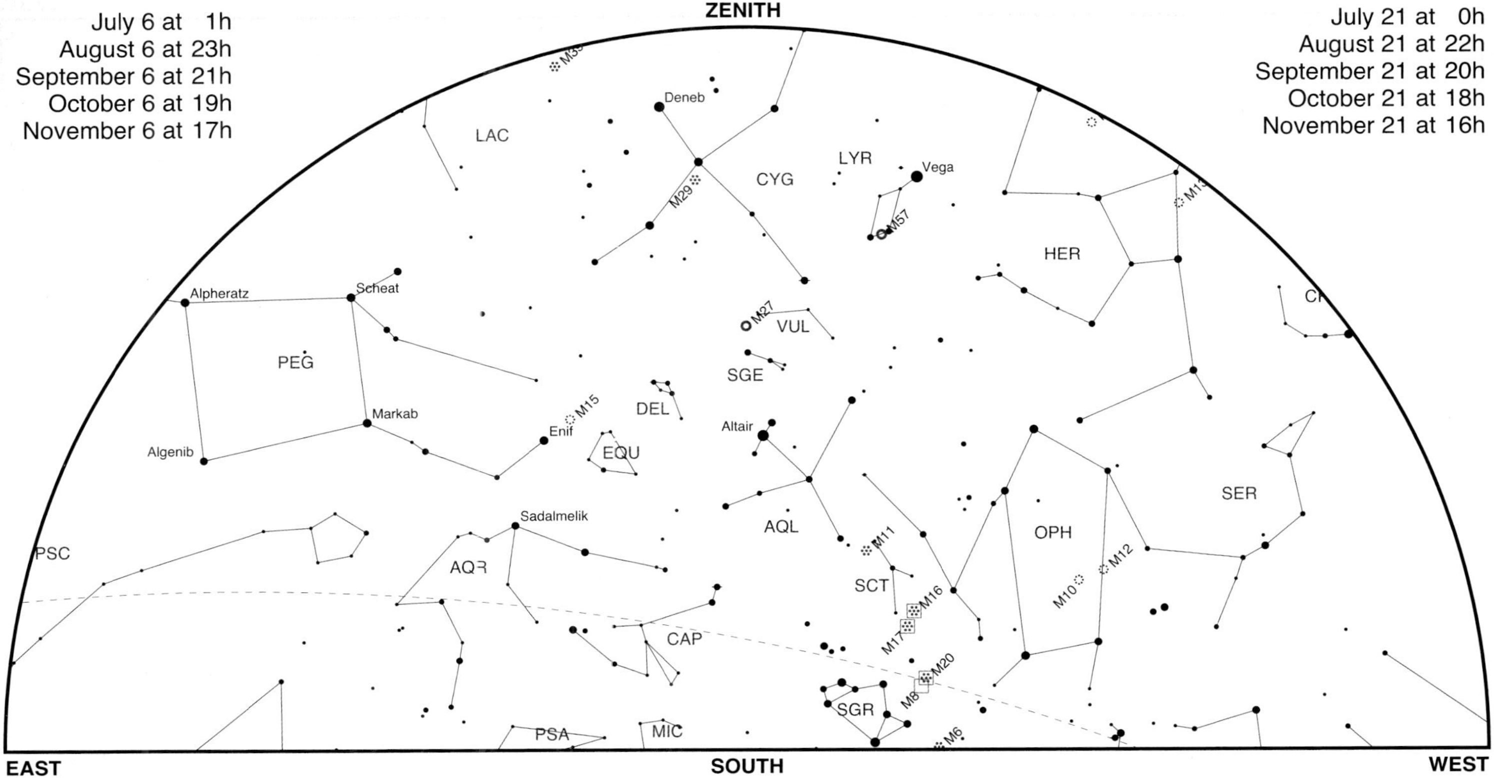

8S

9N

August 6 at 1h
September 6 at 23h
October 6 at 21h
November 6 at 19h
December 6 at 17h

August 21 at 0h
September 21 at 22h
October 21 at 20h
November 21 at 18h
December 21 at 16h

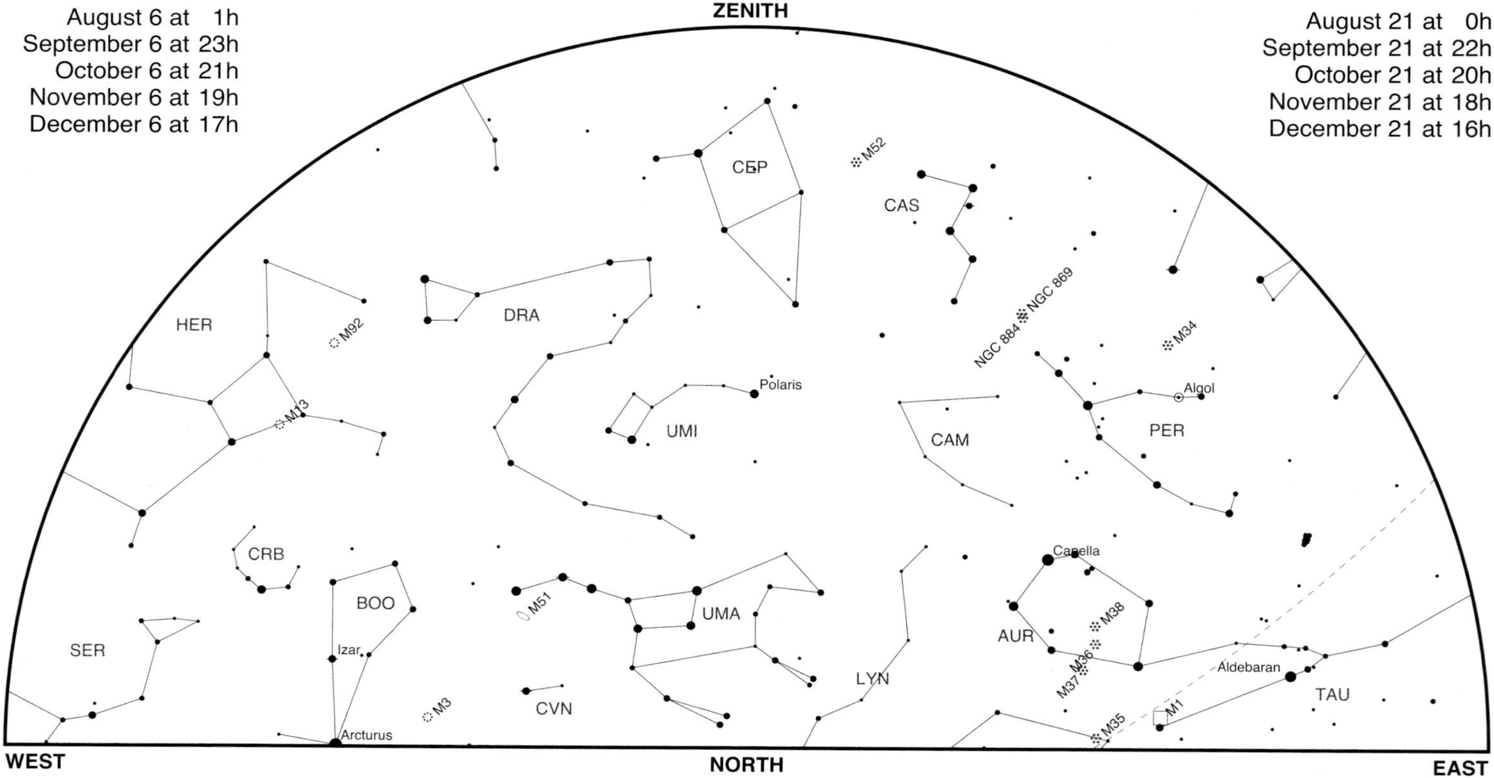

9S

August 6 at 1h
September 6 at 23h
October 6 at 21h
November 6 at 19h
December 6 at 17h

August 21 at 0h
September 21 at 22h
October 21 at 20h
November 21 at 18h
December 21 at 16h

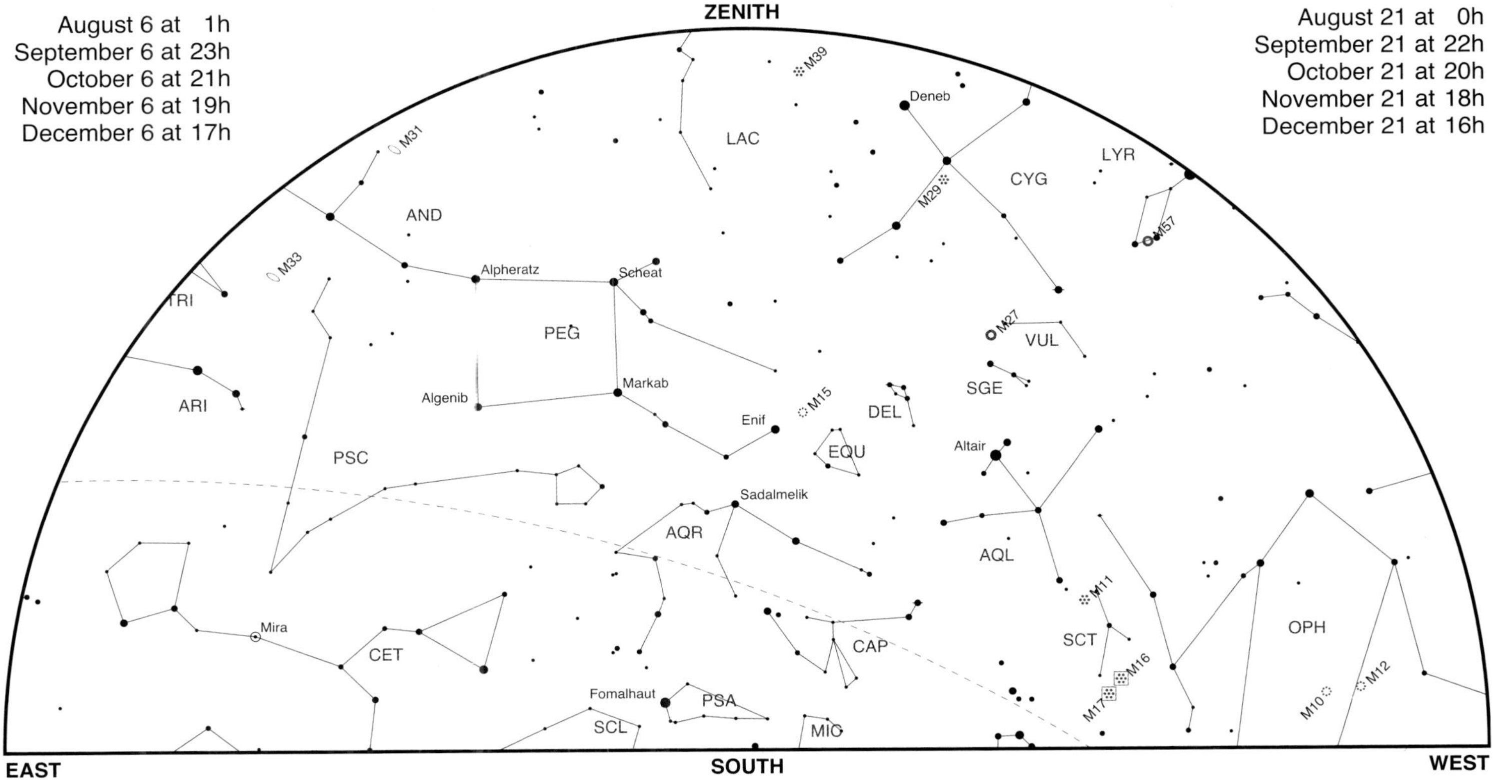

10N

August 6 at 3h
September 6 at 1h
October 6 at 23h
November 6 at 21h
December 6 at 19h

August 21 at 2h
September 21 at 0h
October 21 at 22h
November 21 at 20h
December 21 at 18h

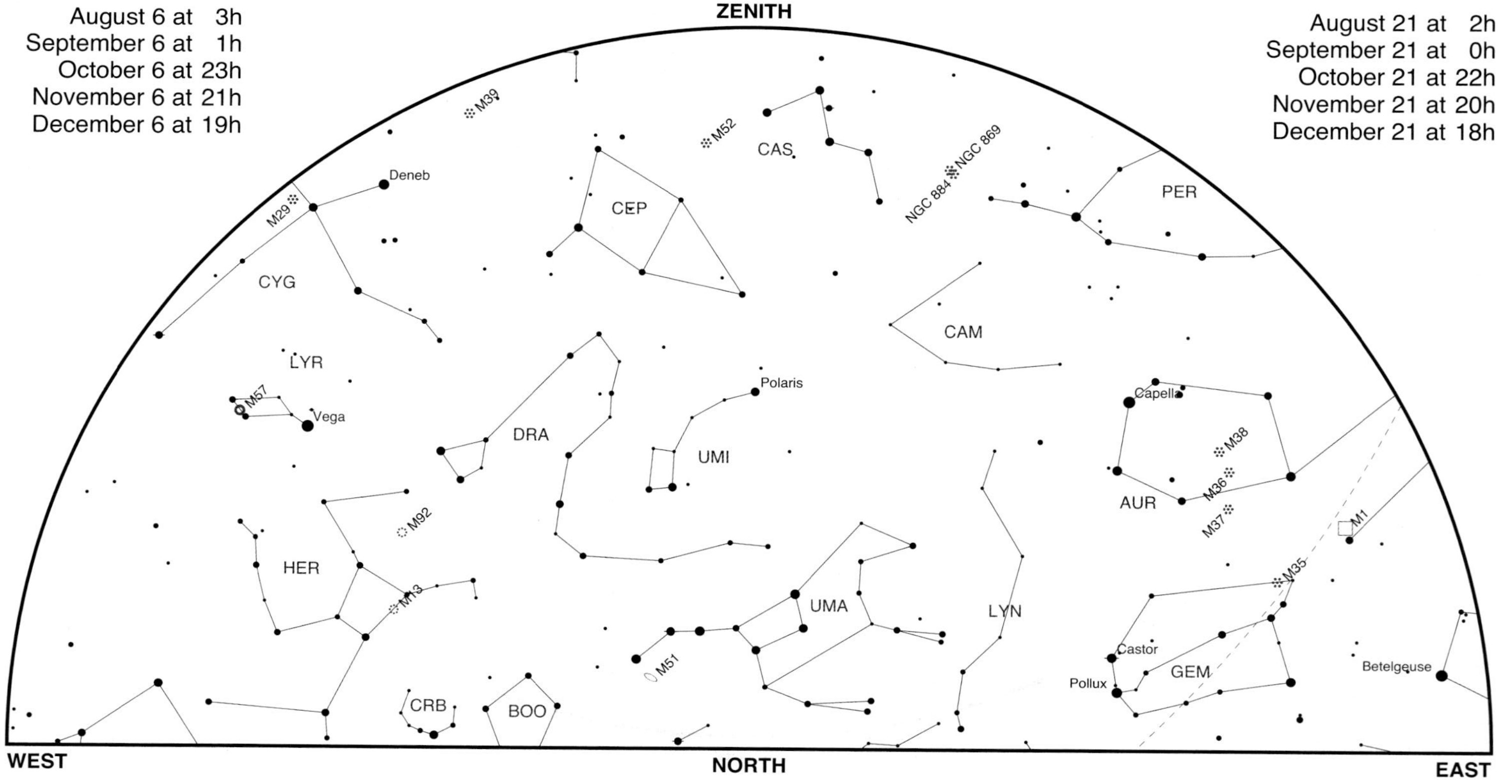

August 6 at 3h
September 6 at 1h
October 6 at 23h
November 6 at 21h
December 6 at 19h

August 21 at 2h
September 21 at 0h
October 21 at 22h
November 21 at 20h
December 21 at 18h

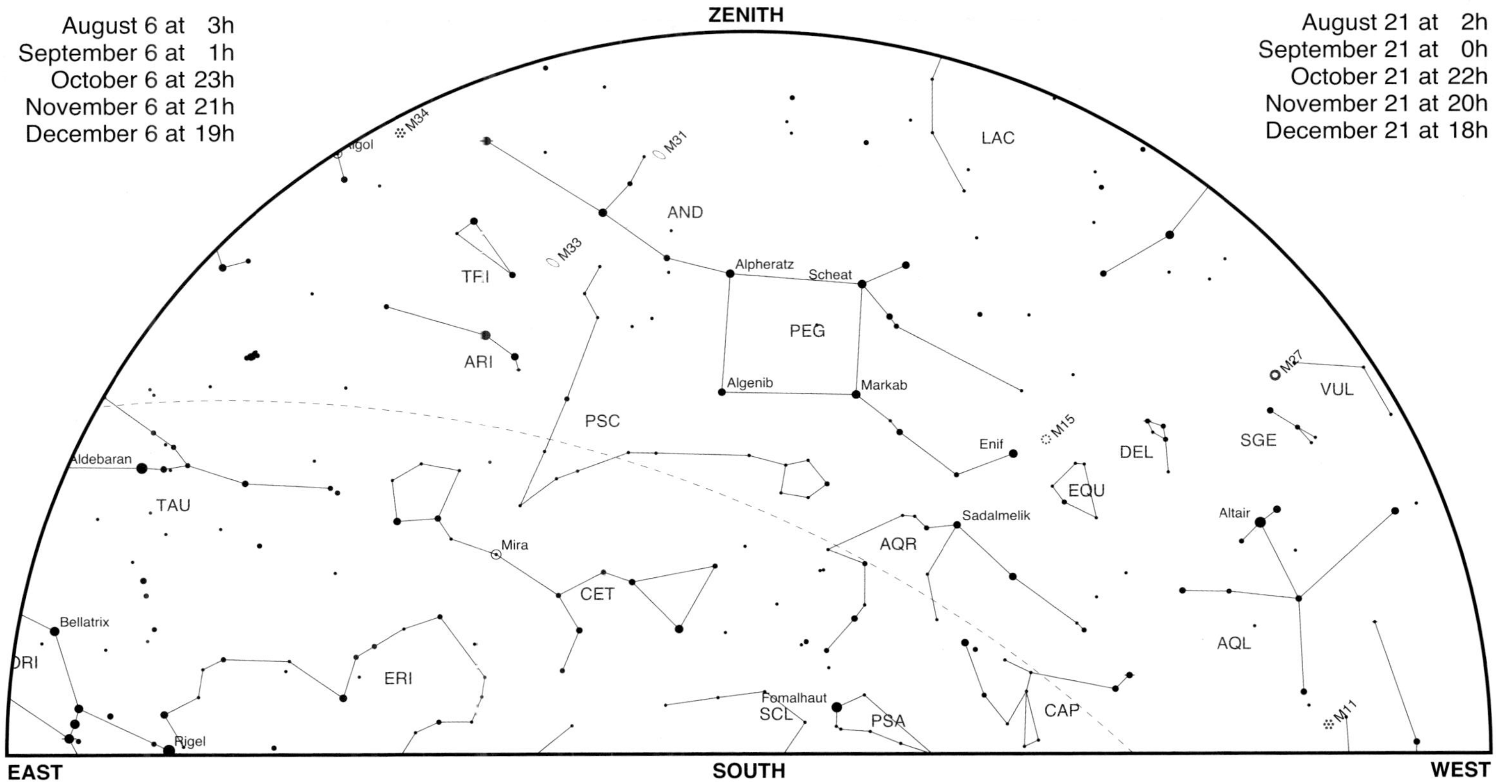

10S

11N

September 6 at 3h
October 6 at 1h
November 6 at 23h
December 6 at 21h
January 6 at 19h

September 21 at 2h
October 21 at 0h
November 21 at 22h
December 21 at 20h
January 21 at 18h

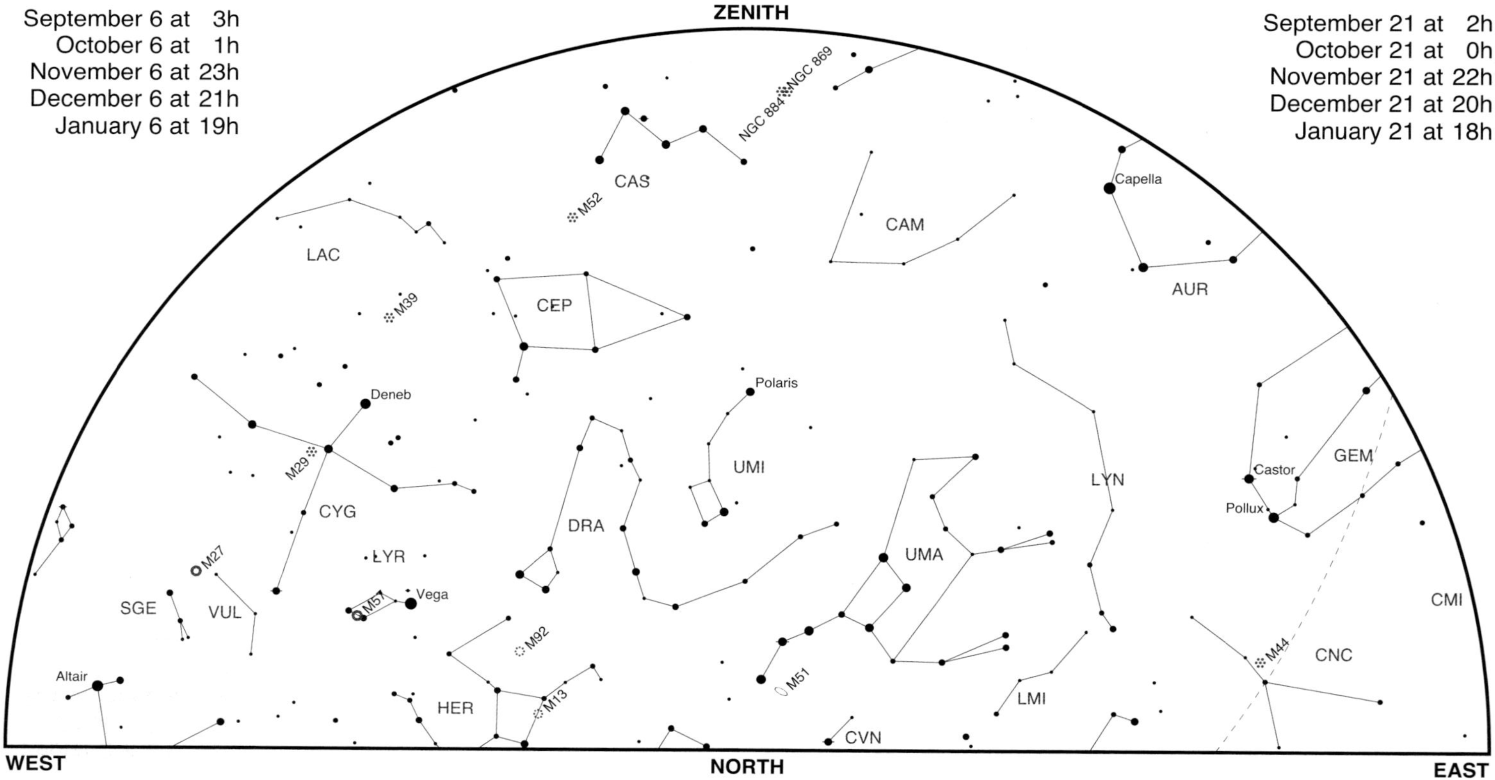

11S

September 6 at 3h
October 6 at 1h
November 6 at 23h
December 6 at 21h
January 6 at 19h

September 21 at 2h
October 21 at 0h
November 21 at 22h
December 21 at 20h
January 21 at 18h

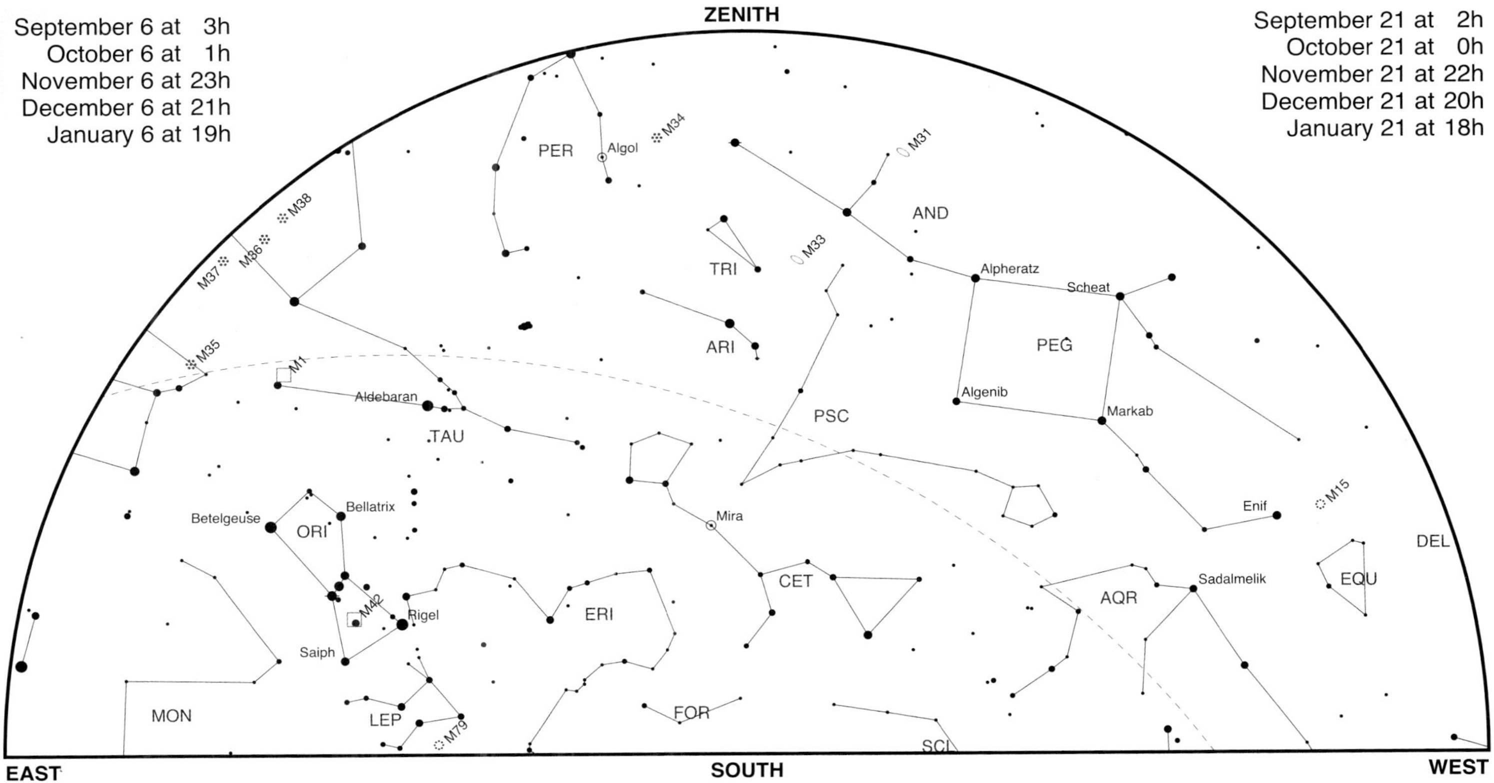

12N

October 6 at 3h
November 6 at 1h
December 6 at 23h
January 6 at 21h
February 6 at 19h

October 21 at 2h
November 21 at 0h
December 21 at 22h
January 21 at 20h
February 21 at 18h

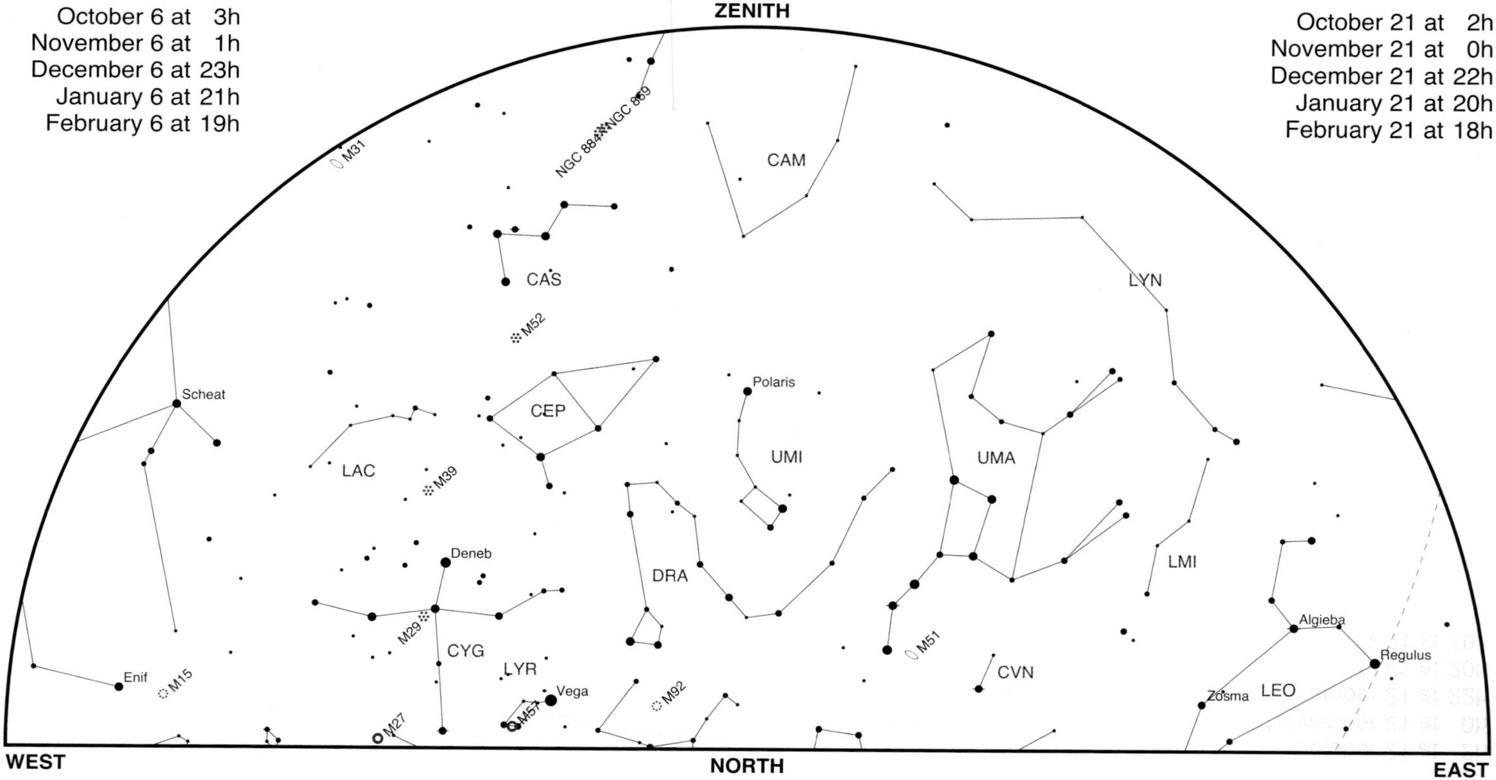

12S

October 6 at 3h
November 6 at 1h
December 6 at 23h
January 6 at 21h
February 6 at 19h

October 21 at 2h
November 21 at 0h
December 21 at 22h
January 21 at 20h
February 21 at 18h

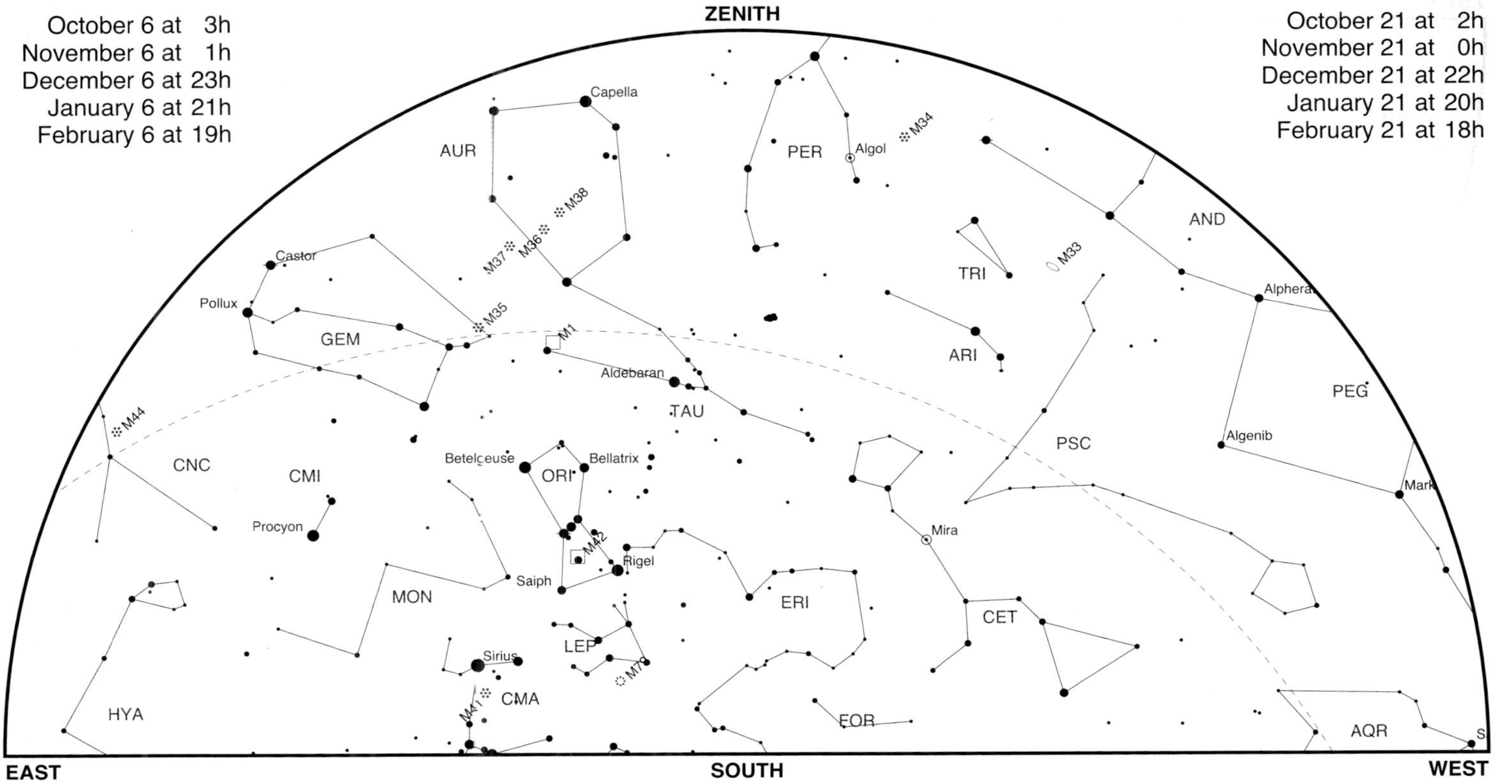

Southern Hemisphere Star Charts

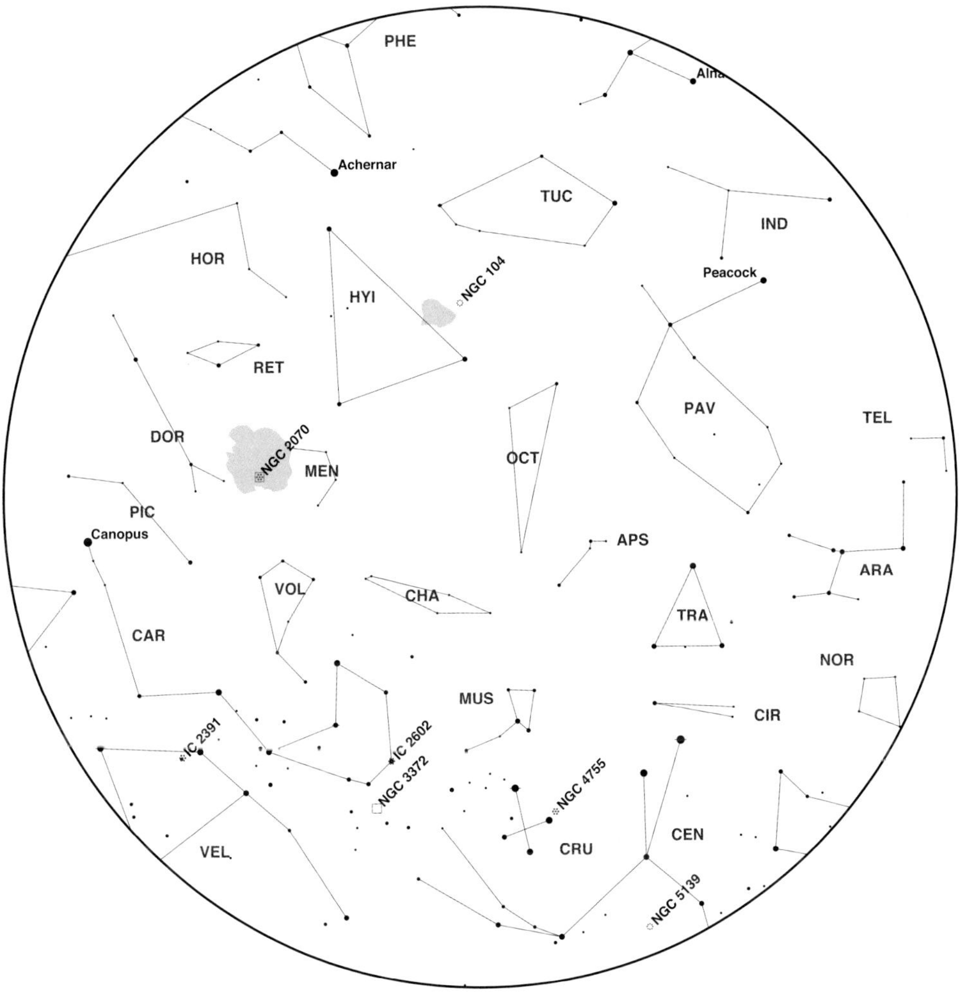

This chart shows stars lying at declinations between − 45 and − 90 degrees. These constellations are circumpolar for observers in Australia and New Zealand.

1N

October 6 at 5h
November 6 at 3h
December 6 at 1h
January 6 at 23h
February 6 at 21h

October 21 at 4h
November 21 at 2h
December 21 at 0h
January 21 at 22h
February 21 at 20h

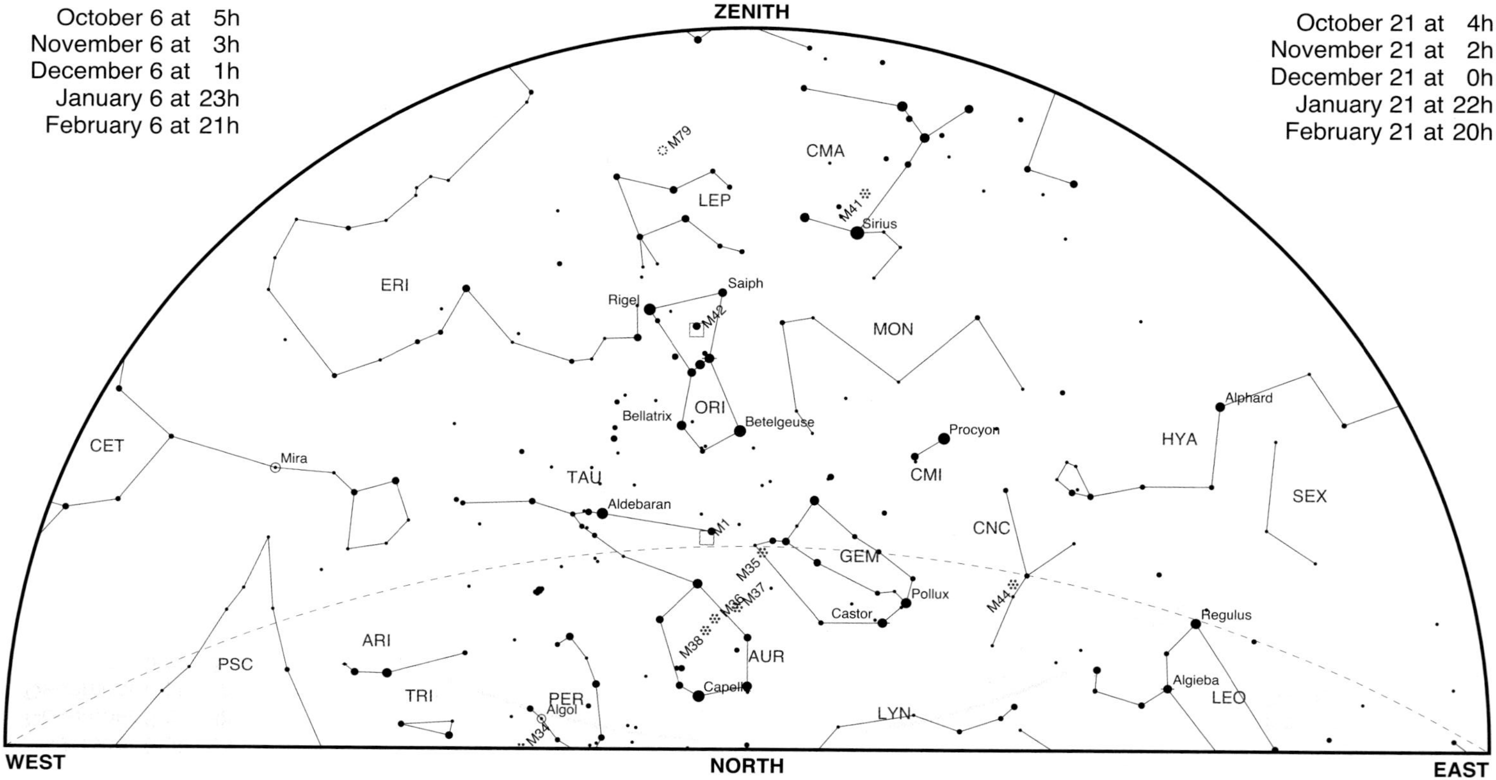

October 6 at 5h
November 6 at 3h
December 6 at 1h
January 6 at 23h
February 6 at 21h

October 21 at 4h
November 21 at 2h
December 21 at 0h
January 21 at 22h
February 21 at 20h

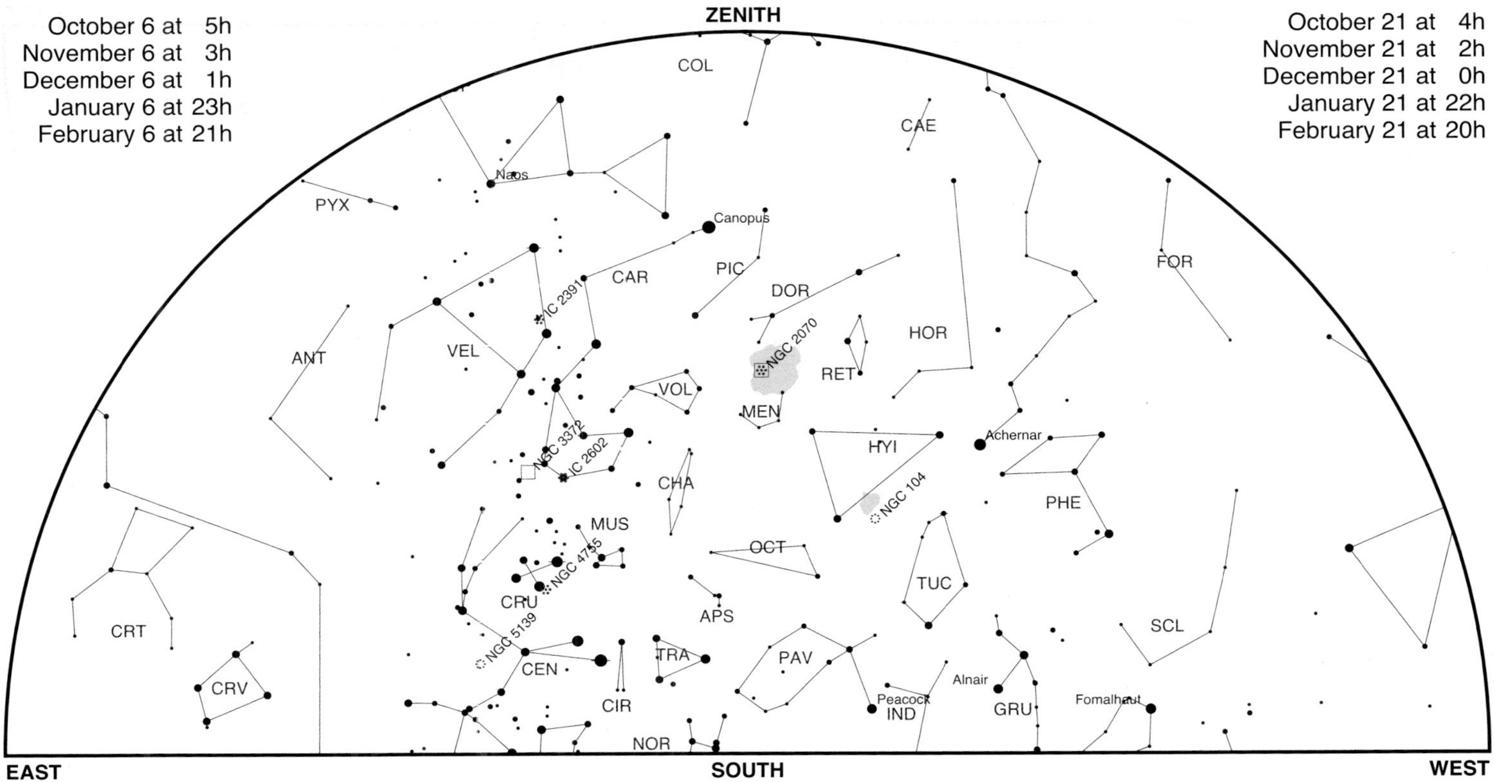

1S

2N

November 6 at 5h
December 6 at 3h
January 6 at 1h
February 6 at 23h
March 6 at 21h

November 21 at 4h
December 21 at 2h
January 21 at 0h
February 21 at 22h
March 21 at 20h

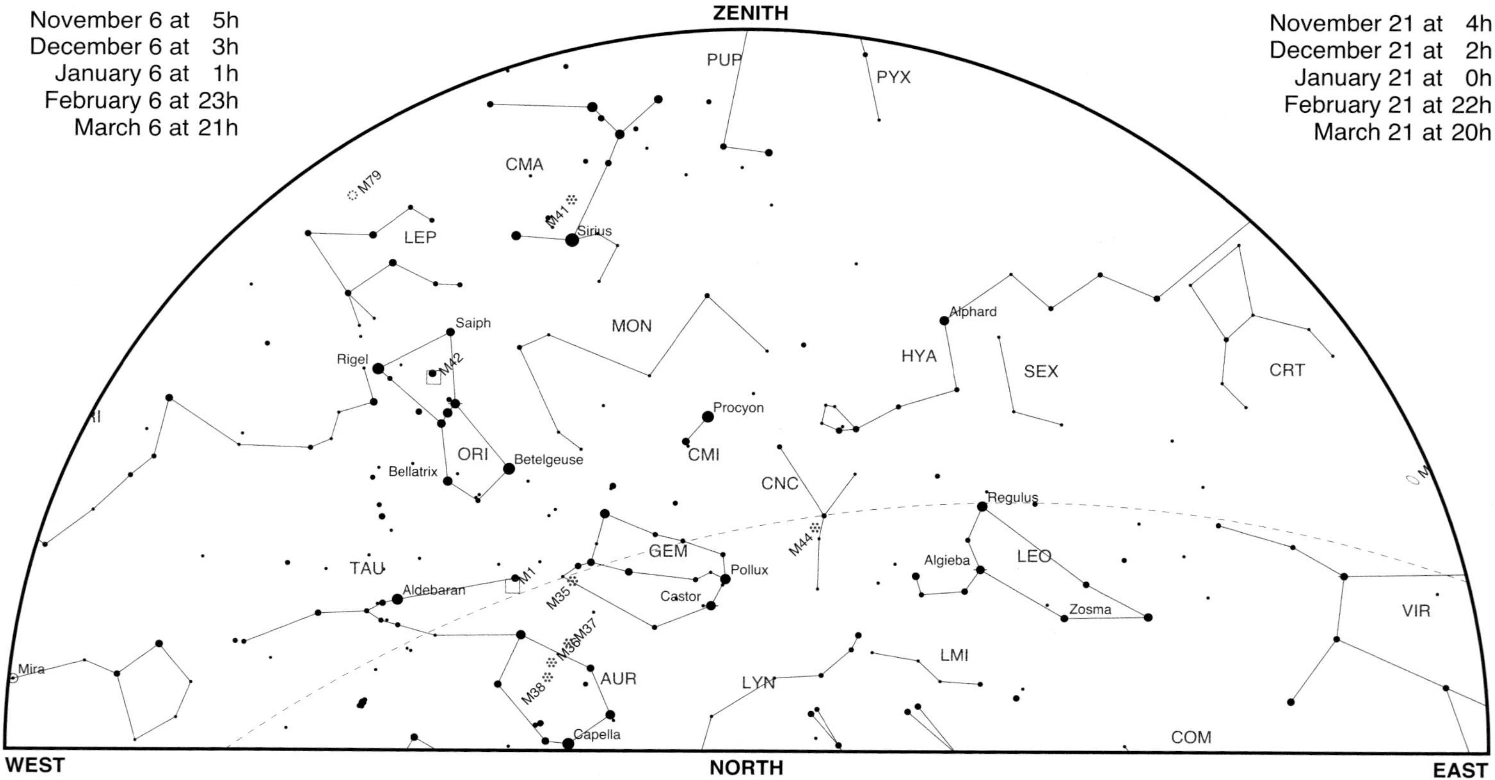

2S

November 6 at 5h
December 6 at 3h
January 6 at 1h
February 6 at 23h
March 6 at 21h

November 21 at 4h
December 21 at 2h
January 21 at 0h
February 21 at 22h
March 21 at 20h

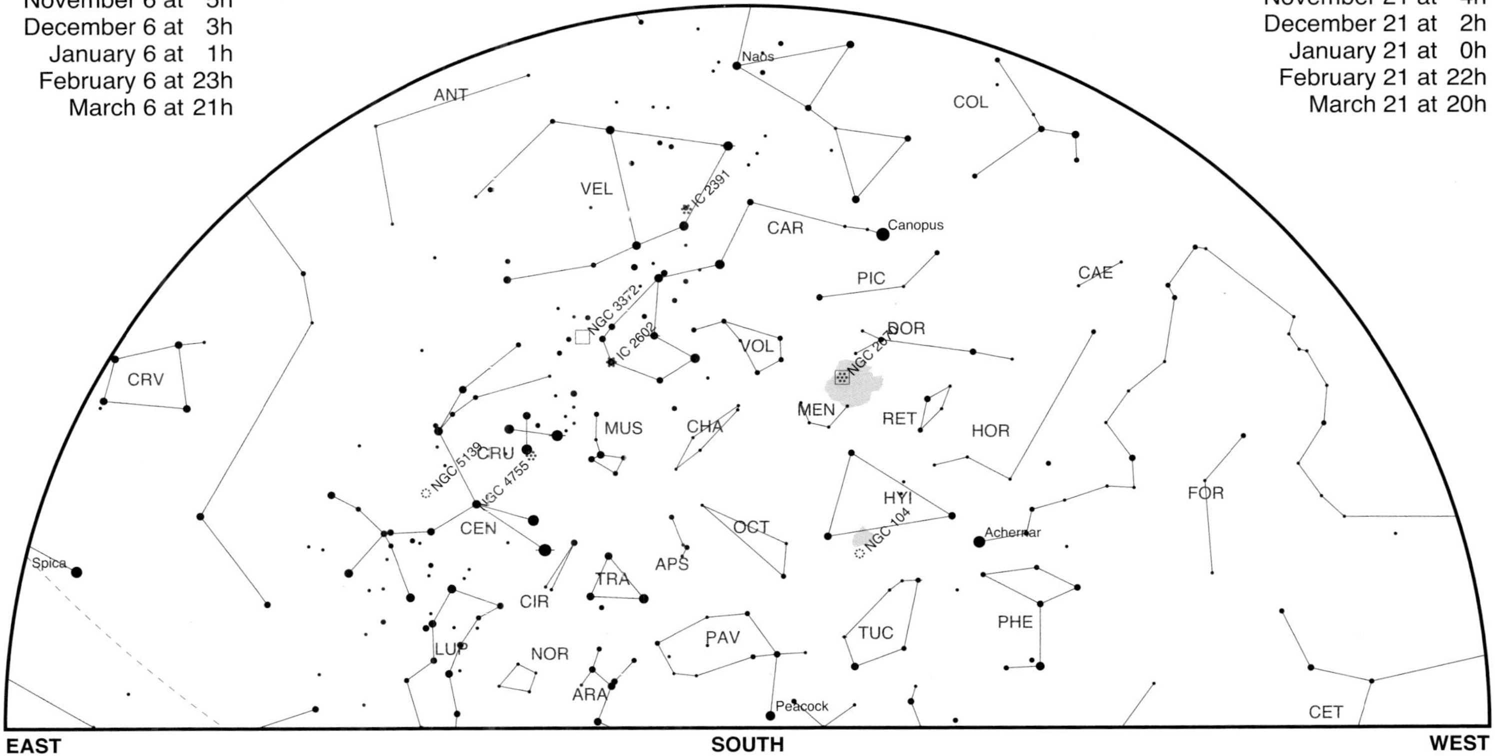

3N

January 6 at 3h
February 6 at 1h
March 6 at 23h
April 6 at 21h
May 6 at 19h

January 21 at 2h
February 21 at 0h
March 21 at 22h
April 21 at 20h
May 21 at 18h

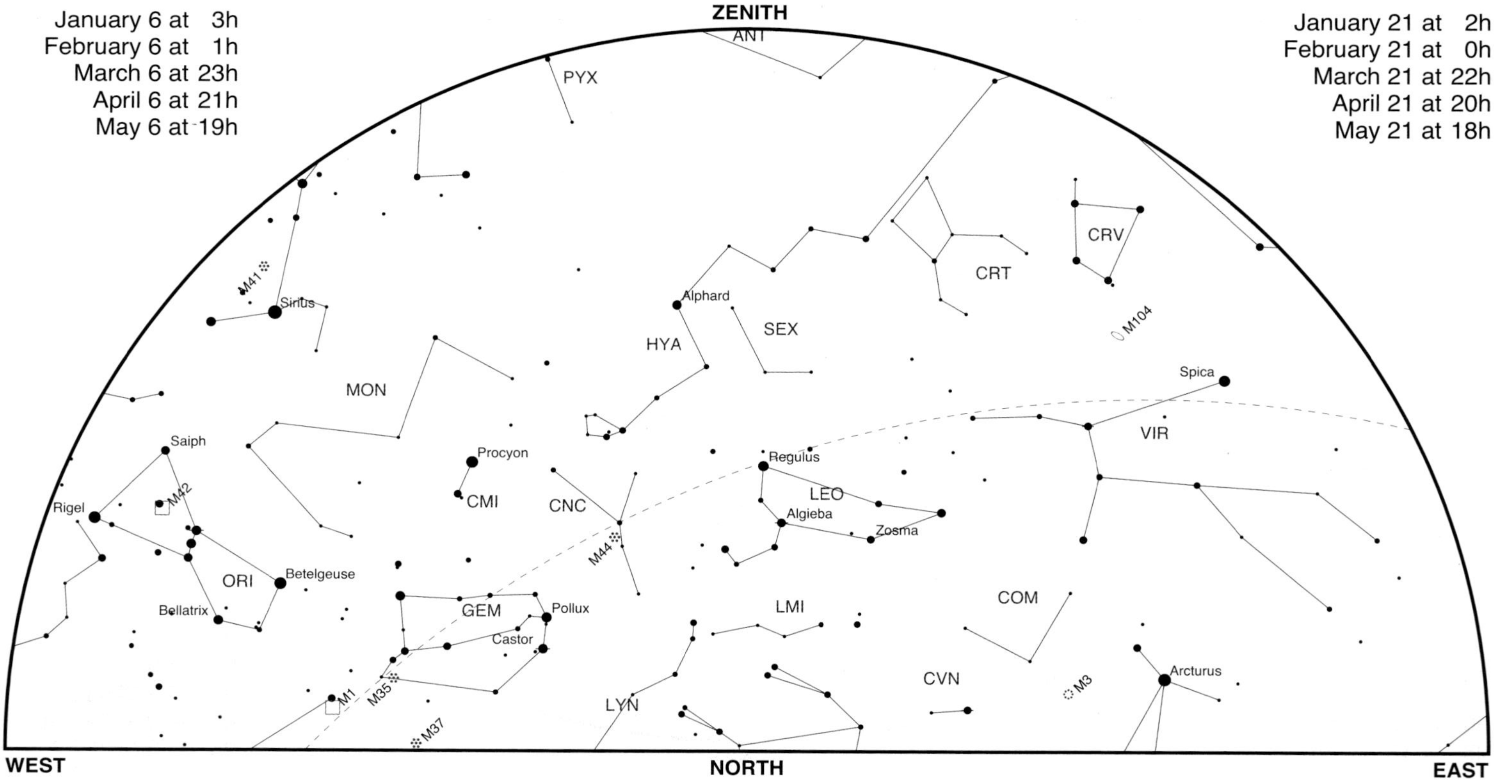

January 6 at 3h
February 6 at 1h
March 6 at 23h
April 6 at 21h
May 6 at 19h

January 21 at 2h
February 21 at 0h
March 21 at 22h
April 21 at 20h
May 21 at 18h

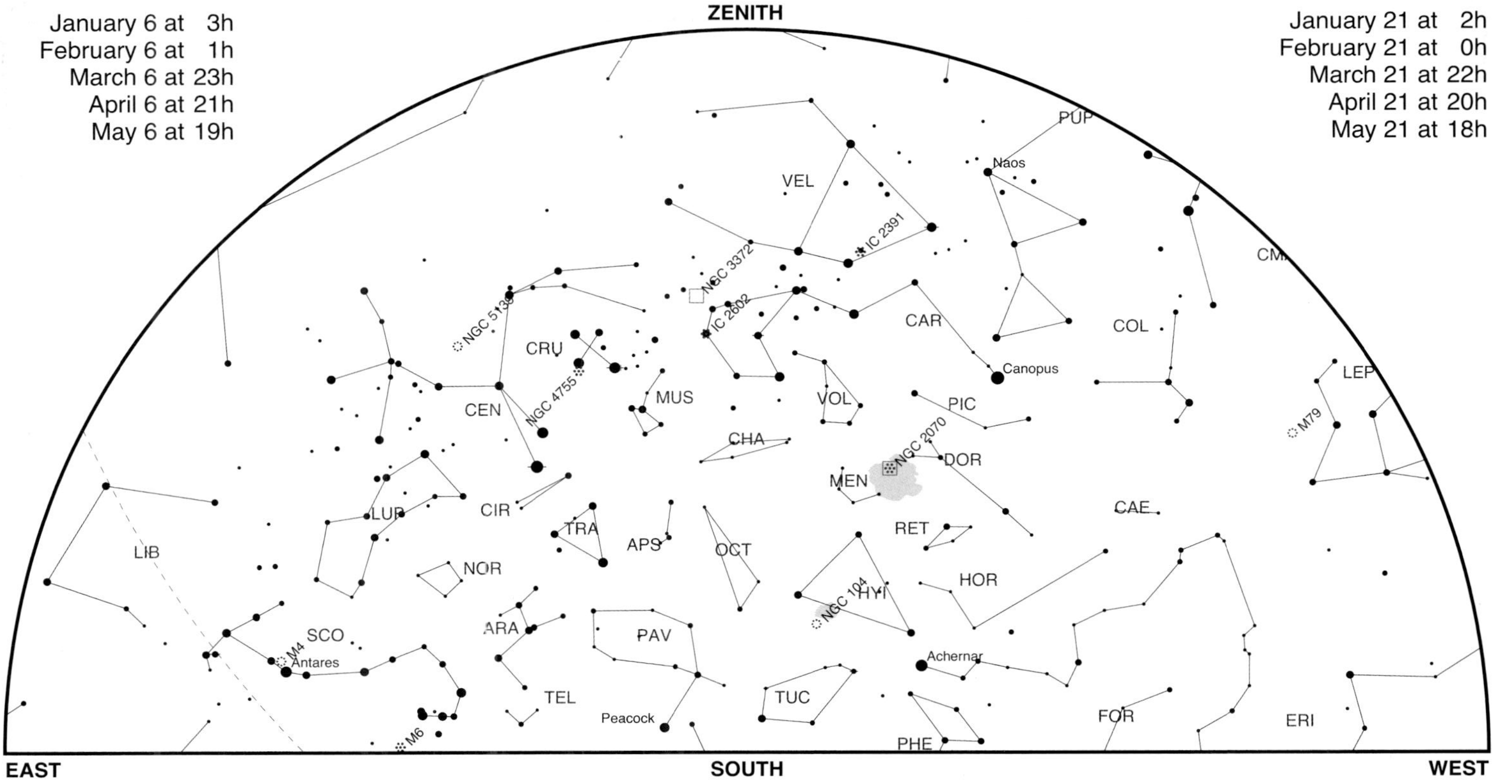

3S

4N

February 6 at 3h
March 6 at 1h
April 6 at 23h
May 6 at 21h
June 6 at 19h

February 21 at 2h
March 21 at 0h
April 21 at 22h
May 21 at 20h
June 21 at 18h

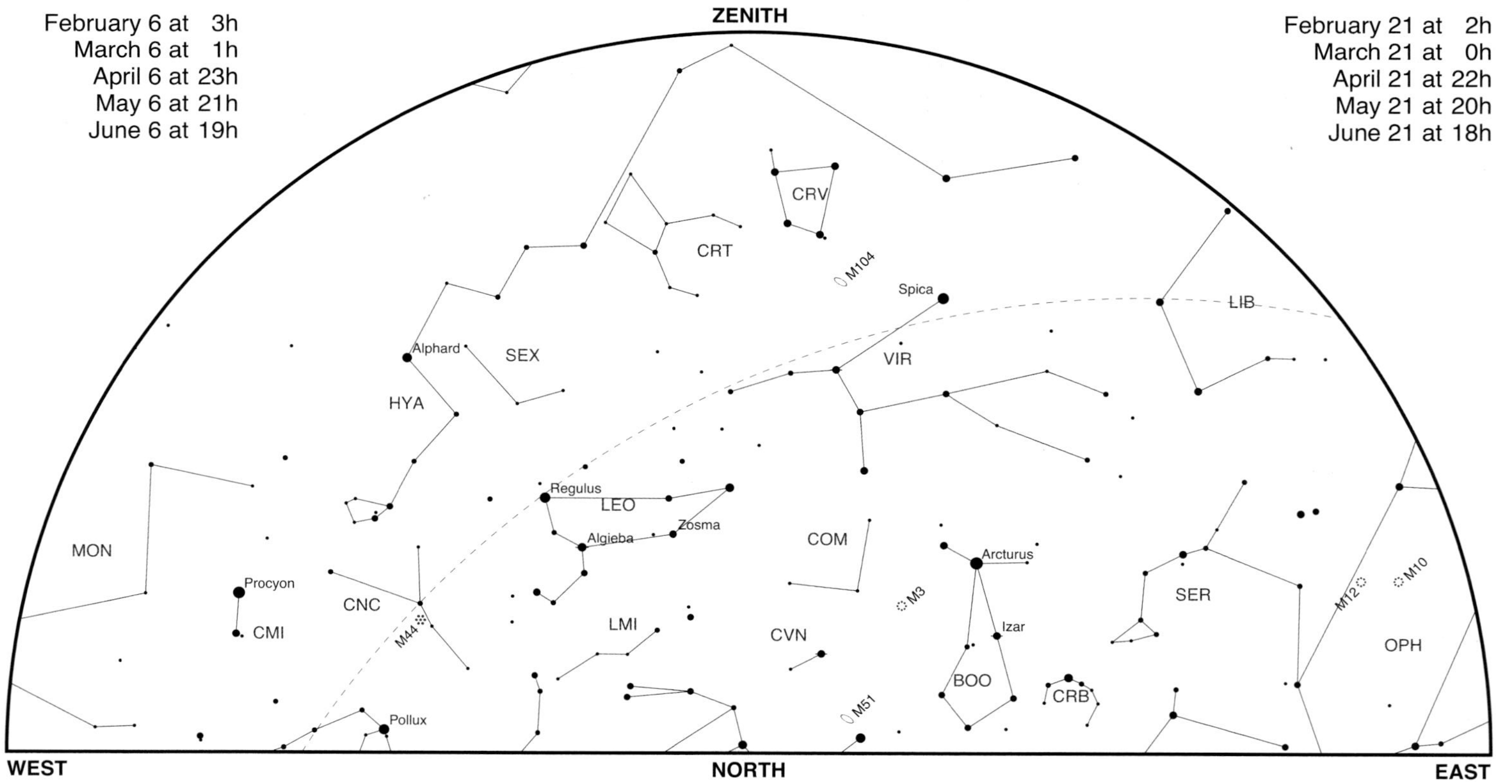

February 6 at 3h
March 6 at 1h
April 6 at 23h
May 6 at 21h
June 6 at 19h

February 21 at 2h
March 21 at 0h
April 21 at 22h
May 21 at 20h
June 21 at 18h

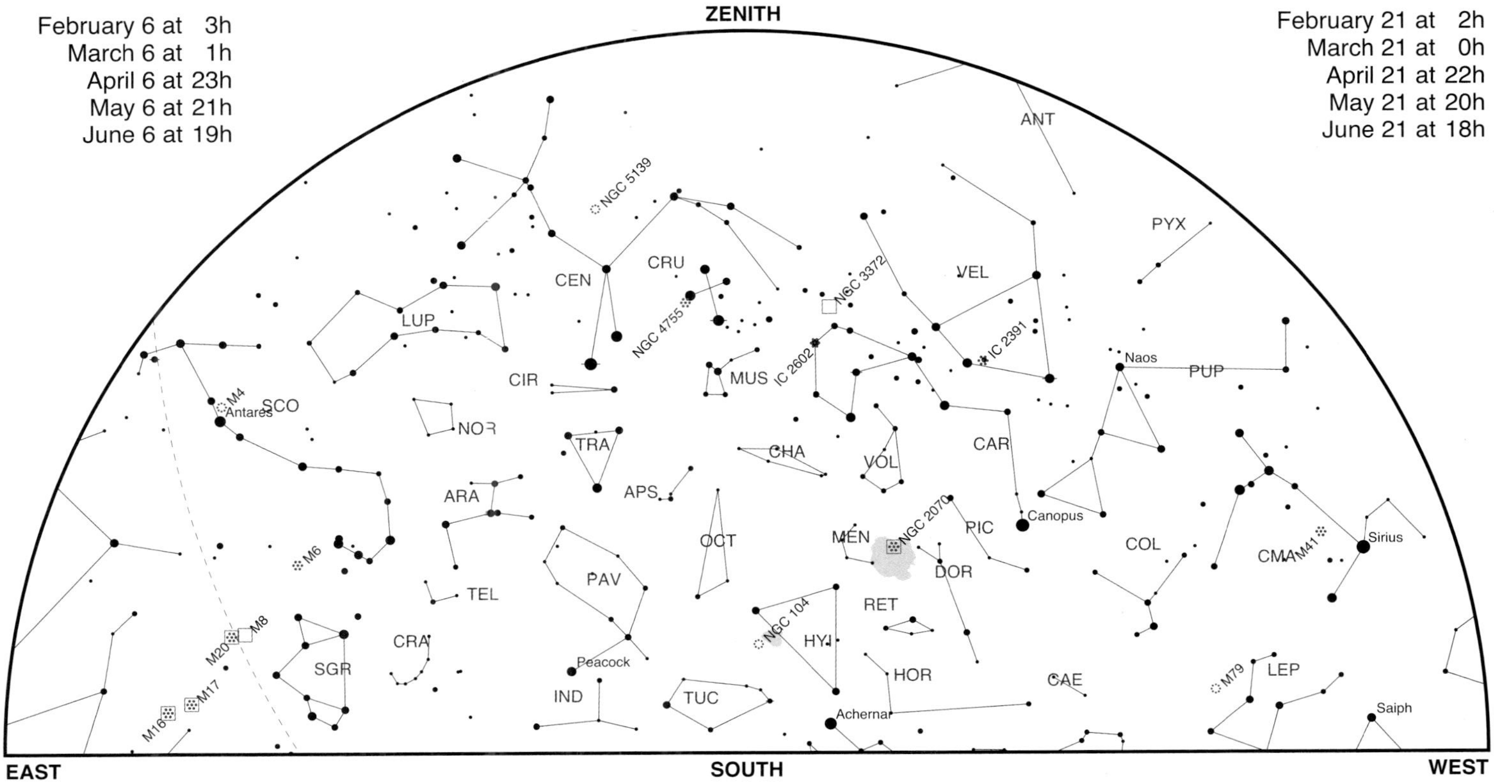

4S

5N

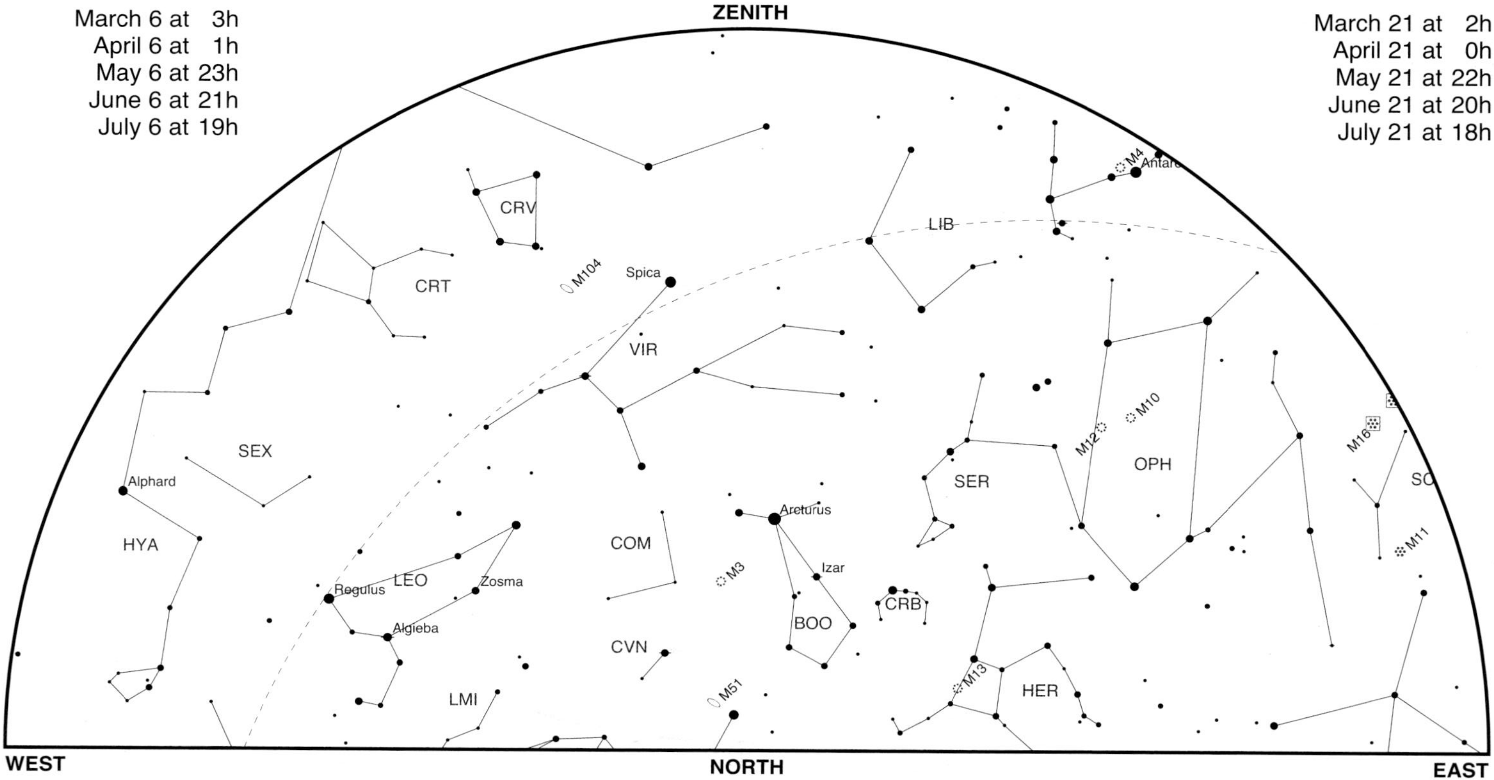
March 6 at 3h
April 6 at 1h
May 6 at 23h
June 6 at 21h
July 6 at 19h
March 21 at 2h
April 21 at 0h
May 21 at 22h
June 21 at 20h
July 21 at 18h
ZENITH
WEST
NORTH
EAST
CRV
CRT
M104
Spica
LIB
M4
Antares
VIR
SEX
Alphard
HYA
LEO
Regulus
Zosma
Algieba
LMI
COM
CVN
M3
M51
Arcturus
Izar
BOO
CRB
SER
M12
M10
OPH
M16
SC
M11
M13
HER

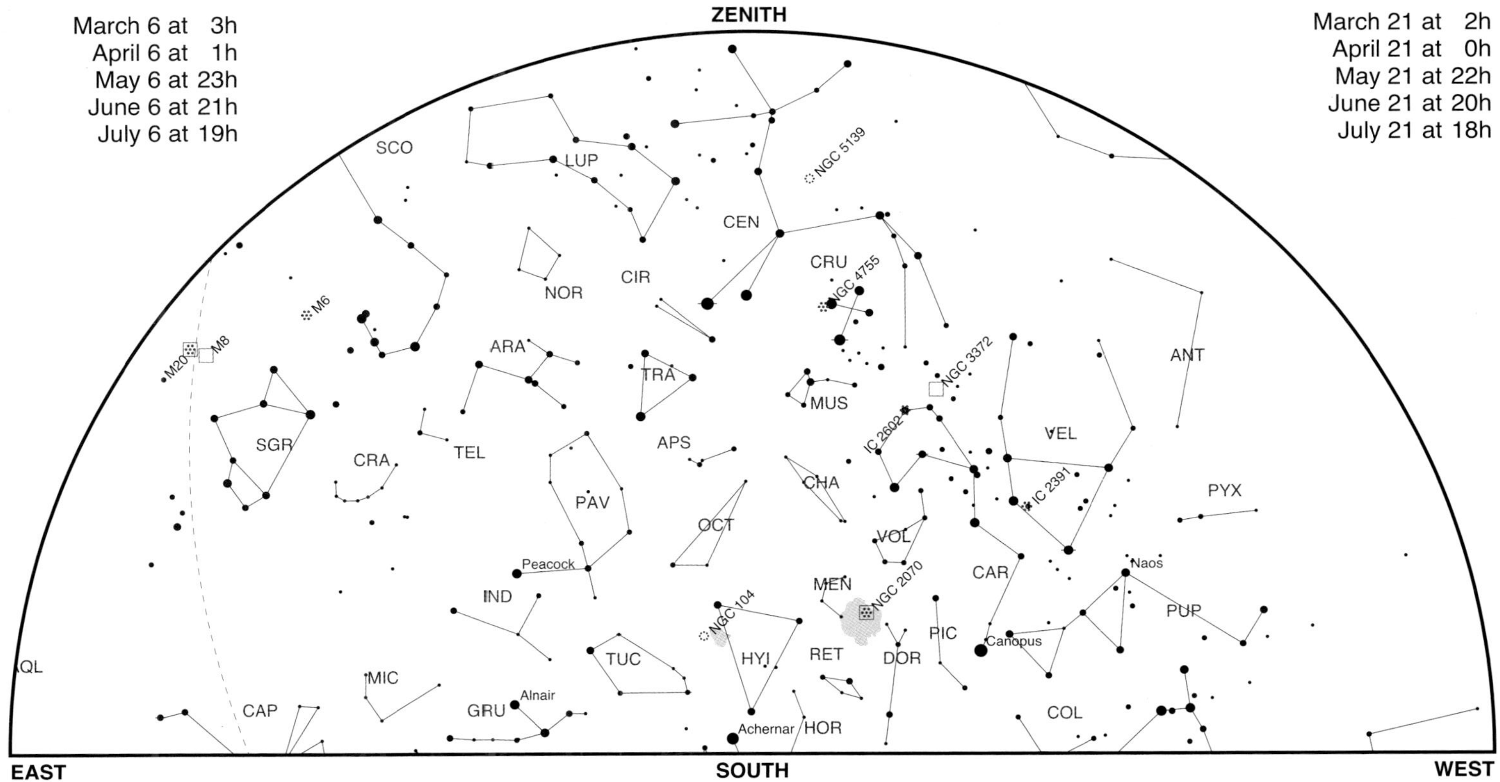
March 6 at 3h
April 6 at 1h
May 6 at 23h
June 6 at 21h
July 6 at 19h
ZENITH
March 21 at 2h
April 21 at 0h
May 21 at 22h
June 21 at 20h
July 21 at 18h
SCO
LUP
NGC 5139
CEN
CRU
NGC 4755
CIR
NOR
M6
M20
M8
ARA
TRA
MUS
NGC 3372
ANT
IC 2602
VEL
SGR
CRA
TEL
APS
CHA
IC 2391
PYX
PAV
OCT
VOL
Peacock
CAR
Naos
IND
MEN
NGC 2070
NGC 104
PUP
PIC
Canopus
TUC
HYI
RET
DOR
QL
MIC
Alnair
CAP
GRU
COL
Achernar
HOR
EAST
SOUTH
WEST

5S

6N

March 6 at 5h
April 6 at 3h
May 6 at 1h
June 6 at 23h
July 6 at 21h

March 21 at 4h
April 21 at 2h
May 21 at 0h
June 21 at 22h
July 21 at 20h

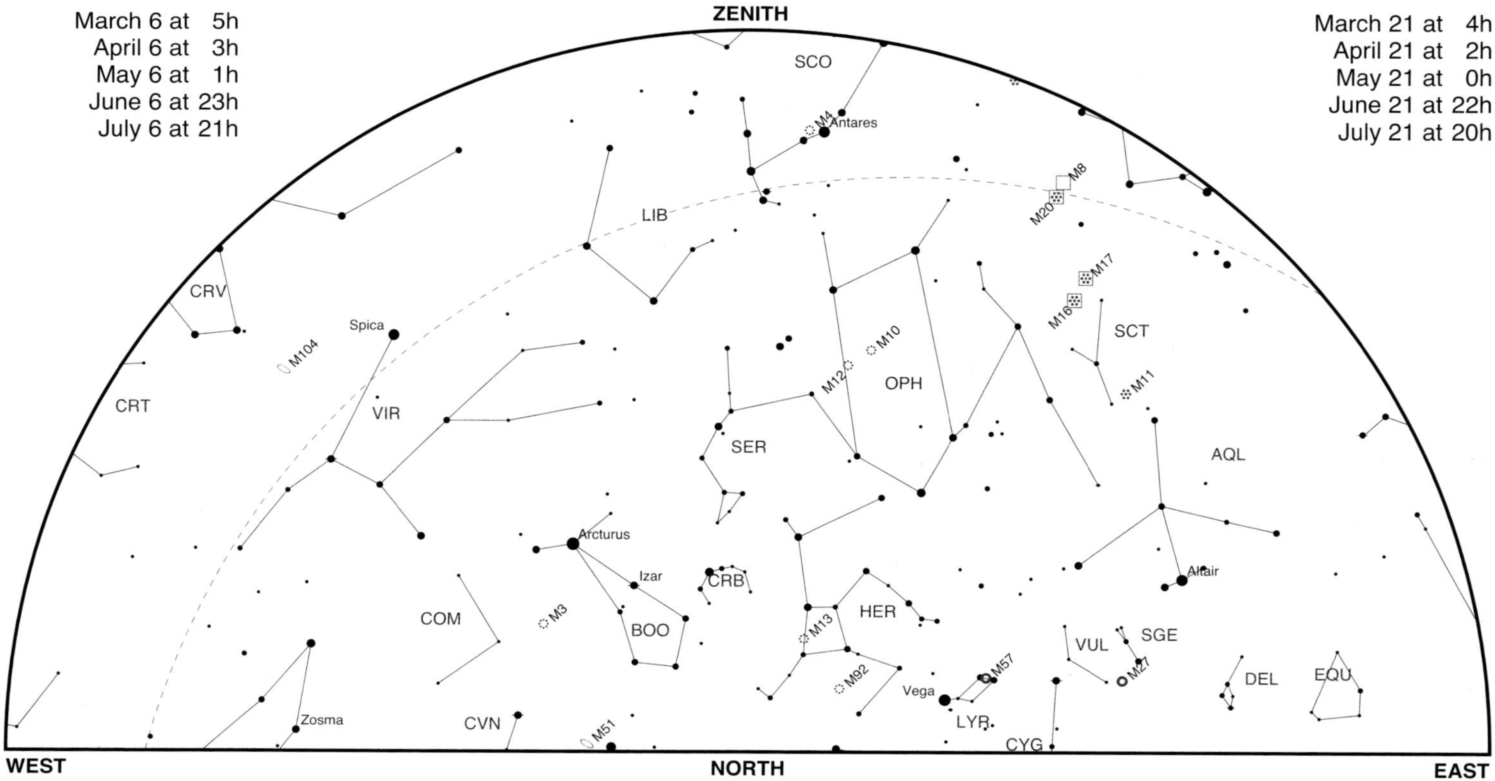

6S

March 6 at 5h	March 21 at 4h
April 6 at 3h	April 21 at 2h
May 6 at 1h	May 21 at 0h
June 6 at 23h	June 21 at 22h
July 6 at 21h	July 21 at 20h

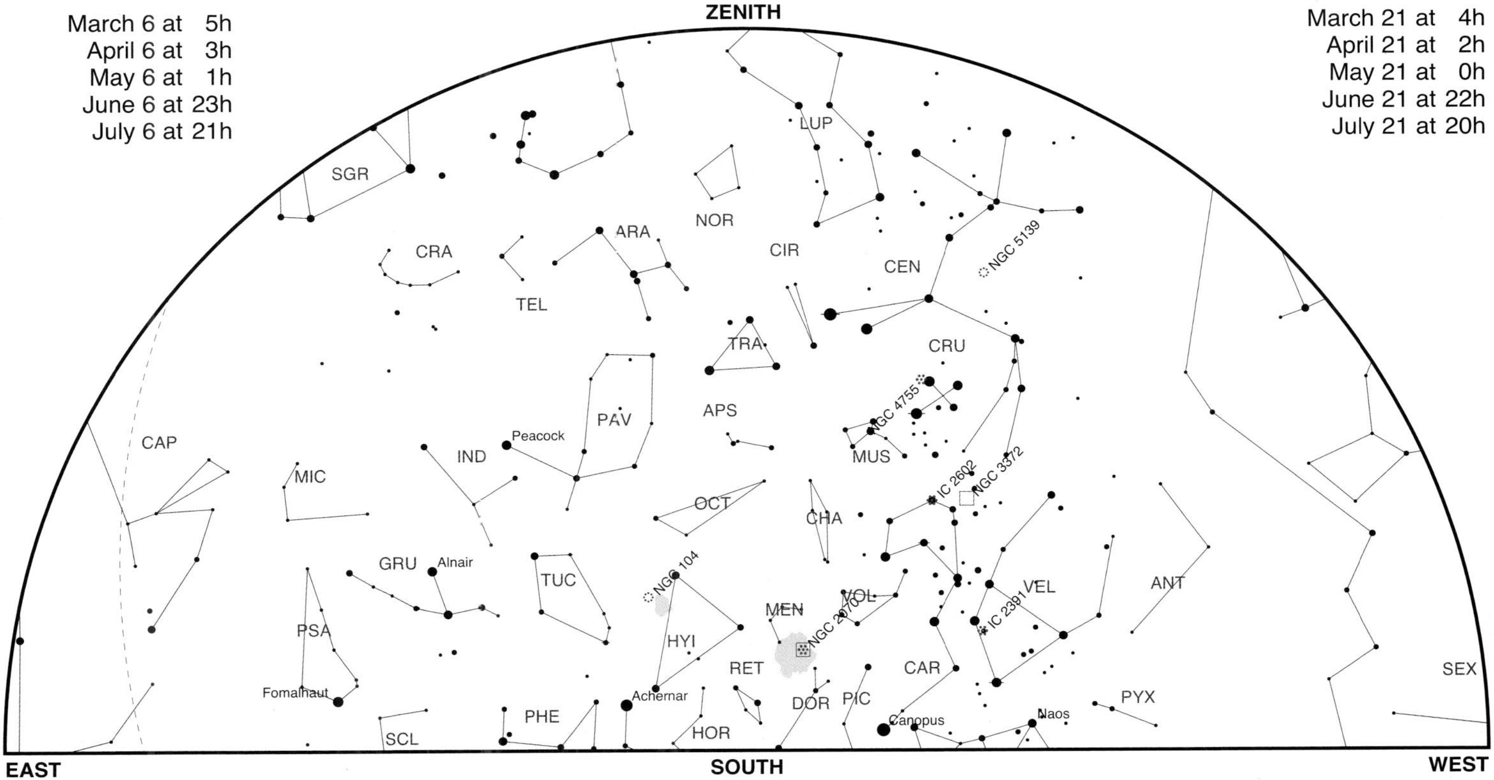

7N

April 6 at 5h
May 6 at 3h
June 6 at 1h
July 6 at 23h
August 6 at 21h

April 21 at 4h
May 21 at 2h
June 21 at 0h
July 21 at 22h
August 21 at 20h

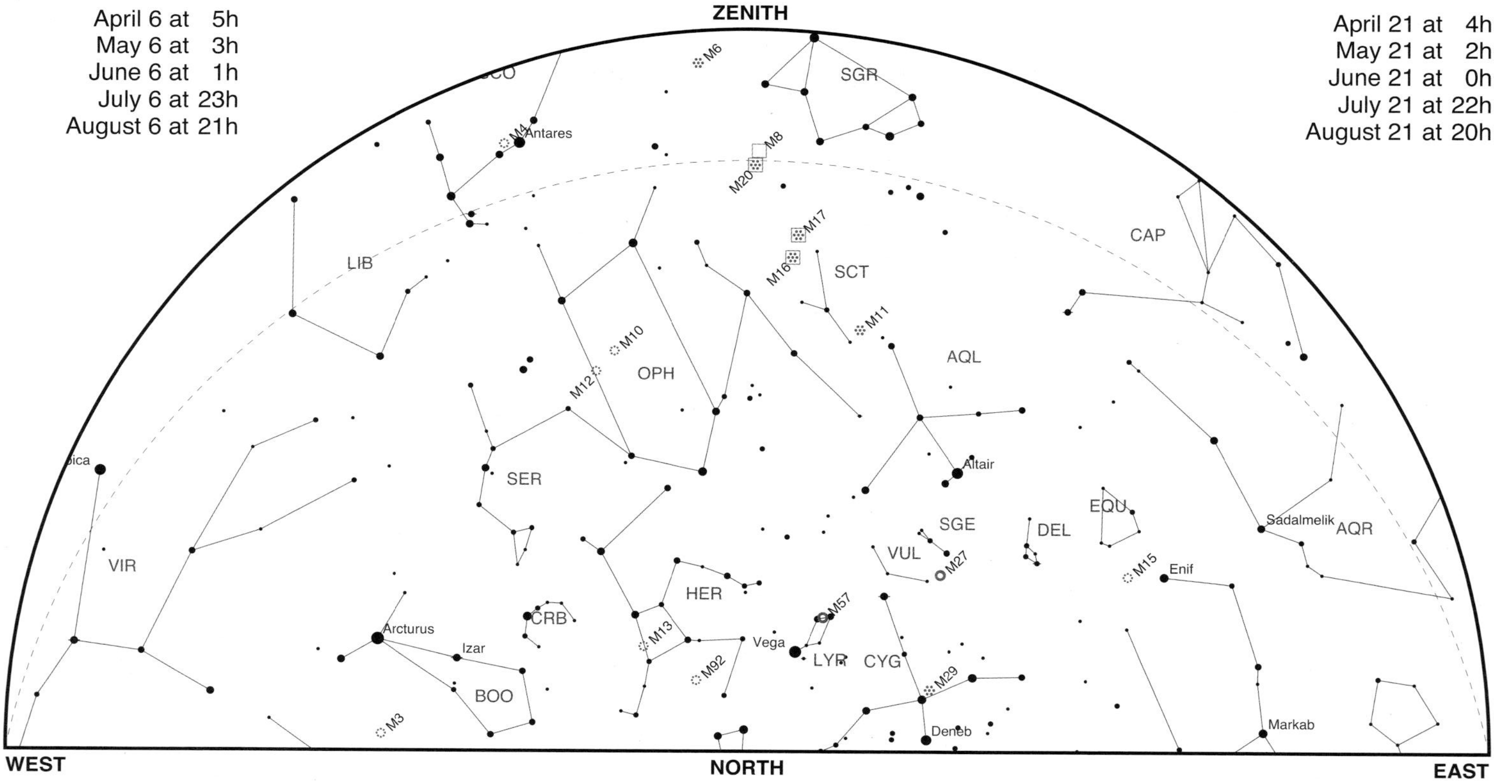

April 6 at 5h
May 6 at 3h
June 6 at 1h
July 6 at 23h
August 6 at 21h

April 21 at 4h
May 21 at 2h
June 21 at 0h
July 21 at 22h
August 21 at 20h

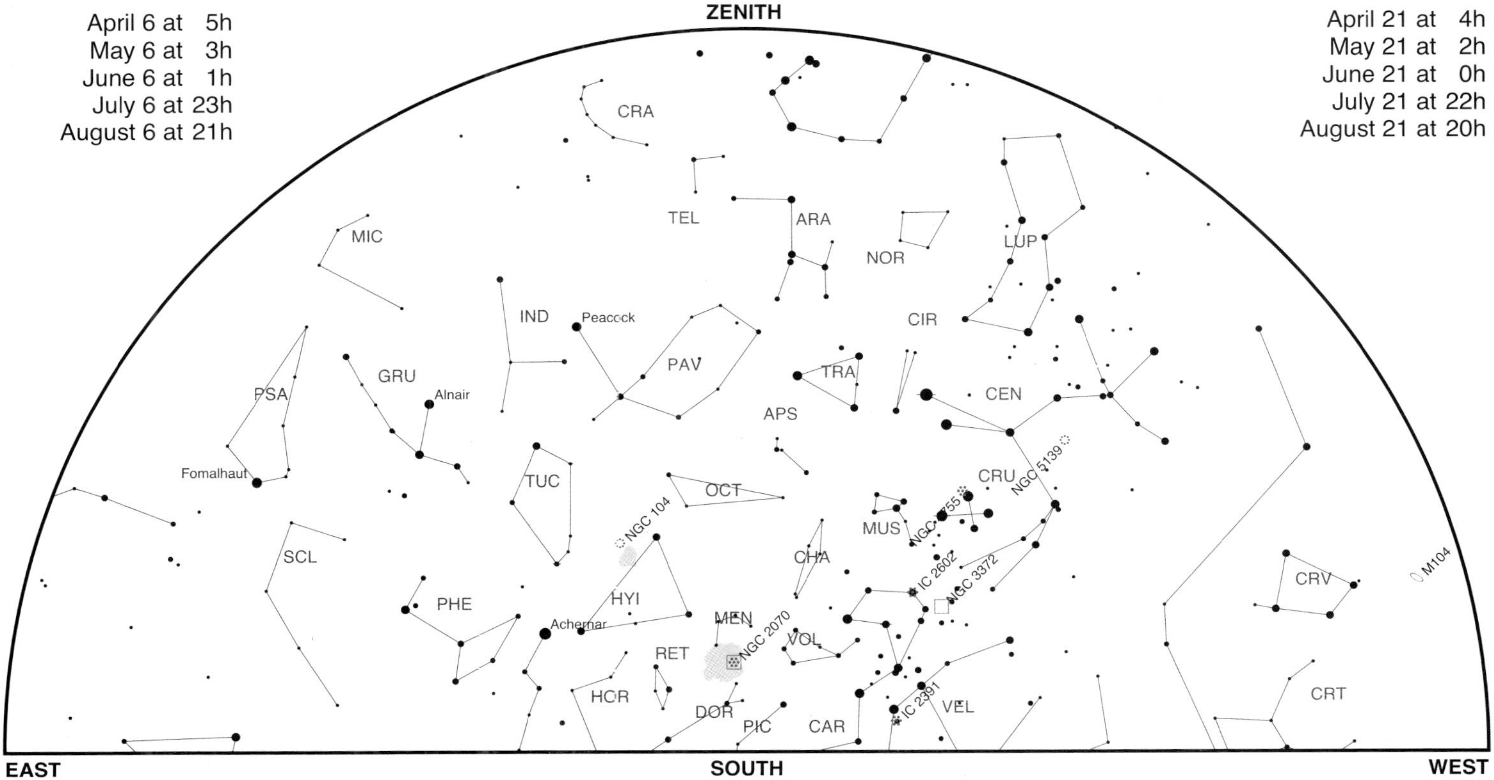

7S

8N

May 6 at 5h
June 6 at 3h
July 6 at 1h
August 6 at 23h
September 6 at 21h

May 21 at 4h
June 21 at 2h
July 21 at 0h
August 21 at 22h
September 21 at 20h

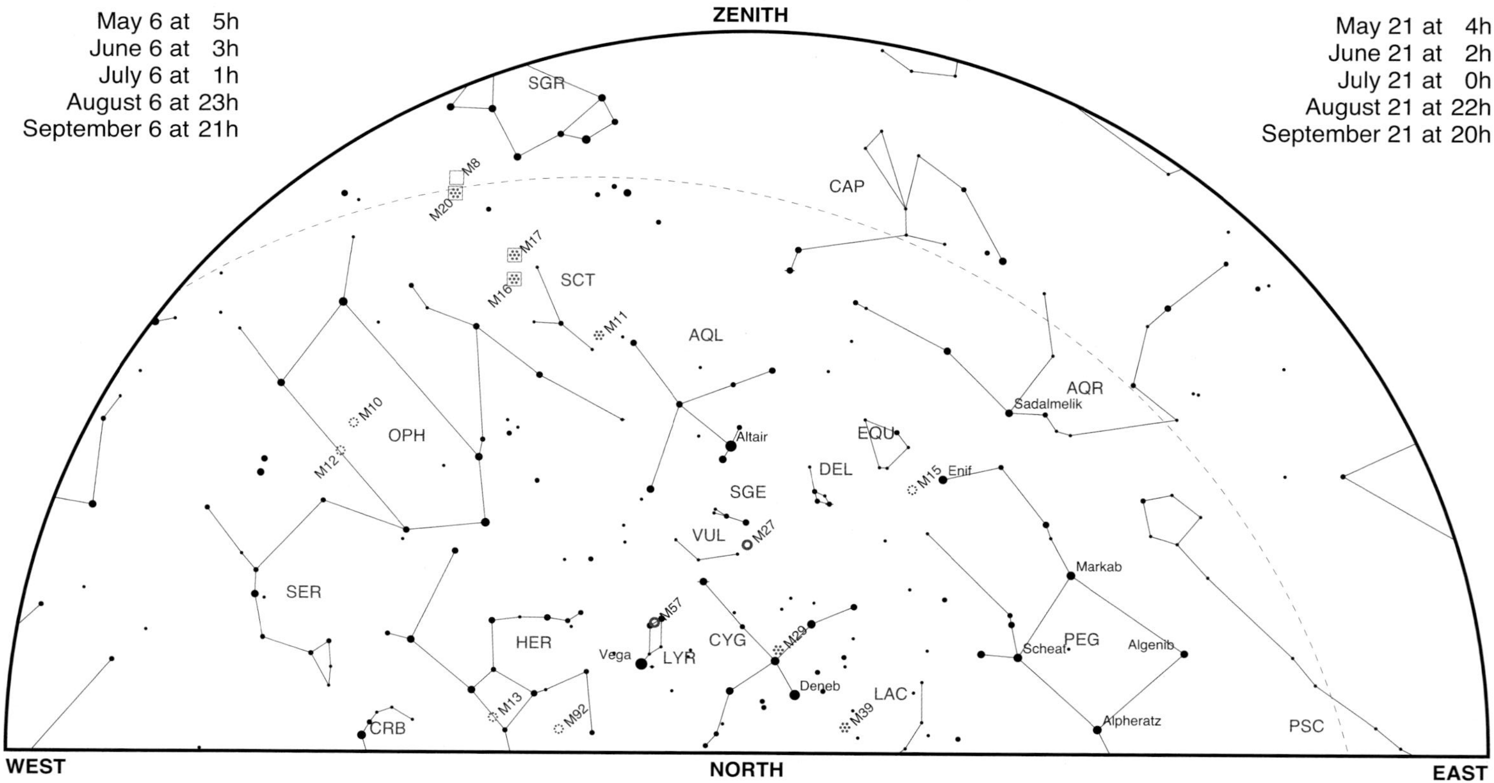

May 6 at 5h
June 6 at 3h
July 6 at 1h
August 6 at 23h
September 6 at 21h

May 21 at 4h
June 21 at 2h
July 21 at 0h
August 21 at 22h
September 21 at 20h

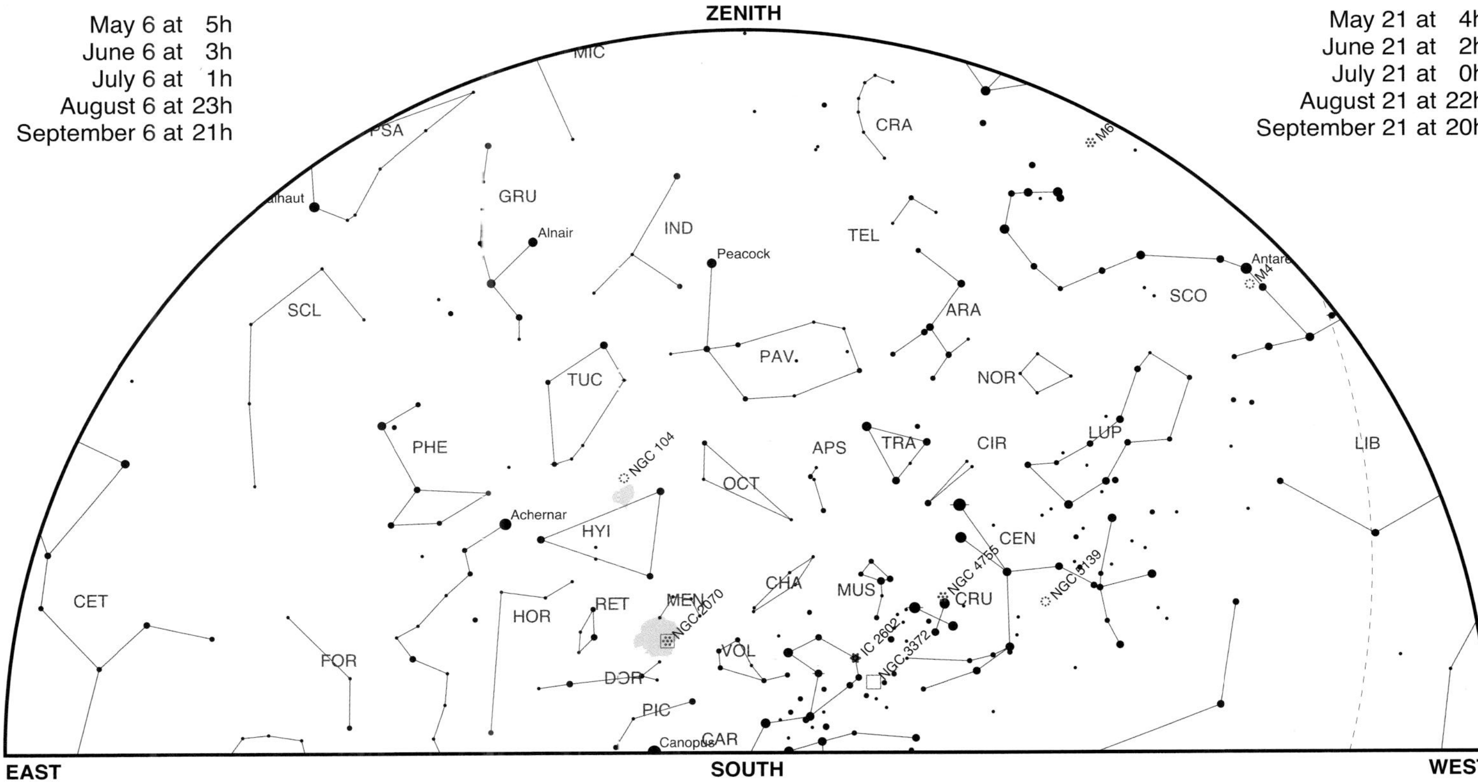

8S

9N

June 6 at 5h
July 6 at 3h
August 6 at 1h
September 6 at 23h
October 6 at 21h

June 21 at 4h
July 21 at 2h
August 21 at 0h
September 21 at 22h
October 21 at 20h

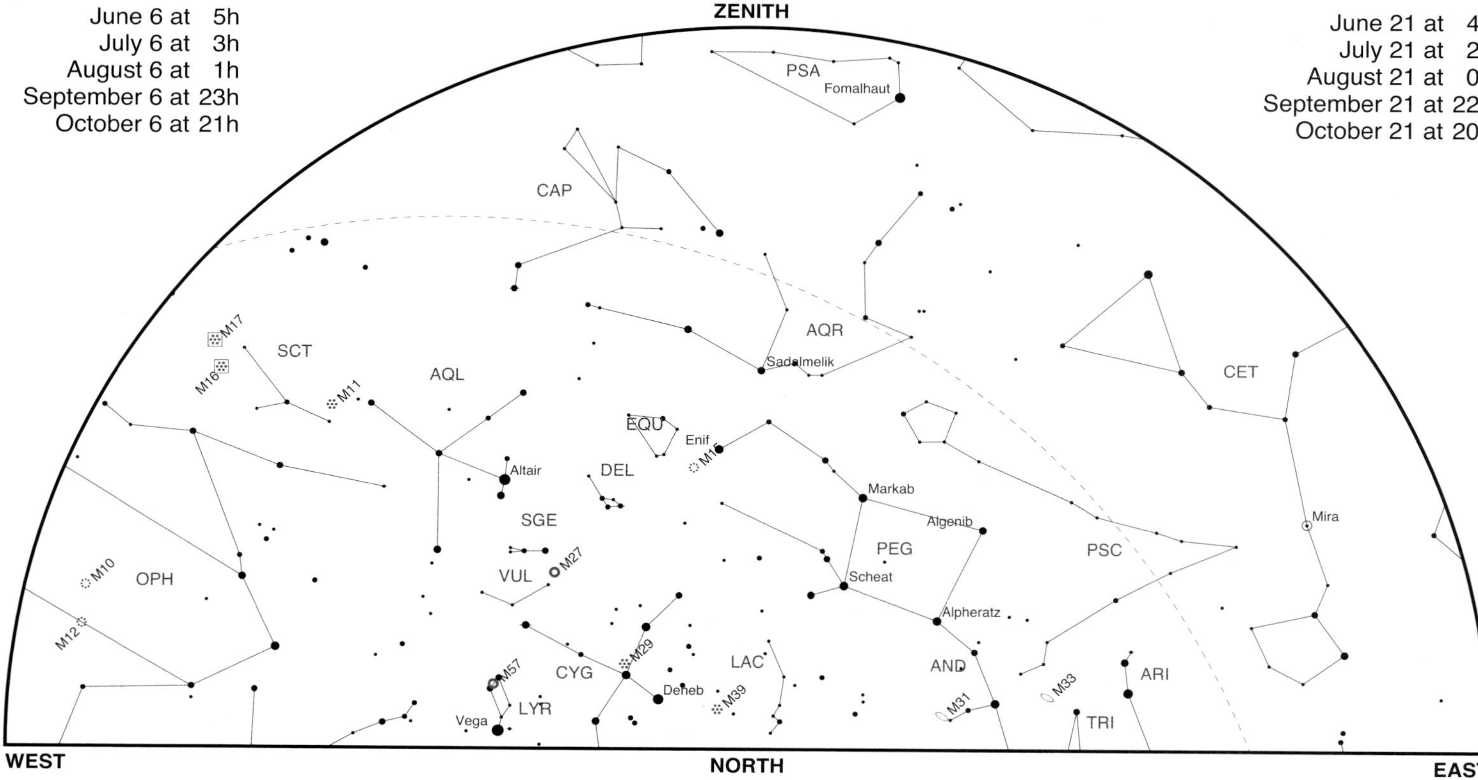

June 6 at 5h
July 6 at 3h
August 6 at 1h
September 6 at 23h
October 6 at 21h

June 21 at 4h
July 21 at 2h
August 21 at 0h
September 21 at 22h
October 21 at 20h

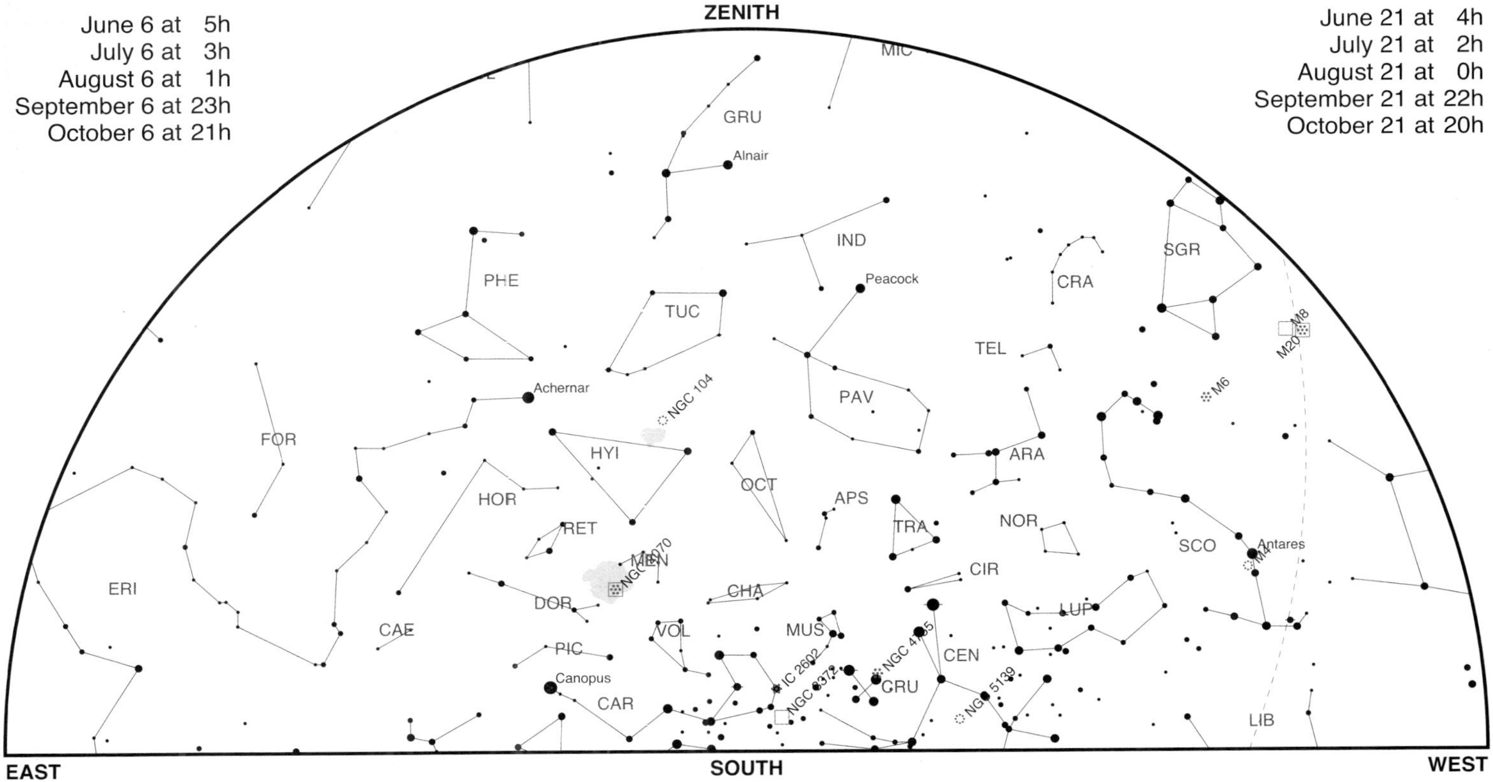

S6

10N

July 6 at 5h
August 6 at 3h
September 6 at 1h
October 6 at 23h
November 6 at 21h

July 21 at 4h
August 21 at 2h
September 21 at 0h
October 21 at 22h
November 21 at 20h

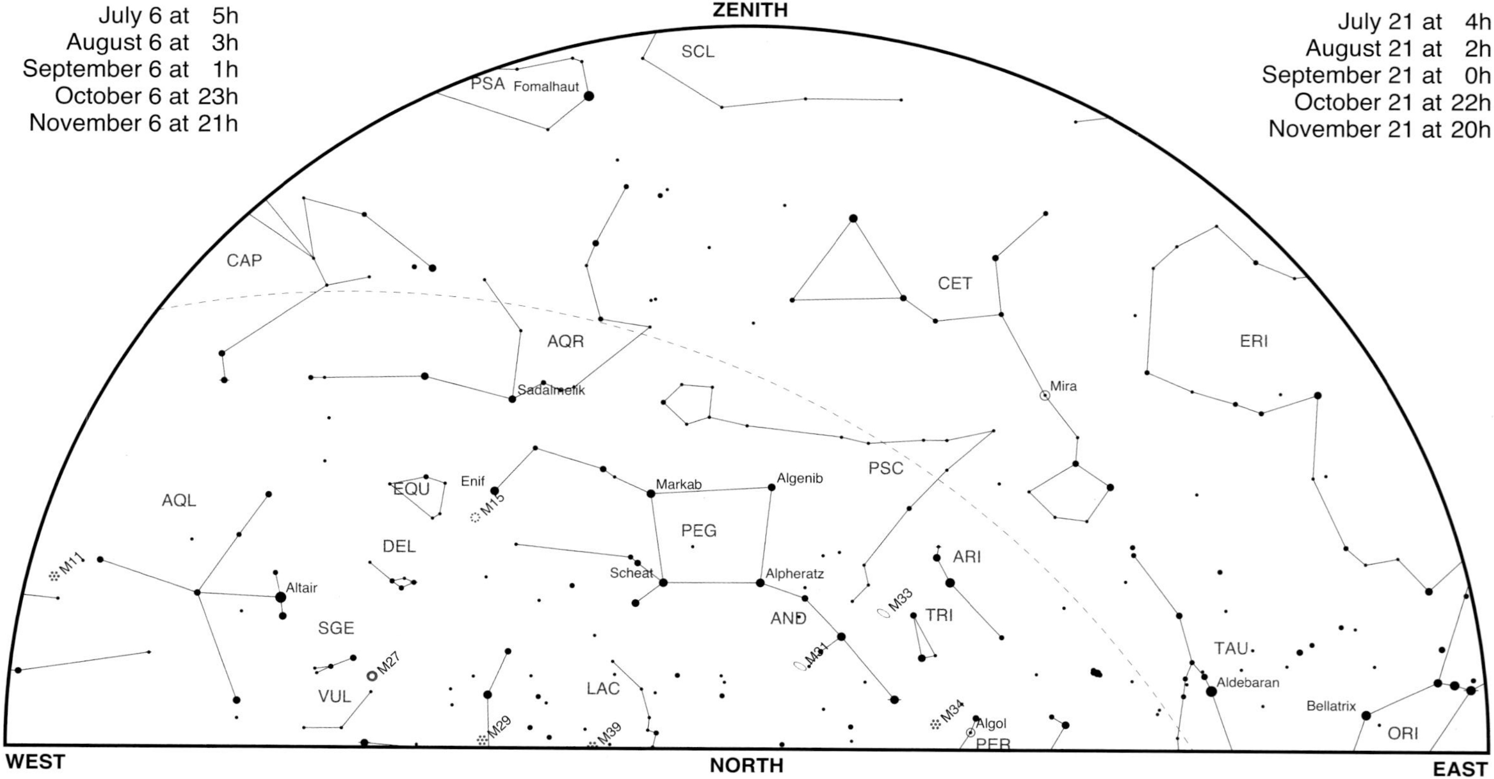

10S

July 6 at 5h
August 6 at 3h
September 6 at 1h
October 6 at 23h
November 6 at 21h

July 21 at 4h
August 21 at 2h
September 21 at 0h
October 21 at 22h
November 21 at 20h

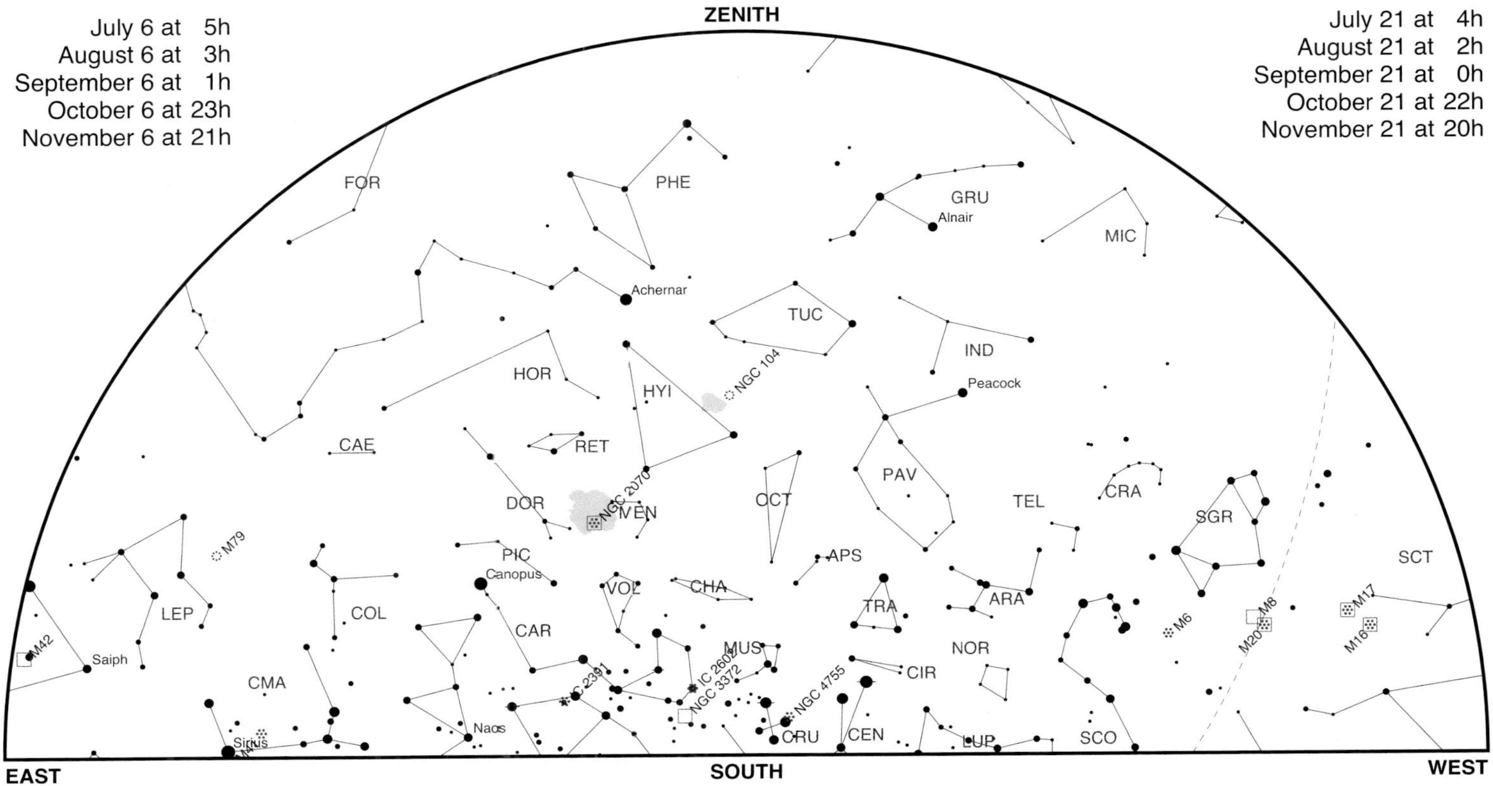

11N

August 6 at 5h
September 6 at 3h
October 6 at 1h
November 6 at 23h
December 6 at 21h

August 21 at 4h
September 21 at 2h
October 21 at 0h
November 21 at 22h
December 21 at 20h

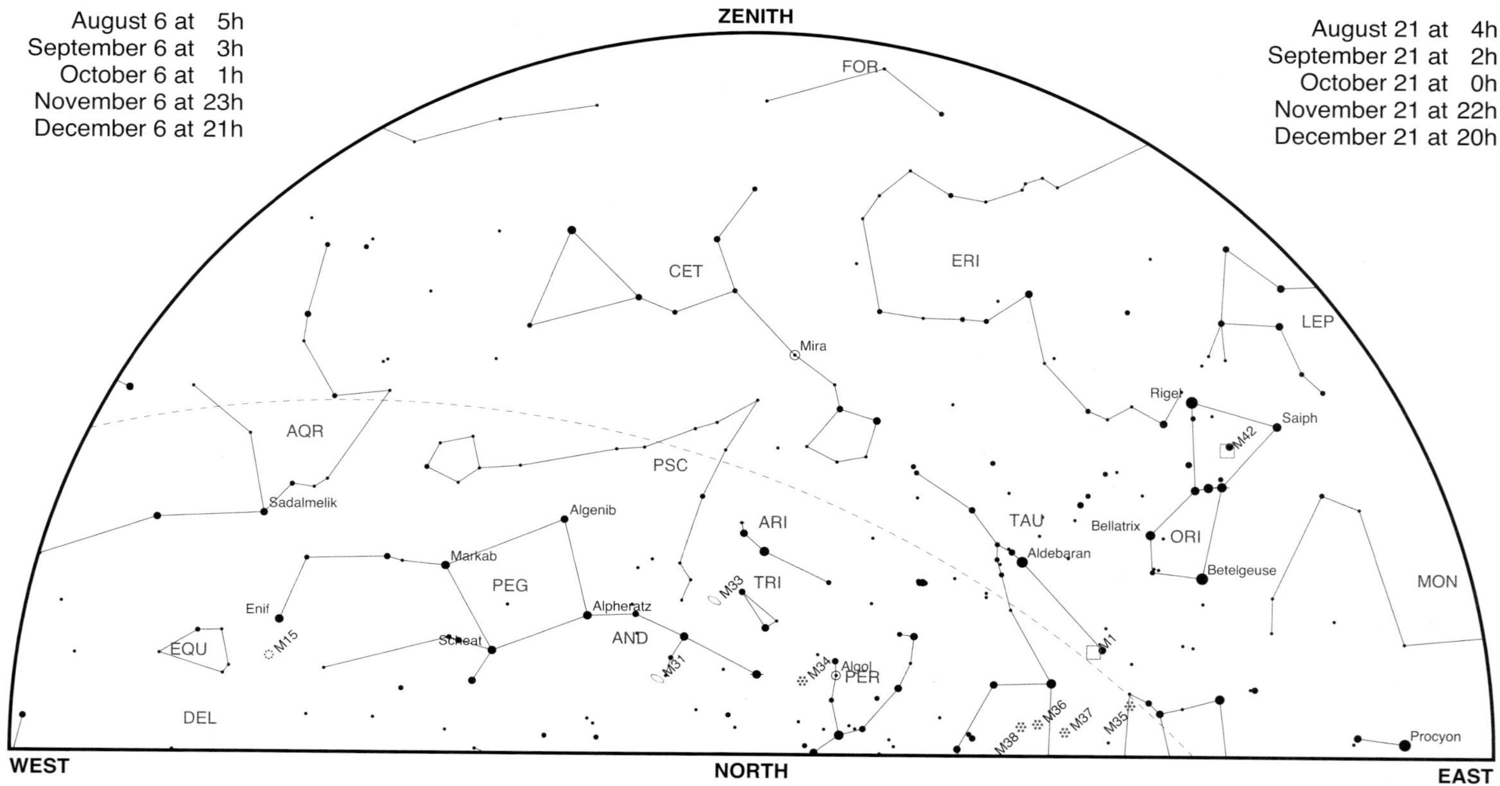

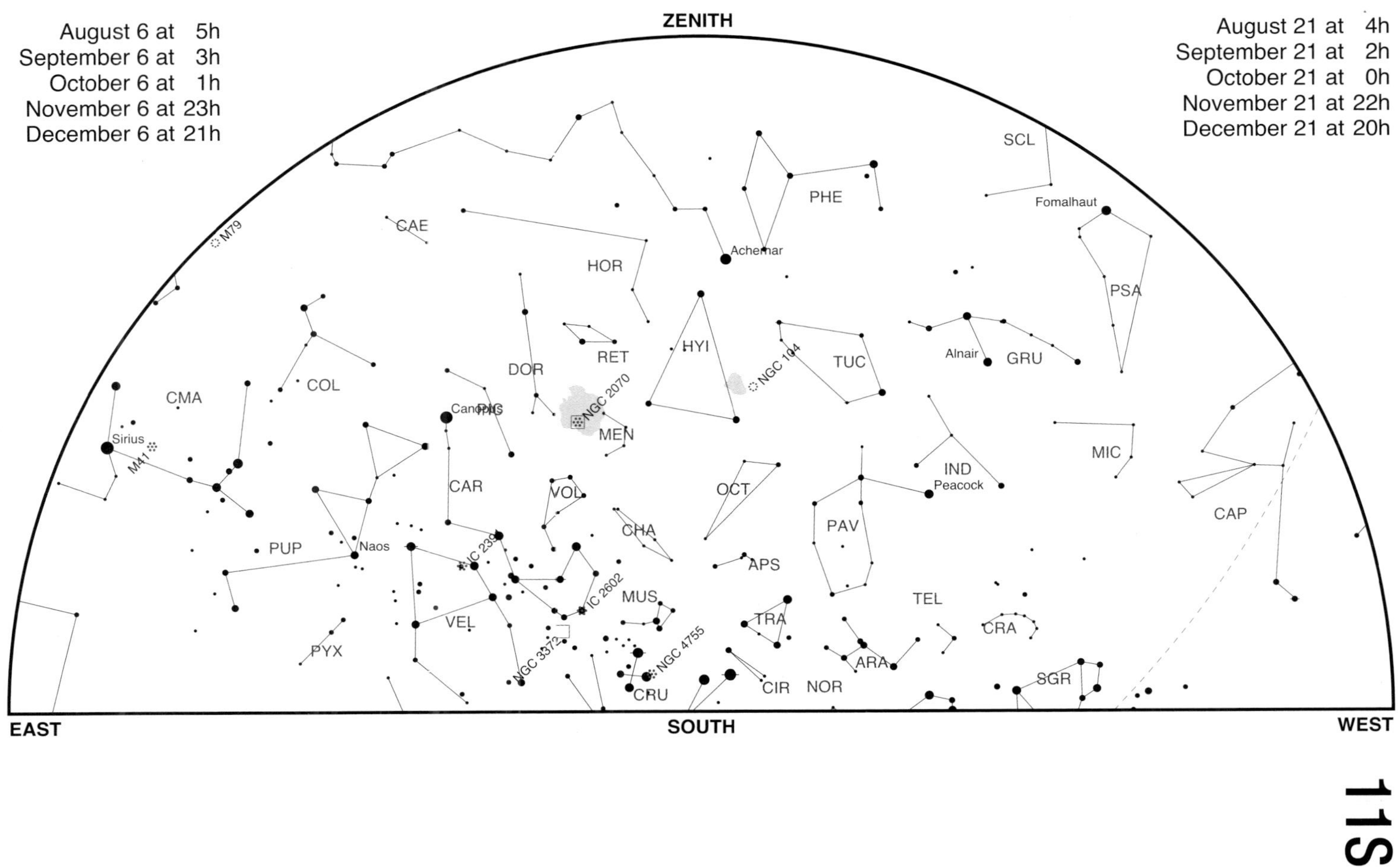

August 6 at 5h
September 6 at 3h
October 6 at 1h
November 6 at 23h
December 6 at 21h
ZENITH
August 21 at 4h
September 21 at 2h
October 21 at 0h
November 21 at 22h
December 21 at 20h
EAST
SOUTH
WEST
11S
CAE
M79
HOR
Achernar
PHE
SCL
Fomalhaut
PSA
CMA
Sirius
M41
COL
DOR
RET
HYI
NGC 104
TUC
Alnair
GRU
Canopus
NGC 2070
MEN
CAR
VOL
OCT
IND
Peacock
MIC
CAP
PUP
Naos
IC 2391
CHA
APS
PAV
IC 2602
MUS
TEL
VEL
TRA
CRA
PYX
NGC 3372
NGC 4755
ARA
CRU
CIR
NOR
SGR

12N

September 6 at 5h
October 6 at 3h
November 6 at 1h
December 6 at 23h
January 6 at 21h

September 21 at 4h
October 21 at 2h
November 21 at 0h
December 21 at 22h
January 21 at 20h

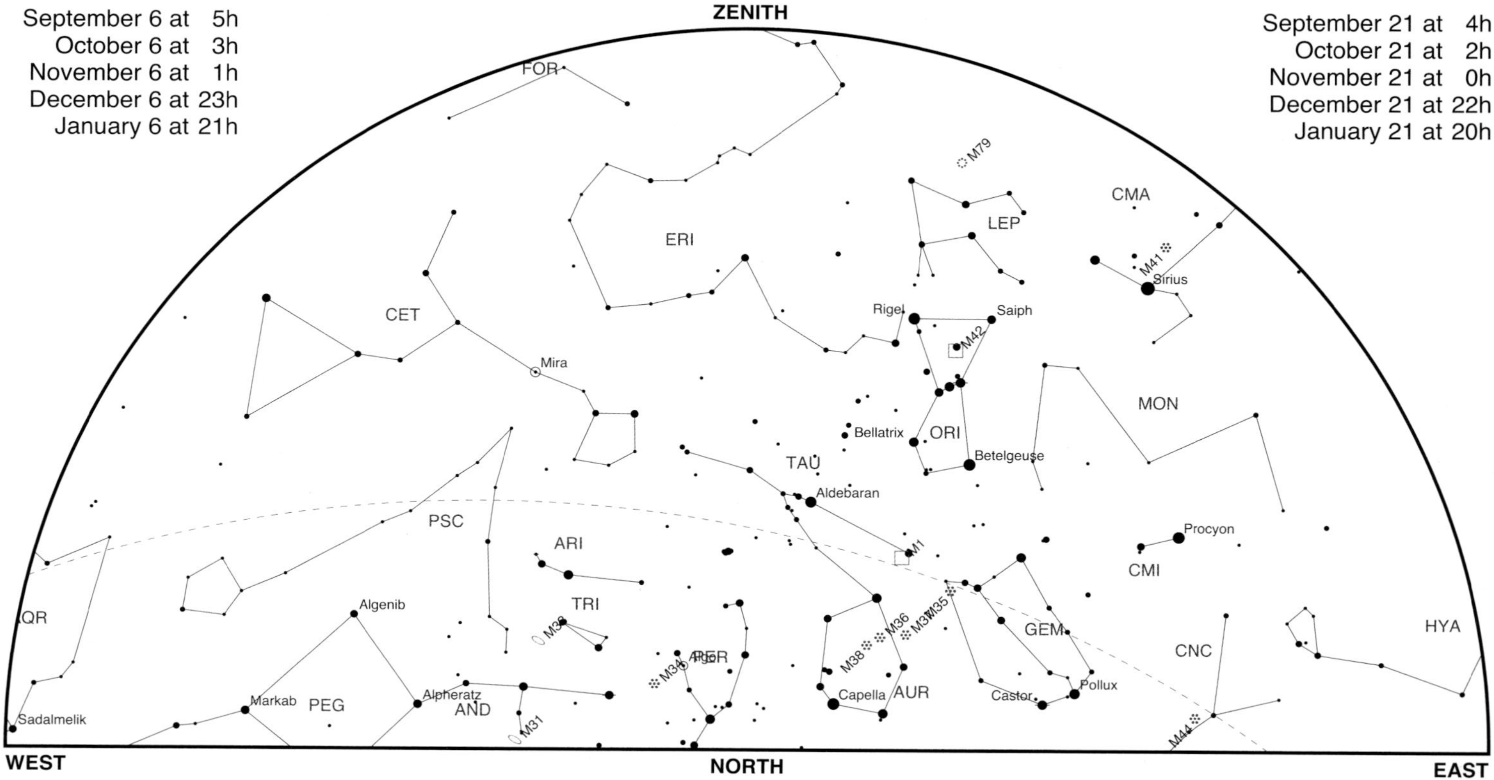

September 6 at 5h
October 6 at 3h
November 6 at 1h
December 6 at 23h
January 6 at 21h

September 21 at 4h
October 21 at 2h
November 21 at 0h
December 21 at 22h
January 21 at 20h

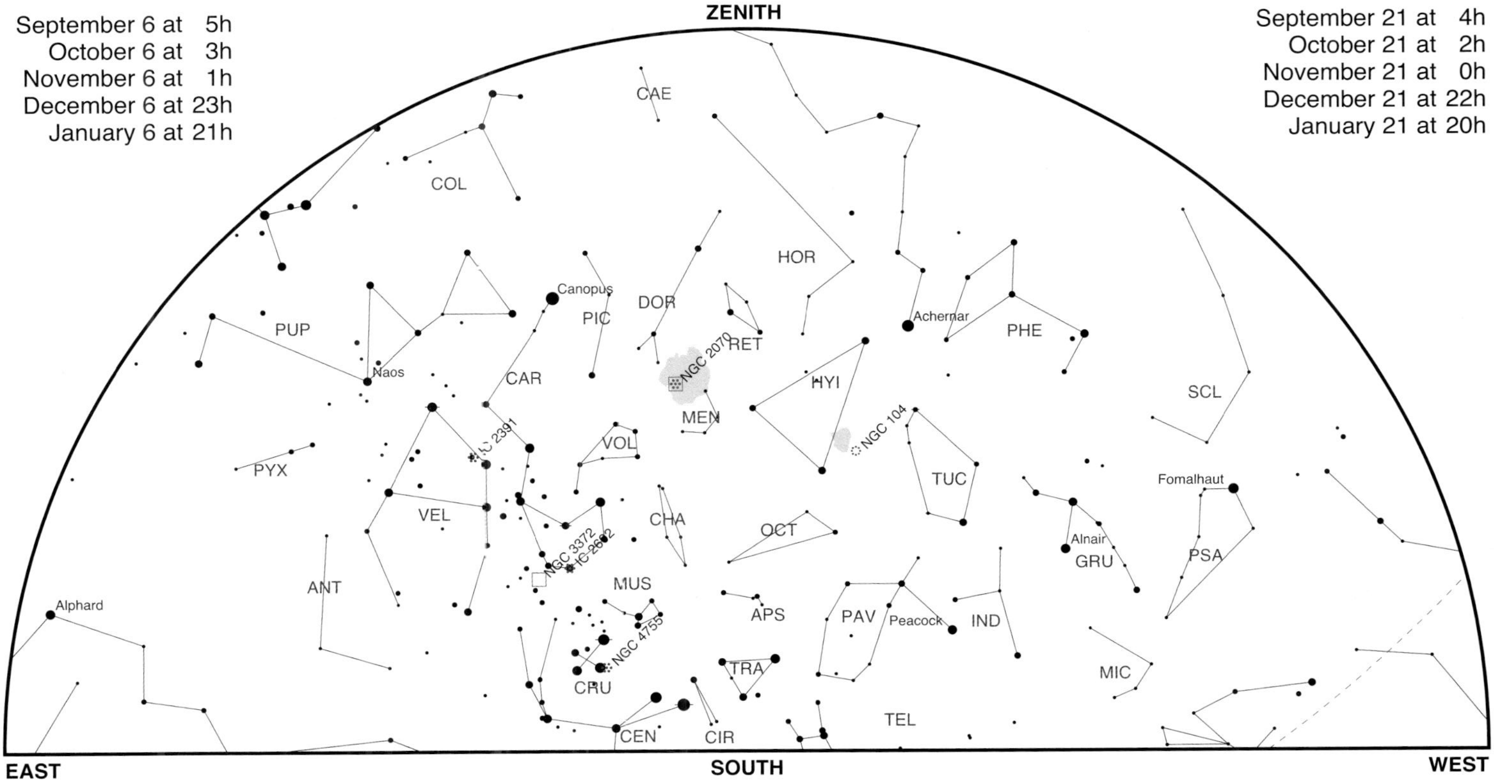

12S

The Solar System

Lunar Phenomena and Eclipses

The dates and circumstances of all lunar phenomena are calculated from the JPL DE406 solar system ephemeris. All dates are given in UT. Positions are geocentric apparent places, referred to the equator and equinox of date.

Phases of the Moon

The apparent shape of the Moon is ever changing; this is called its *phase*. A *lunation* or lunar month typically begins with the unlit New phase. The Moon's illuminated portion becomes larger, progressing through the waxing crescent, first quarter, and waxing gibbous phases before becoming fully lit. After Full Moon, the phase decreases, going through waning gibbous, last quarter, and waning crescent, before returning to New and the beginning of the next lunation. This cycle of phases is called a *synodic month* and its average length is 29.5 days.

BLN	New	First Quarter	Full		Last Quarter
1287	7 January	15 January	**22 January**	2005″	29 January
1288	6 February	14 February	20 February	1973″	28 February
1289	8 March	15 March	22 March	1925″	30 March
1290	6 April	13 April	20 April	1870″	28 April
1291	6 May	13 May	20 May	1820″	28 May
1292	4 June	11 June	19 June	1784″	27 June
1293	4 July	10 July	**18 July**	1766″	26 July
1294	2 August	9 August	17 August	1770″	25 August
1295	31 August	7 September	15 September	1795″	23 September
1296	30 September	7 October	15 October	1839″	22 October
1297	29 October	6 November	14 November	1893″	21 November
1298	28 November	6 December	13 December	1948″	20 December
1299	27 December				

The first column is the Brown Lunation Number.[1] The fifth column gives the apparent diameter of the Full Moon in arc-seconds. Dates of the largest and smallest Full Moons are given in **bold**.

1. Brown, E. W. (1933). The Motion of the Moon, 1923–1931. *Monthly Notices of the Royal Astronomical Society*, **93**(8), 603–619. **doi.org/10.1093/mnras/93.8.603**

Apsides

An *apsis* (plural: apsides) is either the nearest or the farthest point in an elliptical orbit of one body about another. *Periapsis* is the nearest point and *apoapsis* is the farthest point. In the case of Earth, the term *perigee* is used to describe the point where a body (in this case, the Moon) is closest to Earth and *apogee* is where a body is most distant from Earth. The difference between lunar perigee and apogee distances amounts to several tens of thousands of kilometres. The line of apsides of the Moon is not fixed in space but moves approximately 41° eastwards each year, making one full circuit around the Earth in 8.85 years. In contrast, the apsidal precession of Earth around the Sun takes about 112,000 years. The time it takes the Moon to return to the same apsis is just over 27.5 days and is called an *anomalistic month*.

Perigee	Distance (km)	Apogee	Distance (km)
		7 January	406,607
21 January	357,283	3 February	406,189
19 February	361,017	3 March	405,217
19 March	366,436	31 March	404,327
14 April	370,004	**27 April**	404,169
9 May	366,631	25 May	404,800
6 June	361,702	22 June	405,698
4 July	358,283	19 July	406,215
2 August	357,361	15 August	406,085
30 August	359,206	11 September	405,383
27 September	363,453	9 October	404,530
25 October	368,651	6 November	404,181
19 November	369,360	4 December	404,659
16 December	364,027		

Distances are measured from Earth's centre to Moon's centre. Dates of annual maxima/minima are marked in **bold**.

Nodes

The orbit of the Moon is inclined with respect to the ecliptic by just over 5° and a lunar *node* is a point in the Moon's orbit where it intersects the ecliptic plane. The *ascending node* is where the Moon crosses the ecliptic south to north and the *descending node* is where the Moon crosses the ecliptic north to south. The time it takes for the Moon to pass through the same node twice in succession is called a *draconic month*; this is just over 27 days long. These nodes are not fixed in space but precess westward by around 19° each year, making one full circuit every 18.6 years. Solar and lunar eclipses only take place when a New or Full Moon occurs very near to a node crossing. For details on this year's eclipses, see the article *Eclipses* following *Lunar Occultations*.

Ascending		Descending	
10 January		23 January	
6 February	New Moon occurs 6 hours before node crossing: **annular solar eclipse on 6 February**	**20 February**	Full Moon occurs 19 hours after node crossing: **penumbral lunar eclipse on 20–21 February**
6 March		19 March	
2 April		15 April	
29 April		12 May	
26 May		8 June	
22 June		6 July	
19 July	Full Moon occurs 32 hours before node crossing: **penumbral lunar eclipse on 18 July**	**2 August**	New Moon occurs 3 hours before node crossing: **total solar eclipse on 2 August**
16 August	Full Moon occurs 26 hours after node crossing: **penumbral lunar eclipse on 17 August**	29 August	
12 September		26 September	
9 October		23 October	
5 November		19 November	
2 December		16 December	
30 December			

Dates of nodal crossings resulting in eclipses are marked in **bold**.

Lunar Occultations

A *lunar occultation* occurs when the Moon passes in front of a more distant celestial body as seen from Earth, partially or totally obscuring this object and blocking its light. Like total solar eclipses, occultations are seen only at particular times and from particular places, but they have been noticed for centuries; the first known observation dates to 357 BCE when Aristotle spotted Mars disappearing behind the Moon's disk. Occultation data can help astronomers measure the heights and depths of lunar features and improve knowledge about the lunar orbit; such information can be used to detect close companions in multiple star systems and improve the precision of stellar positions. Lunar occultation data was even employed to pinpoint the location of 3C 273, the first quasar to be identified by astronomers.

This year, the Moon regularly occults the first-magnitude star Antares as well as the brightest stars in the Beehive and Pleiades open clusters. Both Mercury and Jupiter are repeatedly obscured: Mercury five times (March may present the best observing opportunity) and Jupiter four times. Venus vanishes behind the Moon's disk twice although the August occultation is unobservable, and Mars is occulted once. Details of individual occultations listed in the table can be found in the corresponding sections of the *Monthly Sky Notes*.

Table of Lunar Occultations in 2027

	Jan	Feb	Mar	Apr	May	Jun	Jul	Aug	Sep	Oct	Nov	Dec
α Scorpii (Antares)	4, 31	28	27	23	21	17	14	11	7	4	1, 28	25
M44 (Beehive / Praesepe)	22	19	18	14	12	8	5	2, 29	26			
M45 (Pleiades)	18	14	13	10	7	3	1, 28	25	21	18	14	12
Mercury	8	8	6					1				28
Venus								2			30	
Mars												29
Jupiter		20	19	15	12							

Occultations of M44 (Praesepe / Beehive) refer to sixth-magnitude star ε Cancri (Meleph).
Occultations of M45 (Pleiades) refer to third-magnitude star η Tauri (Alcyone).

Visit the International Occultation Timing Association (IOTA) at **occultations.org** for more information on lunar and planetary occultations, including observing predictions for the current year.

Eclipses

Solar and lunar eclipses do not happen every month but take place only when a New or Full Moon occurs very near to a node crossing (see the *Nodes* section earlier in this article). The New Moon's shadow covers just a small portion of the Earth's surface so a solar eclipse is visible only from a limited region of our planet. A lunar eclipse, on the other hand, is potentially visible from all locations on the night side of Earth. Depending on the exact location of the observer, the eclipse may already be in progress as the Moon rises (or sets), or the entire eclipse sequence may be visible. There are a minimum of four eclipses in any one calendar year, comprising two solar eclipses and two lunar eclipses. In fact, most years have only four although it is possible to have five (2028), six (2029), or even seven (2038) eclipses during the course of a year. This year there will be five: an annular solar eclipse, three penumbral lunar eclipses, and a total solar eclipse.

All dates and times listed below are given in Universal Time (UT). It is important to note that the times quoted for each event refer to the start, maximum, and ending of the eclipse on a global scale rather than with reference to any specific location.

Annular Solar Eclipse (Saros 131): 6 February

The first eclipse of the year is the annular solar eclipse of 6 February, the annular phase of which begins in the southern Pacific Ocean, crosses Chile and Argentina, and ends on the coast of Africa around Liberia, Côte d'Ivoire, Ghana, Togo, Benin, and Nigeria (including Lagos). The partial phase is visible from most of South America, parts of Antarctica, and the extreme western coast of Africa, including the islands of Cabo Verde and the Canaries.

Time (UT)	Event
12:58	Partial eclipse begins
14:04	Annular eclipse begins
15:56	New Moon
16:00	Maximum eclipse
17:55	Annular eclipse ends
19:02	Partial eclipse ends
21:43	Moon passes through ascending node

Penumbral Lunar Eclipse (Saros 143): 20/21 February

The first of three eclipses of this type in 2027, the penumbral lunar eclipse of 20 February will be visible in whole or part from all but the western edge of North America, the whole of Central and South America, Greenland, Iceland and the Atlantic Ocean, the entirety of Europe (including the UK) and Africa, virtually the whole of Asia and the western regions of Australia.

Time (UT)	Event
04:40	Moon passes through descending node
21:12	Penumbral eclipse begins
23:13	Maximum eclipse
23:23	Full Moon
01:13 (+1 day)	Penumbral eclipse ends

Penumbral Lunar Eclipse (Saros 110): 18 July

The penumbral lunar eclipse of 18 July will be visible in whole or part from eastern Africa, most of Asia, Australia and New Zealand and much of the Pacific Ocean and Antarctica. This eclipse will likely be imperceptible, with less than 1% of the Moon entering the Earth's penumbra.

Time (UT)	Event
15:44	Full Moon
15:54	Penumbral eclipse begins
16:03	Maximum eclipse
16:12	Penumbral eclipse ends
23:33	Moon passes through ascending node

Total Solar Eclipse (Saros 136): 2 August

On 2 August there will be a total solar eclipse, the path of totality of which will commence in the central region of the northern Atlantic Ocean before crossing the southernmost tip of Spain, Gibraltar, the northern coastal regions of Africa, Egypt, the Red Sea, the southern edge of the Arabian peninsula and the Gulf of Aden before travelling over the tip of Somalia (just missing the tiny island of Socotra) and ending its journey in the Indian Ocean. Locations from which a partial eclipse will be seen include the eastern tip of Canada, nearly all of Europe (including the United Kingdom, Iceland, and southern Greenland), most of Africa (including Madagascar), the entire Middle East, and almost all of southern and western Asia (including India).

Time (UT)	Event
07:30	Partial eclipse begins
08:23	Total eclipse begins
10:05	New Moon
10:07	Maximum eclipse
11:50	Total eclipse ends
12:34	Moon passes through descending node
12:43	Partial eclipse ends

Penumbral Lunar Eclipse (Saros 148): 17 August

The penumbral lunar eclipse of 17 August is visible in whole or part from the eastern half of Australia, New Zealand, and the Pacific Ocean, much of Antarctica, virtually the whole of the Americas (Alaska being the notable exception), the Atlantic Ocean, southern Greenland and south western Iceland (including Reykjavik), Portugal, Spain and the westernmost region of Africa.

Time (UT)	Event
05:45 (−1 day)	Moon passes through ascending node
05:24	Penumbral eclipse begins
07:14	Maximum eclipse
07:28	Full Moon
09:03	Penumbral eclipse ends

The *saros* is the period of 6585.32 days (18 years 10.32 or 11.32 days), after which solar or lunar eclipses recur in almost identical circumstances, due to the Earth, Moon and Sun returning to almost the same relative positions. A *saros series* is a series of eclipses which recur at intervals of this period; there are many interleaved saros series running concurrently.

A fuller description/definition of 'saros' can be found in the downloadable glossary available at **yearbookofastronomy.com**, the official *Yearbook of Astronomy* website.

Eclipse circumstances, with maps, are available from a variety of online sources, including **EclipseWise.com** (Fred Espenak), **ssp.imcce.fr/forms** (Institut de mécanique céleste et de calcul des éphémérides), and **aa.usno.navy.mil** (United States Naval Observatory).

All eclipse circumstances are taken from **EclipseWise.com**, Fred Espenak's web site dedicated to solar and lunar eclipse predictions and information.

Meteor Showers

Brian Jones and Lynne Marie Stockman

A *meteor* or shooting star is the streak of light left behind when a bit of space debris enters the Earth's atmosphere at high speed. On any given dark night, an observer may see several meteors.

Sporadic meteors appear at any time and from any direction. Meteor showers, on the other hand, seem to emanate from a particular part of the sky (called the *radiant*) only at a particular time of the year. Showers occur when the Earth intercepts a stream of cometary or asteroidal cast-offs.

Meteor showers are typically named after the constellation in which the radiant appears. The relative strength of the shower is indicated by the *Zenithal Hourly Rate* (ZHR) which is the theoretical number of meteors an observer would see if the radiant was located directly overhead and observing conditions were perfect. However, do not expect to see this many meteors in an hour!

The accompanying table details the major – and a few of the minor – meteor showers that will occur this year.

Observing Notes

Moonlight can wash out all but the brightest of meteors, and for most showers, the best viewing occurs after local midnight, so the phase of the Moon is important. The location of the radiant in the sky also determines where on Earth the shower is most easily seen. Visit the International Meteor Organization (IMO) at **imo.net** for more information on meteor showers, including observing predictions for this year.

Favourable

The limited light from the waning crescent Moon provides good conditions for viewing the **Quadrantids** in early January. The meteors are of medium speed, with the shower being slightly richer in bright meteors than average. The radiant is circumpolar above 41° north latitude and reaches a useful height after midnight. The speedy **η-Aquariids** in early May are best seen from points south of 40° north latitude. The radiant culminates after sunrise so the best viewing opportunities are just before dawn. Moonlight is irrelevant to the **β-Taurids** in late June as they are a daytime shower, detectable in radio wavelengths. At the end of July comes

Meteor Showers in 2027

Shower	Peak	Activity	ZHR	Parent	Moon	Notes
Quadrantids	3/4 Jan	Late-Dec to mid-Jan	80	2003 EH_1 (asteroid)	Waning crescent	Radiant located in Boötes.
April Lyrids	22/23 Apr	Mid- to late-April	18	C/1861 G1 Thatcher	Waning gibbous	
π-Puppids	23/24 Apr	Mid- to late-April	Varies	26P/Grigg-Skjellerup	Waning gibbous	
η-Aquariids	5/6 May	Mid-April to late-May	50	1P/Halley	New	
June ε-Ophiuchids	19/26 Jun	Late-June	Low	300P/Catalina	Waning gibbous	
β-Taurids	25/28 Jun	Mid-June to early-July	25	2P/Encke	Last Quarter	Daytime radio shower.
Southern δ-Aquariids	30/31 Jul	Mid-Jul to late-Aug	25	96P/Machholz	Waning crescent	
Perseids	12/13 Aug	Mid-Jul to late-Aug	100	109P/Swift–Tuttle	Waxing gibbous	
August δ-Capricornids	16/17 Aug	Mid-August	Low	45P/Honda-Mrkos-Pajdušáková	Full	
κ-Cygnids	16/17 Aug	August	3	Not Known	Full	Episodic shower with a period of 7 years. Latest enhanced activity was in 2020/21.
Draconids	7/8 Oct	Early-October	5	21P/Giacobini–Zinner	First Quarter	Also known as the Giacobinids.
Orionids	21/22 Oct	Early-Oct to early-Nov	20	1P/Halley	Waning gibbous	
Southern Taurids	5 Nov	Mid-Sep to mid-Nov	7	2P/Encke	Waxing crescent	Early peak occurs on 13/14 October.
Northern Taurids	12 Nov	Mid-Oct to mid-Dec	5	2004 TG_{10} (asteroid)	Waxing gibbous	2004 TG_{10} may be a fragment of 2P/Encke.
Leonids	17/18 Nov	November	15	55P/Tempel–Tuttle	Waning gibbous	
Puppid-Velids	7 Dec	Early- to mid-Dec	10	Not Known	First Quarter	Also known as the γ-Velids.
Geminids	14 Dec	Early- to mid-Dec	150	3200 Phaethon (asteroid)	Full	
Ursids	22/23 Dec	Mid- to late-Dec	10	8P/Tuttle	Waning crescent	

the **Southern δ-Aquariid** meteor shower. These objects are of average speed and brightness, and are best detected from the southern hemisphere where the radiant rises before midnight. October brings the slow-moving **Draconids**. The radiant is circumpolar above 44° north latitude and highest before midnight. However, because of the First Quarter Moon, the best observations may come after midnight. The **Southern Taurids** of early November are also slow moving, with the shower being richer in bright meteors than average. This shower is observable from all latitudes although the northern hemisphere is favoured. The waxing crescent Moon should present no problems later in the evening. Southern and tropical latitudes are favoured for viewing the **Puppid-Velid** meteor shower. These objects are of medium speed and tend to be rather fainter than average. The radiant is highest towards dawn so the First Quarter Moon is no obstacle. Another December shower, the **Ursids**, is a northern shower, with the radiant circumpolar above 14° north latitude. Like the Puppid-Velids, the radiant of the Ursids is highest before dawn so the waning crescent Moon should not affect observations.

Unfavourable

Moonlight seriously interferes with observing opportunities for the **April Lyrids**, the **June ε-Ophiuchids**, the **August δ-Capricornids**, the **κ-Cygnids**, the **Leonids**, and the **Geminids**. The κ-Cygnid meteor shower is episodic, peaking every seven years, and no enhanced activity is forecast for this year anyway. However, next year will be a different story.

Worth a try?

The **π-Puppids** peak around the same time as the April Lyrids. These slow, bright meteors are best spotted from the southern hemisphere. The radiant is very low after midnight so better views occur earlier in the evening. Thus, the earliest observations may not be unduly affected by light of the waning gibbous Moon. The famous August **Perseids** are best seen from mid-northern latitudes where the radiant reaches a useful height before midnight. Observations before dawn may be possible after the waxing gibbous Moon sets but extended twilight could be an issue. October brings the **Orionids** which are known for their speed. The radiant is at a useful altitude by midnight but with a waning gibbous Moon, the best observations will be in the evening. Finally, the slow but bright **Northern Taurids** of November are observable from all latitudes but light from the waxing gibbous Moon will prove to be a major obstacle until well after midnight.

Comets

Brian Jones and Lynne Marie Stockman

Comets are small icy bodies originating in the far reaches of the solar system and even beyond. The nucleus of a comet ranges in size from hundreds of metres to kilometres in diameter and is comprised of rock, dust, ice, and various frozen gases. When a comet approaches perihelion (closest approach to the Sun), its nucleus heats up and it begins outgassing, releasing the dust and gas on the surface to produce a fuzzy atmosphere or coma. This coma can be many Earth diameters in width. Solar radiation and solar wind interact with the coma to push out a tail. Cometary tails come in all sizes but the longest have been known to stretch one astronomical unit from the nucleus.

Comets usually have highly elliptical orbits, with periods ranging from a few years to many millennia. Some comets have parabolic or hyperbolic orbits, however, and make only one pass around the Sun. Short-period comets are thought to originate in the Kuiper Belt which lies beyond Neptune. Amongst the subtypes of short-period comets are *Encke-type* (orbit lies completely within that of Jupiter) and *Jupiter-family* (period of less than 20 years and a relatively low inclination). Long-period comets come from further afield, in the Oort Cloud, whereas intermediate-period comets such as the *Halley-type* (period between 20 and 200 years with any inclination) may originate in either place. Some comets may even arrive from outside the solar system. A comet becomes extinct when it passes by the Sun so many times that it loses most or all of its volatile gases, and can no longer form a coma or tail. Such an object may then resemble an asteroid. A comet may also disintegrate if it strays too close to the Sun. This is sometimes the fate of so-called *sungrazing comets*. If the Earth intercepts the debris field left by the earlier passage of a comet, it will experience a meteor shower.

There are hundreds of known comets with more being discovered all the time. The difficulty in preparing this section for the *Yearbook of Astronomy* is that the choice of comets to highlight is limited to those which are known at the time of writing (two or three years ahead of publication) to return to perihelion this year. For the latest comet discoveries, consult such authorities as the IAU Minor Planet Center, the British Astronomical Association (BAA) Comet Section at **people.ast.cam.ac.uk/~jds** or Seiichi Yoshida's Home Page at **aerith.net**

Geocentric right ascensions and declinations tabulated in this section are referred to the mean equator and equinox of J2000. This allows them to be plotted directly on modern star charts such as *Sky Atlas 2000.0* by Wil Tirion and Roger Sinnott.

Orbital and physical parameters are derived from the JPL Solar System Dynamics Small-Body Database at **ssd.jpl.nasa.gov/tools/sbdb_lookup.html#**

Positional and magnitude data are taken from the IAU Minor Planet & Comet Ephemeris Service at **minorplanetcenter.net/iau/MPEph/MPEph.html**

Data in the following tables were correct as of 10 January 2026.

2P/Encke

Discovery	17 January 1786 by Pierre François André Méchain (1744–1804)		
Name	Named after Johann Franz Encke (1791–1865) who first calculated its orbit in 1819 and correctly predicted its return in 1822		
Meteor showers	β-Taurids (daytime, mid June to early July) Taurids (late September to early December)		
Classification	Encke-type comet	**Orbital eccentricity**	0.85
Aphelion	4.10 au	**Orbital inclination to ecliptic**	11.4°
Perihelion	0.34 au	**Orbital period (sidereal)**	3.31 years
Next perihelion	10 February 2027	**Closest approach to Earth**	3 February 2027

This image of 2P/Encke was captured by NASA's *MESSENGER* spacecraft on 17 November 2013. (NASA/Johns Hopkins University Applied Physics Laboratory/Carnegie Institution of Washington/Southwest Research Institute)

Comet 2P/Encke approaches the Sun from north of the ecliptic. It is found in the constellation of Pegasus at the opening of the year before moving into Aquarius soon after and then into Capricornus as it rounds the Sun. It is at its closest to Earth on 3 February when it is 0.6 au distant but it will be only 10° from the Sun at that time. 2P/Encke begins this apparition as an evening sky object, visible after sunset in the southwest (northern hemisphere) or west (southern hemisphere). It dives south of the ecliptic and passes into the morning sky in early February where it is then best seen from southern vantage points. Maximum brightness estimates vary wildly, from the Minor Planet Center's optimistic fourth magnitude to JPL's more cautious thirteenth, but it is likely that the comet will not be visible to the naked eye. 2P/Encke is the prototype Encke-type comet.

Date	RA			Declination			Magnitude	Constellation
	h	m	s	°	′	″		
1 January 2027	22	32	28	+03	30	43	+11.0	Pegasus
15 January 2027	22	29	10	+01	34	40	+8.9	Aquarius
1 February 2027	21	48	55	−07	25	08	+4.8	Aquarius
15 February 2027	20	38	25	−21	48	37	+4.4	Capricornus
1 March 2027	20	52	25	−23	40	00	+8.2	Capricornus
15 March 2027	21	23	22	−21	58	24	+11.0	Capricornus

45P/Honda–Mrkos–Pajdušáková

Discovery	3 December 1948 by Minoru Honda at Kurashiki, Japan		
Name	Named after Minoru Honda (1913–1990), Antonín Mrkos (1918–1996), and L'udmila Pajdušáková (1916–1979) who independently discovered the comet		
Meteor showers	August δ-Capricornids		
Classification	Jupiter-family comet	Orbital eccentricity	0.82
Aphelion	5.52 au	Orbital inclination to ecliptic	4.3°
Perihelion	0.53 au	Orbital period (sidereal)	5.26 years
Next perihelion	30 August 2027	Closest approach to Earth	21 July 2027

Comet **45P/Honda-Mrkos-Pajdušáková** never strays far from the ecliptic. At closest approach, it is just over 0.5 au away from Earth and 68° from the Sun, but the comet is unlikely to brighten past eighth magnitude during this apparition. It is visible in the morning sky, and is best seen from the southern hemisphere and equatorial regions where the constellations of Cetus, Aries, Taurus, Orion, Gemini, Cancer, and Leo rise high in the sky before sunrise.

This image of 45P/Honda–Mrkos–Pajdušáková was captured on the early morning of 10 February 2017, in bright moonlight and during the last few hours before dawn. (Wikimedia Commons/CajunAstro/Attribution 2.0 Generic license)

Date	RA			Declination			Magnitude	Constellation
	h	m	s	°	′	″		
1 July 2027	00	31	31	−02	33	44	+14.7	Cetus
15 July 2027	02	12	00	+06	03	12	+12.7	Cetus
1 August 2027	04	53	16	+16	26	19	+10.6	Taurus
15 August 2027	06	48	33	+19	08	58	+9.2	Gemini
1 September 2027	08	34	42	+17	27	39	+8.5	Cancer
15 September 2027	09	46	44	+13	45	21	+10.1	Leo

300P/Catalina

Discovery	6 May 2005 by the Catalina Sky Survey, University of Arizona, Tucson, Arizona, USA		
Name	Named after the Catalina Sky Survey, a NASA-funded project dedicated to discovering and tracking Near-Earth Objects		
Meteor showers	June ε-Ophiuchids (possible)		
Classification	Jupiter-family comet	**Orbital eccentricity**	0.69
Aphelion	4.57 au	**Orbital inclination to ecliptic**	5.7°
Perihelion	0.83 au	**Orbital period (sidereal)**	4.43 years
Next perihelion	13 September 2027	**Closest approach to Earth**	23 October 2027

Comet **300P/Catalina** first appears in the evening sky, racing south through the zodiacal constellations of Virgo and Libra. It passes through perihelion just inside Earth's orbit before approaching our planet in October when it closes to about 0.2 au. The comet will be around 90° away from the Sun at that point and is not expected to get any brighter than thirteenth magnitude. It is best seen from the southern hemisphere where it will be visible for much of the night from October onwards.

Date	RA			Declination			Magnitude	Constellation
	h	m	s	°	′	″		
1 September 2027	14	05	11	−15	37	41	+14.4	Virgo
15 September 2027	14	47	22	−22	58	40	+13.7	Libra
1 October 2027	16	00	18	−34	10	37	+13.2	Lupus
15 October 2027	18	21	01	−45	05	05	+12.9	Corona Australis
1 November 2027	22	39	34	−33	22	35	+13.5	Piscis Austrinus
15 November 2027	00	15	42	−15	36	33	+14.8	Cetus

104P/Kowal 2

Discovery	27 January 1979 by Charles Thomas Kowal at Palomar Observatory, California, USA		
Name	Named after its discoverer Charles Thomas Kowal (1940–2011)		
Meteor showers	None identified (does not cross Earth's orbit)		
Classification	Jupiter-family comet	**Orbital eccentricity**	0.67
Aphelion	5.32 au	**Orbital inclination to ecliptic**	5.7°
Perihelion	1.07 au	**Orbital period (sidereal)**	5.74 years
Next perihelion	12 October 2027	**Closest approach to Earth**	23 September 2027

Comet **104P/Kowal 2** is visible in the morning sky from both hemispheres. It traverses the constellations of Taurus, Orion, Gemini, Cancer, Leo, and finally Sextans between September and December, with southern-latitude observing sites favoured later in the apparition. The comet never comes very close to Earth this time and is 69° away from the Sun at its closest approach to our planet. Comet 104P/Kowal 2 is a very faint object and may not brighten past sixteenth magnitude although JPL suggests that it may reach +13.

An image of 104P/Kowal 2 taken on 12 February 2022 by the Zwicky Transient Facility (ZTF), installed on the 48-inch Samuel Oschin Telescope at Palomar Observatory. (NASA/JPL-Caltech/ZTF)

Date	RA			Declination			Magnitude	Constellation
	h	m	s	°	′	″		
1 September 2027	05	32	08	+22	54	57	+16.8	Taurus
15 September 2027	06	44	36	+21	05	53	+16.4	Gemini
1 October 2027	08	03	21	+16	42	22	+16.2	Cancer
15 October 2027	09	04	04	+11	44	39	+16.2	Cancer
1 November 2027	10	06	14	+05	31	57	+16.5	Sextans
15 November 2027	10	48	42	+00	51	38	+16.8	Sextans

Minor Planets

Brian Jones and Lynne Marie Stockman

A *minor planet* is a body orbiting the Sun, and which is neither a planet, satellite, nor a comet. Many of these bodies are found in the main asteroid belt between Mars and Jupiter with the largest being the dwarf planet 1 Ceres, the first asteroid discovered. Others, called *Trojans*, congregate at a planet's L_4 and L_5 Lagrange points, places of gravitational equilibrium between the Sun, planet, and minor planet. Some asteroids cross the orbits of the inner rocky planets. A *Near-Earth Object* (NEO) has an orbit which takes it inside that of Mars. If an NEO larger than 140 metres in diameter crosses Earth's orbit, it is additionally classified as a *Potentially Hazardous Asteroid* (PHA).

Other minor planets inhabit the outer reaches of the solar system. *Centaurs* orbit the Sun between Jupiter and Neptune and cross the orbits of at least one of the outer gas giants. They typically display the characteristics of both asteroids and comets. *Trans-Neptunian Objects* (TNOs) have semi-major axes greater than that of Neptune with the dwarf planet 134340 Pluto being the first TNO discovered. TNOs are divided into several subgroups, including *Kuiper Belt Objects* and *Oort Cloud Objects*.

Over a million minor planets have been discovered since 1801, but with the occasional exception of 4 Vesta, none are bright enough to be seen with the naked eye. Detailed star charts and precise coordinates are necessary in order to find these faint planetoids, but even the brightest of these tiny worlds will appear only as a pinpoint of light in a telescope. The following is a small selection of minor planets that come to opposition this year, with the brightest objects well-placed for observation.

Geocentric right ascensions and declinations tabulated in this section are referred to the mean equator and equinox of J2000. This allows them to be plotted directly on modern star charts such as *Sky Atlas 2000.0* by Wil Tirion and Roger Sinnott.

Orbital and physical parameters are derived from the JPL Solar System Dynamics Small-Body Database at **ssd.jpl.nasa.gov/tools/sbdb_lookup.html#**

Positional and magnitude data is taken from the Minor Planet & Comet Ephemeris Service at **minorplanetcenter.net/iau/MPEph/MPEph.html**

Data in the following tables were correct as of 10 January 2026.

1 Ceres

Discovery	1 January 1801 by Giuseppe Piazzi (1746–1826) at Palermo Astronomical Observatory, Sicily		
Name	Named after the Roman goddess of agriculture		
Classification	Main-belt asteroid	**Orbital eccentricity**	0.080
Aphelion	2.99 au	**Orbital inclination to ecliptic**	10.6°
Perihelion	2.55 au	**Orbital period (sidereal)**	4.60 years
Opposition	7 January 2027 in the constellation of Gemini		

1 Ceres is the only dwarf planet (as of this writing) located in the main asteroid belt between Mars and Jupiter. Although some authorities do not classify it as an asteroid, only as a dwarf planet, it is unquestionably the largest object in this region. It was visited by the *Dawn* spacecraft in 2015 and the retired orbiter is still circling the dwarf planet. Despite its size, 1 Ceres is still too faint to be seen by the naked eye as it is usually seventh magnitude or fainter, so a telescope is always needed to view it. This miniature world comes to opposition this year on 7 January when it is 1.63 au distant from Earth. The New Moon of that date ensures that there will be no lunar light interference for any observer seeking out this distant world.

Date	**RA**			**Declination**			**Magnitude**	**Constellation**
	h	**m**	**s**	°	′	″		
1 December 2026	07	44	03	+26	00	34	+7.6	Gemini
15 December 2026	07	38	13	+27	26	22	+7.3	Gemini
1 January 2027	07	24	13	+29	18	25	+6.9	Gemini
15 January 2027	07	09	49	+30	37	26	+6.9	Gemini
1 February 2027	06	54	26	+31	38	44	+7.3	Gemini
15 February 2027	06	47	08	+31	59	55	+7.6	Gemini

6 Hebe

Discovery	1 July 1847 by Karl Ludwig Hencke (1793–1866) at Driesen, Brandenburg (now Drezdenko, Poland)		
Name	Named after the Greek goddess of youth		
Classification	Main-belt asteroid	**Orbital eccentricity**	0.20
Aphelion	2.92 au	**Orbital inclination to ecliptic**	14.7°
Perihelion	1.93 au	**Orbital period (sidereal)**	3.78 years
Opposition	17 February 2027 in the constellation of Leo		

6 Hebe is another main-belt asteroid. It is one of the brightest asteroids in the sky but never reaches naked-eye visibility. Analysis of its light curve suggests that it is angular in shape, rather than spherical, and the minor planet is known to have a high density, suggesting that it is not a gravitationally-bound rubble pile like so many other denizens of the asteroid belt. Ninth-magnitude 6 Hebe reaches opposition in Leo on 17 February when it is 1.69 au from Earth. The waxing gibbous Moon may provide some light interference.

Date	RA			Declination			Magnitude	Constellation
	h	m	s	°	′	″		
15 January 2027	10	32	15	+10	21	04	+9.8	Leo
1 February 2027	10	22	30	+12	55	36	+9.4	Leo
15 February 2027	10	10	32	+15	22	09	+9.1	Leo
1 March 2027	09	57	50	+17	39	34	+9.4	Leo
15 March 2027	09	47	19	+19	27	11	+9.8	Leo
1 April 2027	09	40	26	+20	45	12	+10.2	Leo

3548 Eurybates

Discovery	19 September 1972 by Cornelius Johannes van Houten (1920–2002) and Ingrid van Houten-Groeneveld (1921–2015) at Leiden University, Netherlands, from plates taken at Palomar Observatory, California, USA		
Name	Named after a Greek herald who served in the Trojan War		
Classification	Jupiter Trojan	**Orbital eccentricity**	0.091
Aphelion	5.69 au	**Orbital inclination to ecliptic**	8.1°
Perihelion	4.74 au	**Orbital period (sidereal)**	11.91 years
Opposition	11 May 2027 in the constellation of Libra		

The Jupiter trojans are two groups of asteroids clustered around Jupiter's L_4 (60° ahead in the orbit) and L_5 (60° behind) Lagrange points. They are typically named after characters in Greek mythology who took part in the Trojan War. **3548 Eurybates** is located in the leading camp and is the largest member of the Eurybates family, a group of asteroids formed from collisions with the original asteroid. It has a small satellite, Queta, which was discovered from *Hubble Space Telescope* images. 3548 Eurybates is scheduled to be visited by NASA's *Lucy* space probe (**science.nasa.gov/mission/lucy**) in August 2027, about three months after the eighteenth-magnitude asteroid reaches opposition for terrestrial astronomers.

Date	RA			Declination			Magnitude	Constellation
	h	m	s	°	′	″		
1 April 2027	15	26	29	−19	22	32	+17.5	Libra
15 April 2027	15	21	34	−19	15	45	+17.3	Libra
1 May 2027	15	14	04	−19	00	05	+17.1	Libra
15 May 2027	15	06	47	−18	41	53	+17.0	Libra
1 June 2027	14	58	24	−18	18	57	+17.3	Libra
15 June 2027	14	52	53	−18	03	34	+17.4	Libra

134340 Pluto

Discovery	18 February 1930 by Clyde Tombaugh (1906–1997) at Lowell Observatory, Flagstaff, Arizona		
Name	Named after the Roman god of the underworld		
Classification	Trans-Neptunian Object	**Orbital eccentricity**	0.25
Aphelion	49.56 au	**Orbital inclination to ecliptic**	17.1°
Perihelion	29.62 au	**Orbital period (sidereal)**	249.09 years
Opposition	29 July 2027 in the constellation of Capricornus		

Like 1 Ceres, **134340 Pluto** was once considered a fully-fledged planet, but it was reclassified as a 'dwarf planet' by the International Astronomical Union in 2006. It has five known satellites, the largest of which is Charon. Pluto and Charon are tidally locked, and their barycentre (centre of mass) actually lies outside Pluto. Pluto and its moons were visited by the *New Horizons* spacecraft when it sped by in 2015. At fifteenth-magnitude, it is a difficult object for amateurs although the Moon will be a waning crescent in late July and provide little to no light interference. Opposition occurs this year on 29 July when the dwarf planet will be a whopping 34.81 au distant from Earth (and getting further by the day).

Date	RA			Declination			Magnitude	Constellation
	h	m	s	°	′	″		
15 June 2027	20	40	33	−23	10	50	+15.1	Capricornus
1 July 2027	20	39	24	−23	17	43	+15.1	Capricornus
15 July 2027	20	38	10	−23	24	03	+15.0	Capricornus
1 August 2027	20	36	33	−23	31	34	+15.0	Capricornus
15 August 2027	20	35	13	−23	37	09	+15.0	Capricornus
1 September 2027	20	33	45	−23	42	43	+15.1	Capricornus

(137108) 1999 AN_{10}

<table>
<tr><td>Discovery</td><td colspan="3">13 January 1999 by LINEAR at White Sands Missile Range, Socorro, New Mexico, USA</td></tr>
<tr><td>Name</td><td colspan="3">Unnamed</td></tr>
<tr><td>Classification</td><td>Apollo asteroid, NEO, PHA</td><td>Orbital eccentricity</td><td>0.56</td></tr>
<tr><td>Aphelion</td><td>2.28 au</td><td>Orbital inclination to ecliptic</td><td>39.9°</td></tr>
<tr><td>Perihelion</td><td>0.64 au</td><td>Orbital period (sidereal)</td><td>1.76 years</td></tr>
<tr><td>Closest Approach</td><td colspan="3">~07:11 UT on 7 August 2027</td></tr>
</table>

(137108) 1999 AN_{10} will make an unusually close approach to Earth on 7 August 2027, passing 390,000 kilometres (just over a lunar distance) from the Earth's centre. This Near-Earth Object may be a kilometre in size or more, and is classified as a Potentially Hazardous Asteroid. It should be bright enough ($m < 8.0$) to be visible in binoculars and small telescopes from 07:00 to 12:00 UT. Closest approach is at 07:11 UT, but greatest brightness ($m = +7.5$) is at 09:00 UT. The Moon is a waxing crescent on 7 August and should not interfere with observations of this fast-moving object.

The event occurs during daylight for European observers. Regions in the western half of the United States and Canada are best placed to observe the entire close pass during full darkness, although observers in the Eastern and Central time zones will be able to see the early stages before their skies begin to brighten. Hawaii will see the last couple of hours of the pass as Pegasus rises just before midnight.

A chart depicting the close approach and passage of asteroid (137108) 1999 AN_{10} – cutting across Aries and Pisces, and skirting through Andromeda towards the north-western reaches of Pegasus – appears on the following page.

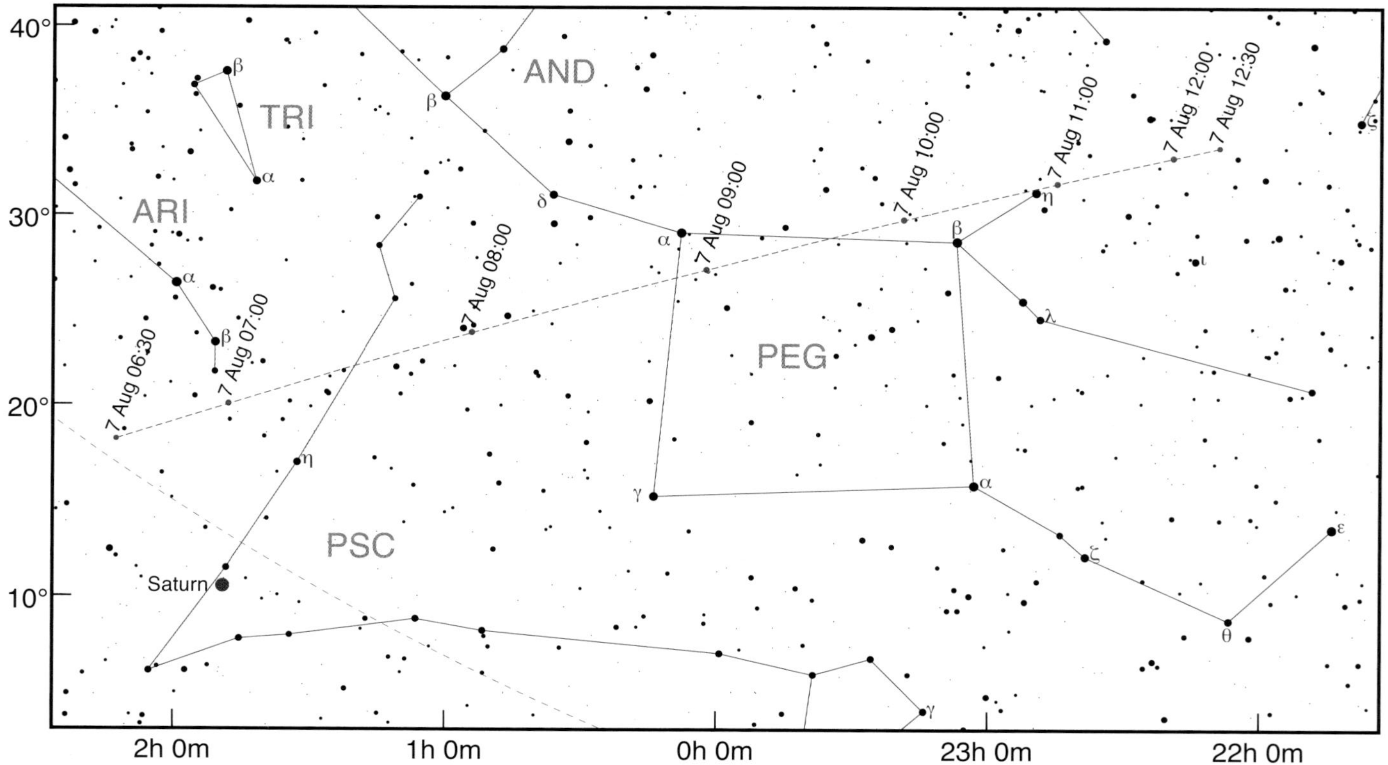

The close approach of asteroid (137108) 1999 AN_{10} on 7 August 2027. The asteroid is magnitude 7.5 at its brightest at 09:00 UT, and brighter than 8.1 along the whole of the track depicted. Stars are shown to magnitude 7.0. The asteroid track is geocentric. Parallax may shift its observed position by up to 0.5°. (David Harper)

Planets

Lynne Marie Stockman

Mercury

The nearest planet to the Sun is at superior conjunction on the first day of the year, leading into the first evening apparition of 2027. This western appearance, the first of three, favours northern observers but the second evening apparition, in May and June, is even better. Those seeking this elusive planet from the southern hemisphere will enjoy their best views of **Mercury** in the west later in the year, from mid-August to mid-October. Mercury also puts in three morning appearances. The first one, from mid-February through to the end of April, is the best dawn presentation for astronomers in the southern hemisphere, whilst the last morning apparition from mid-October to mid-December provides early risers in northern temperate latitudes their best opportunity to catch this planet before sunrise. Mercury is occulted by the Moon three months running – January, February, March – and again in August and December. It also glides past all of the planets in the sky except Mars. However, all of these planetary match ups take place less than 20° from the Sun and some of them may be difficult, if not impossible, to see.

Best Apparitions	52° N Latitude	35° S Latitude
Morning	19 October–11 December	18 February–28 April
Evening	28 April–23 June	11 August–19 October

Apparition diagrams showing the position of Mercury above the eastern horizon at sunrise and the western horizon at sunset may be found throughout the *Monthly Sky Notes*.

Venus

Venus is the morning star at the beginning of 2027. For southern hemisphere observers, it continues to gain altitude above the eastern horizon throughout January, but it is already on the descent for early birds living in northern temperate latitudes. The bright planet's evening apparition begins after superior conjunction in mid-August, with the southern hemisphere again favoured. Close encounters with Neptune (April), Saturn (May), and Mars (November) take place over 25° from the Sun but conjunctions with Uranus (June), Mercury (July, August, and October), and Jupiter (also August) occur too close to the Sun to be easily visible.

The Moon mostly keeps its distance but Venus is occulted twice this year, in early August (morning sky) and again in late November (evening sky). Venus begins the year at its brightest, magnitude −4.5, before dimming to 'only' −3.9 as it approaches conjunction with the Sun, and then recovering to −4.0 by the end of the year.

Apparition diagrams showing the position of Venus above the eastern horizon at sunrise and the western horizon at sunset may be found in the January and August *Sky Notes* respectively.

Mars

Rising mid-evening as 2027 commences, **Mars** is in the constellation of Leo and passes within a degree of three Messier galaxies – M95, M96, and M105 – in late January. Opposition takes place in mid-February and the red planet remains an evening sky object for the rest of the year. Mars passes into the constellation of Virgo in early July, then Libra in September, Scorpius and Ophiuchus in October, and finally Sagittarius in mid-November. The first-magnitude planet teams up with a much-brighter Venus low in the west in late November, then passes just 0.1° south of the Lagoon Nebula (M8) a few days later. Mars also passes by the globular clusters M22 and M28 in early December but the planet is drawing near to the Sun by that point. The Moon never approaches closer than 3° until the end of November, with a lunar occultation taking place in December.

An apparition diagram showing the position of Mars above the western horizon at sunset may be found in the September *Sky Notes*, and a finder chart showing the position of Mars throughout the first six months of 2027 may be found at the end of this article.

Jupiter

Jupiter begins the year in Leo, retrograding back into the constellation of Cancer in mid-March before direct motion returns it to Leo in mid-May and eventually Virgo in late November. The largest planet in the solar system is rising by mid-evening at the beginning of 2027, with opposition taking place in mid-February. July sees the gas giant pass less than a degree from fifth-magnitude star ν Leonis and first-magnitude Regulus (α Leonis) but Jupiter is getting low in the west by that point. Conjunction occurs at the end of August. Afterwards a morning sky object, Jupiter is found close to fourth-magnitude star σ Leonis in late October. The Moon occults Jupiter four times this year, in February, March, April, and May.

A finder chart showing the position of Jupiter throughout 2027 may be found at the end of this article.

Saturn

The ringed planet is found in the non-zodiacal constellation of Cetus at the start of the year, and moves into Pisces in late February, where it remains for the duration of

2027; it is slowly leaving Neptune behind after the two planets' triple conjunction in 2025–6. **Saturn** is found in the evening sky as the year gets underway but is setting ever earlier on its way to conjunction in early April. As a morning sky object, Saturn is best viewed from the southern hemisphere where its close encounter with Venus in early May takes place in dark skies. The rings open to their maximum tilt for the year in early August before slightly closing up again. However, the rings are still open to nearly 13° when Saturn is at opposition in mid-October. The planet passes less than a degree away from the fifth-magnitude star μ Piscium twice, in May and October, but the Moon stays over 6° away from the planet all year.

A finder chart showing the position of Saturn throughout 2027 may be found at the end of this article.

Uranus

Sixth-magnitude **Uranus** continues it course across the constellation of Taurus, passing by the fifth-magnitude Bayer-lettered stars ω Tauri, κ^1 Tauri, and κ^2 Tauri, and the fifth- and sixth-magnitude Flamsteed-numbered stars 51 Tauri, 53 Tauri, and 56 Tauri in its travels. Uranus is visible in the evening as the year commences, having reached opposition in late November 2026. It sets after midnight until late March for observers in northern latitudes but sinks into the evening twilight by May, undergoing conjunction near the end of the month. Afterwards it appears in the east, more easily detected in the southern hemisphere winter skies than the all-night astronomical twilight of northern temperate latitudes. Opposition occurs at the end of November. The Moon keeps its distance, never approaching closer than 4° but given the faintness of the green ice giant, choose a moonless night when observing Uranus.

A finder chart showing the position of Uranus throughout 2027 may be found at the end of this article.

Neptune

The most distant planet in our solar system is found in Pisces, a constellation it will inhabit until April 2028 when it dips into neighbouring Cetus. **Neptune** opens the year in the evening sky and is best seen from the northern hemisphere, but the eighth-magnitude ice giant is lost to view by March, undergoing conjunction late in the month. Reappearing at dawn, the planet's April rendezvous with Mercury and Venus are best viewed in the dark autumnal skies of the southern hemisphere. Neptune returns to the evening sky around mid-year and reaches opposition at the end of September. The Moon never comes closer than about 5° this year and the planet does not pass by any bright stars.

A finder chart showing the position of Neptune throughout 2027 may be found at the end of this article.

Mars

January to June 2027

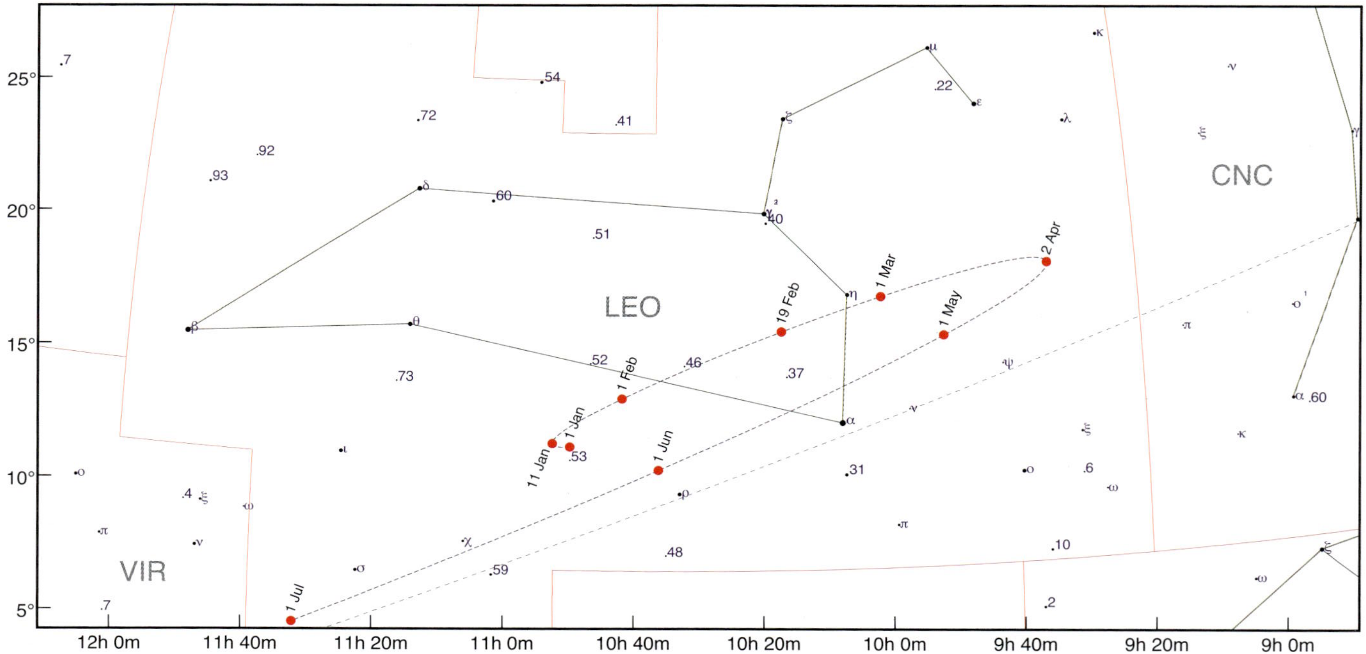

Background stars are shown to magnitude +5.5. (David Harper)

Jupiter
January to December 2027

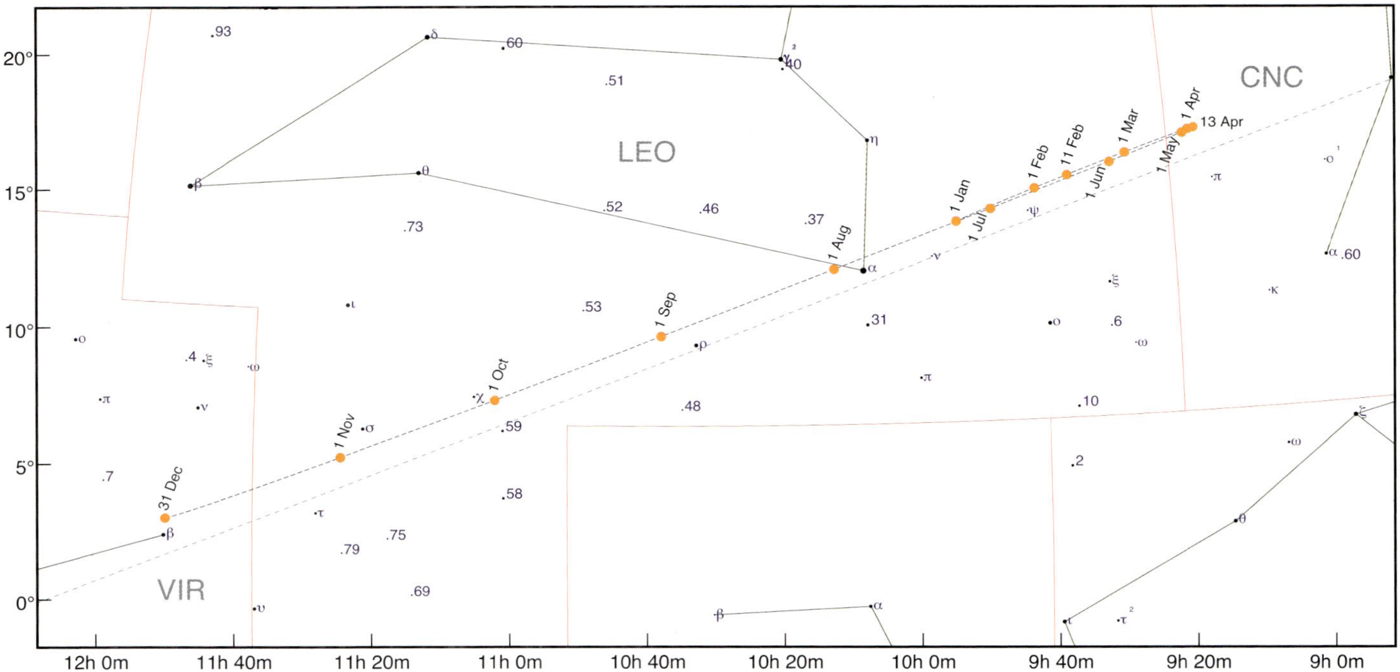

Background stars are shown to magnitude +5.5. (David Harper)

Saturn
January to December 2027

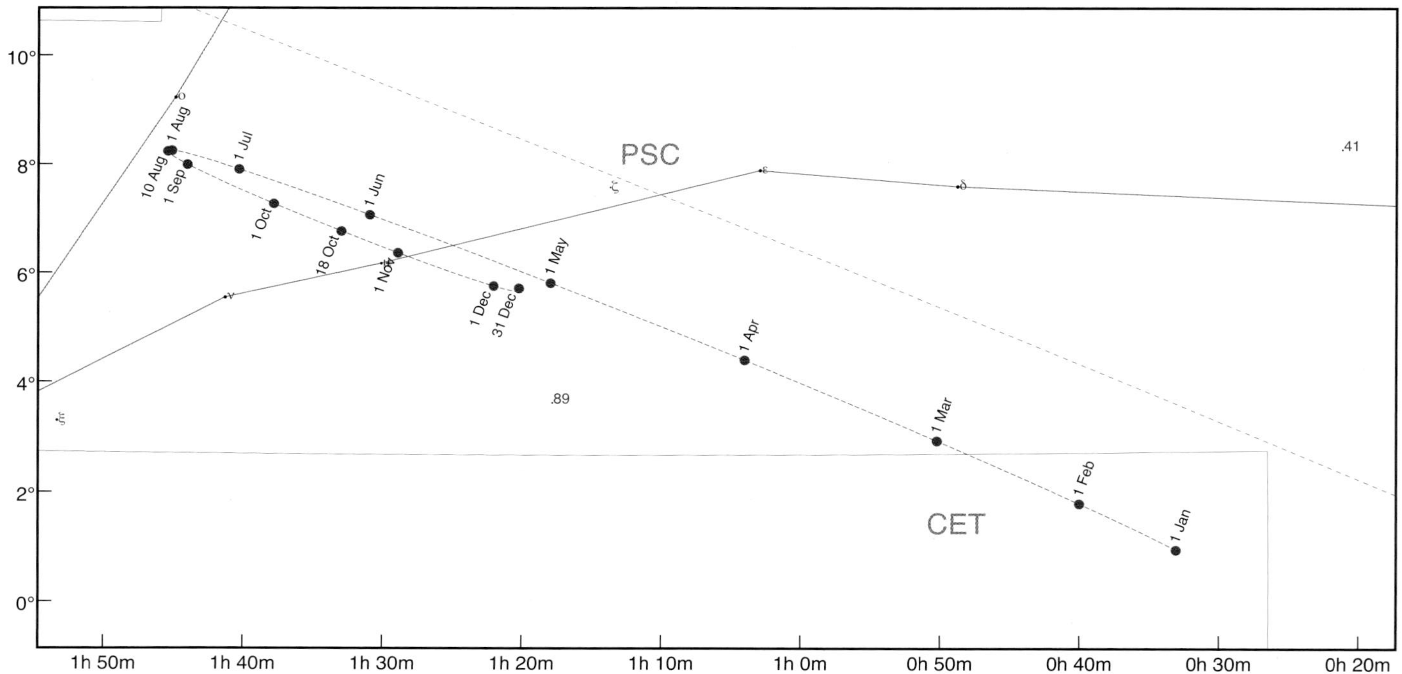

Background stars are shown to magnitude +5.5. (David Harper)

Uranus

January to December 2027

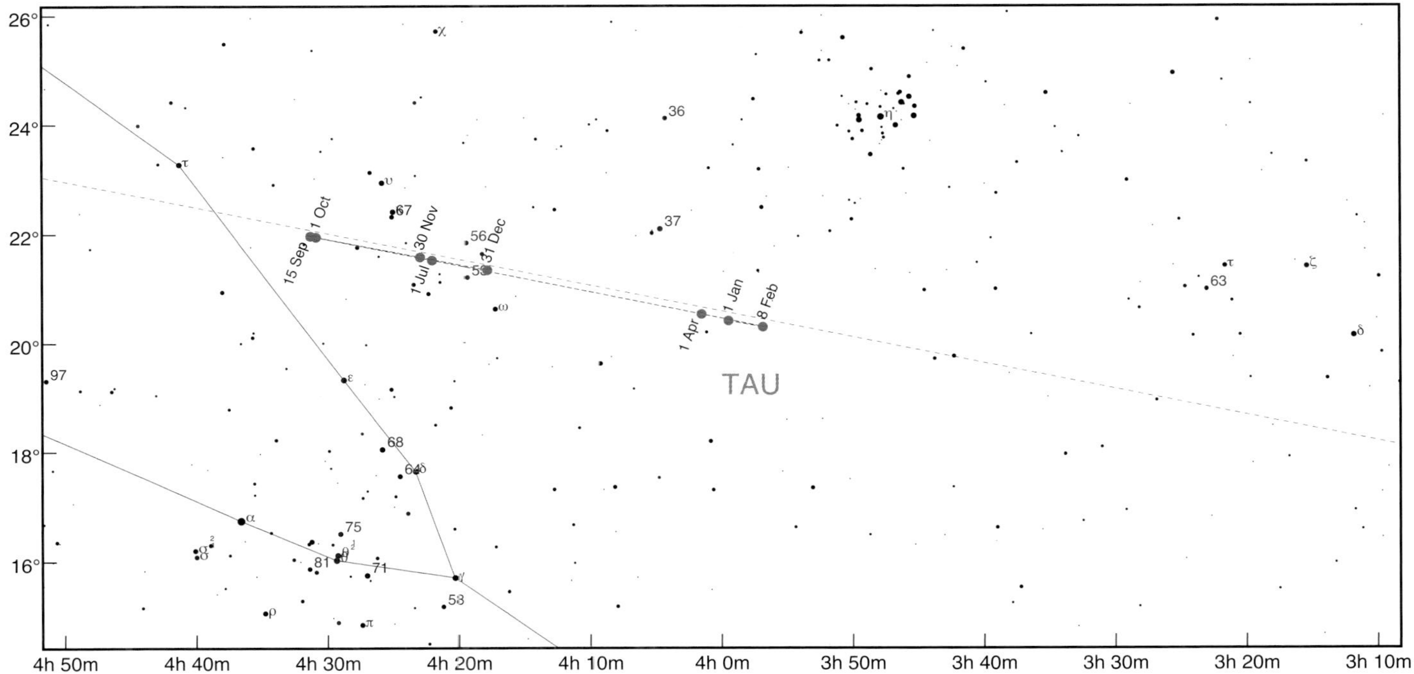

Background stars are shown to magnitude +8.0. (David Harper)

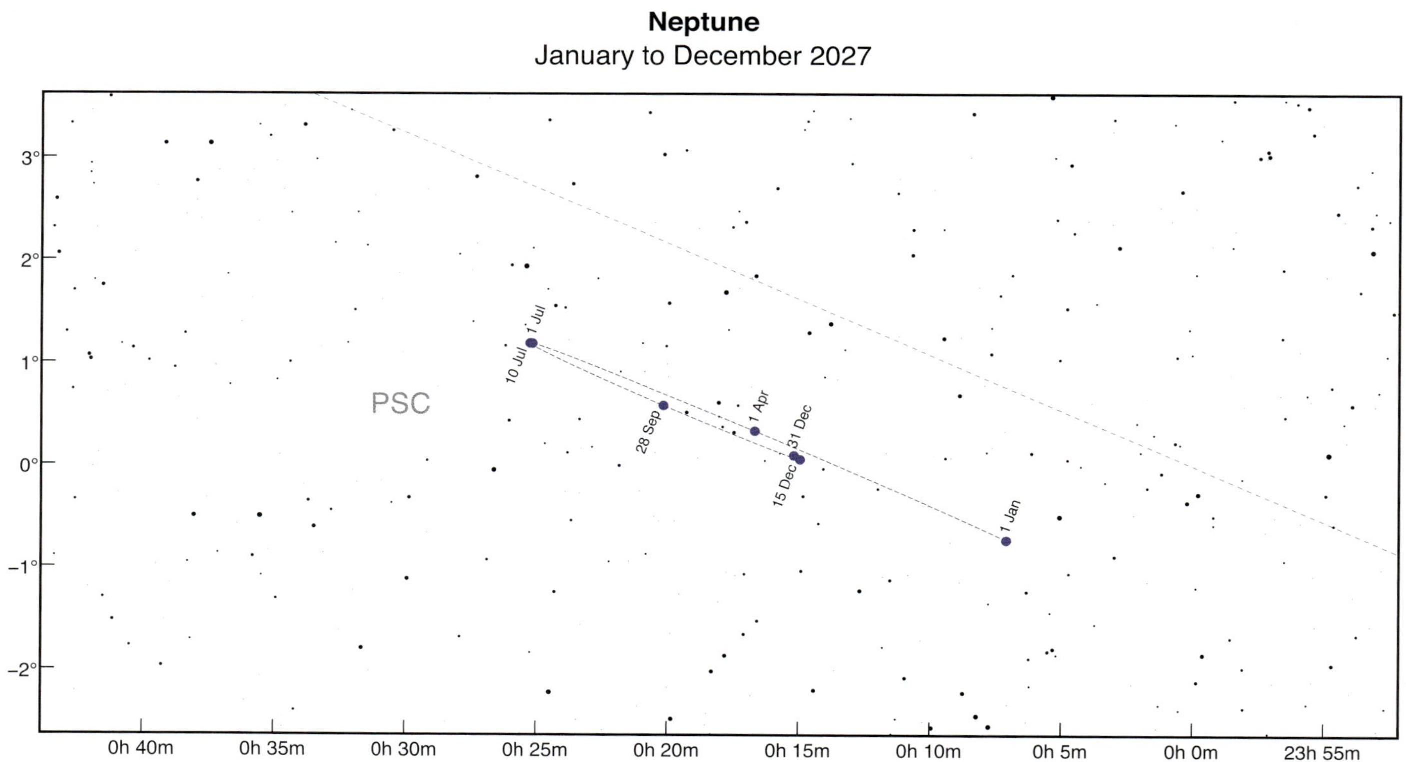

Background stars are shown to magnitude +10.0. (David Harper)

Monthly Sky Notes and Articles

Morning Apparition of Venus

24 October 2026 to 12 August 2027

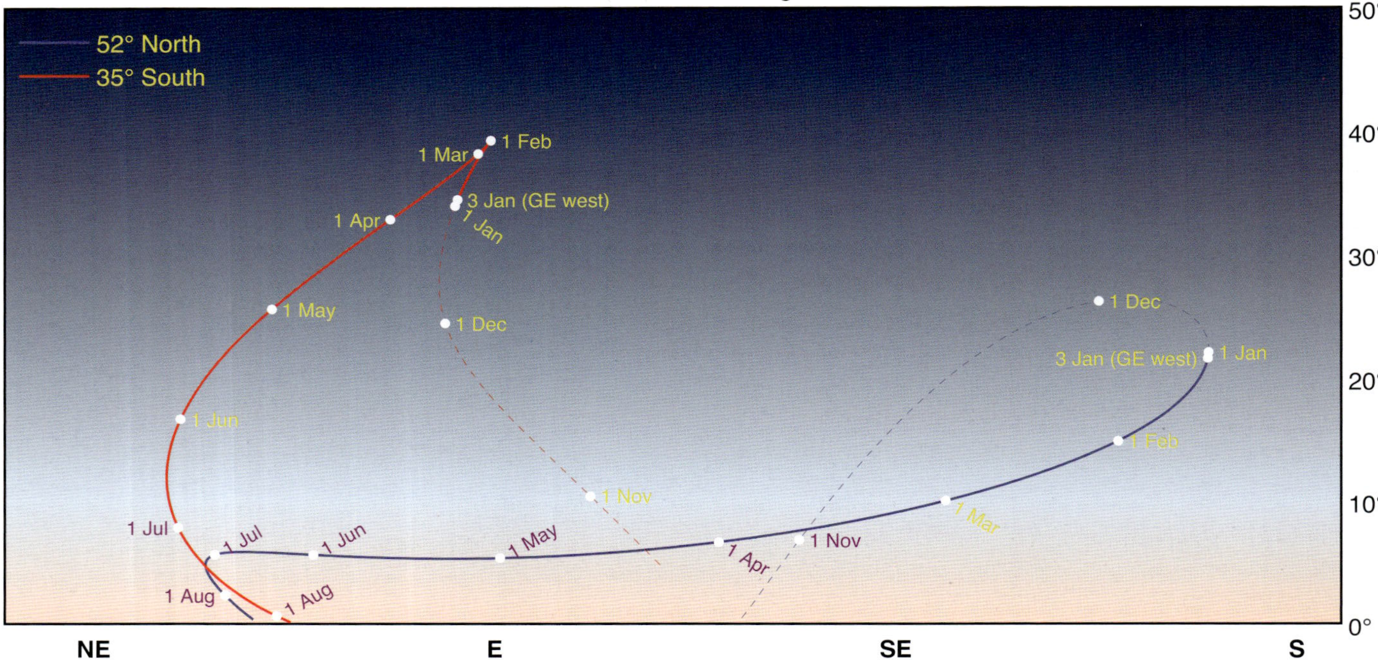

January

1	Saturn	Minimum geocentric ring tilt: −6.4°	88°
1	Mercury	Superior Conjunction (morning → evening)	2°
3	Earth	Perihelion: 0.9833 au	
3	Venus	Greatest Elongation West (morning)	47°
3/4	Earth	Quadrantids (ZHR 80)	
4	Moon, α Scorpii (Antares)	0.3° apart: occultation	35°
7	1 Ceres	Opposition	172°
7	Moon	New	3°
8	Moon, Mercury	0.5° apart: occultation	5°
15	Moon	First Quarter	90°
18	Moon, M45 (Pleiades)	1.1° apart: occultation	122°
22	Moon	Full	178°
22	Moon, M44 (Beehive)	28″ apart: occultation	173°
24	Mars, M96	0.3° apart	146°
26	Mars, M105	0.5° apart	147°
27	Mars, M95	0.8° apart	149°
29	Moon	Last Quarter	90°
31	Moon, α Scorpii (Antares)	0.4° apart: occultation	62°

Dates are based on Universal Time (UT). Occultations of M44 (Beehive/Praesepe) refer to sixth-magnitude star ε Cancri (Meleph). Occultations of M45 (Pleiades) refer to third-magnitude star η Tauri (Alcyone). Peak activity dates and ZHRs for meteor showers are estimates. The last column gives the approximate elongation from the Sun of the event.

Mercury is at superior conjunction on the first day of the year and spends the rest of the month in the west after sunset. This apparition favours northern temperate observers although the next evening apparition in May and June is slightly better. A lunar occultation by a very young crescent Moon takes place on 8 January but both objects are very close to the Sun at the time.

Venus adorns the morning sky for the first eight months of 2027, with greatest elongation west (47.0°) taking place on the third day of January. It is higher

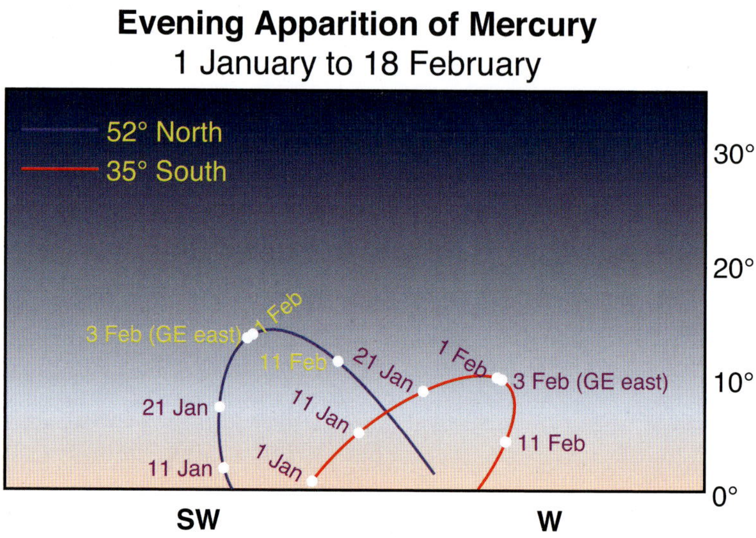

in the sky as seen from the southern hemisphere than from the north, and will continue to climb above the eastern horizon all month for those favoured observers. For early risers in northern temperate latitudes, however, the morning star is already on the descent. The planet, which appears as a waxing crescent when seen through a telescope, is at its brightest at the beginning of the year: magnitude −4.5.

Earth swoops in to perihelion, its closest approach to the Sun, on 3 January. The Sun's disk is 1952″ across on this day, which is 3.4% larger than appears when Earth is at aphelion in July. (*Warning: Do not observe the Sun directly without proper equipment!*) The waning crescent Moon provides minimal interference for observing the Quadrantids, the first major meteor shower of the year, which coincidentally peak around the same time as perihelion.

The **Moon** continues its relentless occultation schedule, obscuring the first-magnitude star Antares twice this month and the open clusters known as the Pleiades and the Beehive once each. However, the light of the Full Moon will completely obliterate any observations of the faint Beehive Cluster on 22 January. Mercury also falls victim to the Moon but having just completed superior conjunction, is extremely close the Sun when it is occulted. The Moon reaches its most distant apogee of the year on the seventh, followed by its nearest perigee 14 days later.

The Full Moon of the following day exhibits the largest apparent angular size, a phenomenon often termed a 'Supermoon' by the popular media.

Mars enters 2027 in direct motion but reverses course in the constellation of Leo over 10–11 January. This retrograde motion bring the red planet back into close proximity to three prominent galaxies – M96, M105, and M96 – between 24 January and 27 January although you will need a telescope to see these objects as they are all ninth or tenth magnitude. Mars rises mid-evening.

Jupiter is in retrograde in Leo, rising an hour or so before Mars. The waning gibbous Moon passes 1.4° south of the gas giant on 23–24 January but it will get even closer next month, resulting in a lunar occultation.

Saturn is located in the north-west corner of the constellation of Cetus, near the border with Pisces, which makes it an evening sky object this month. It sets before midnight. At the beginning of January, the planet's rings are at their minimum opening aspect for the year; the tilt will widen until August. This month, Saturn is at magnitude +0.8.

Uranus retrogrades into the year in the constellation of Taurus. The sixth-magnitude planet is best viewed (on a dark night!) from the northern hemisphere where it does not set until well after midnight.

Neptune is found in not far from Saturn in the sky, having completed a triple conjunction (in right ascension) with the ringed planet last year. Viewing conditions for this eighth-magnitude ice giant are somewhat better in northern latitudes where the planet does not set until mid-evening.

The Trapezium Cluster

Jonathan Powell

Orion the Hunter makes for one of the most impressive constellations in the winter sky which along with its easily recognisable pattern of stars, hosts a number of celestial treasures. Two of Orion's stars, Betelgeuse and Rigel, sit in the top ten list of brightest stars that reside within the 88 designated constellations that can be observed in both the northern and southern hemisphere respectively. Nestled beneath Orion's Belt, which is comprised of three further bright stars, is the impressive diffuse nebula Messier 42. Also known as the Orion Nebula, this famous object has at its very heart an open cluster of stars known as the Trapezium Cluster.

The Orion Nebula – believed to be the cosmic fire of creation by the Maya civilisation, one of the most famous of classical societies of Mesoamerica – is a well-researched stellar nursery and the closest star-forming region to Earth. Its discovery as a diffuse nebula is widely credited to French astronomer Nicolas-Claude Fabri de Peiresc in 1610, with his observing partner Joseph Gaultier de la Vallette having also observed the nebula that same year.

The first published observation of the Orion Nebula was made in 1619 by Swiss Jesuit mathematician and astronomer Johann Baptist Cysat of Lucerne, making a comparison between the gaseous cloud and a bright comet that he had observed in 1618, describing how the nebula appeared in his telescope. There were three comets visible to the naked eye in 1618, the brightest of these being referred to as 'The Great Comet', C/1618 W1, which sported a tail measuring up to 90 degrees long.

The French astronomer Charles Messier observed the nebula in 1769 with the first edition of his catalogue published in 1774; the nebula numbered 42 on his list of deep sky objects, hence the designation M42. At the time of observing, Messier also noted the three stars of the Trapezium. In the *New General Catalogue of Nebulae and Clusters of Stars*, compiled by Danish astronomer John Louis Emil Dreyer and published in 1888, Dreyer afforded the Orion Nebula the tag of NGC 1976.

The Trapezium Cluster lies within the heart of the Orion Nebula. Also known as the Orion Trapezium Cluster – or by its Bayer designation (after the German celestial cartographer Johann Bayer) Theta^1 Orionis – the cluster is 1,600 light years from Earth. The focal point of the cluster is its four brightest stars which form

Focusing in on the central region of the Trapezium Cluster, the *Hubble Space Telescope* captures the sparkling imagery of this interstellar nursery nestled in Orion, the Hunter's M42 nebula. (Image ©ESA)

an asterism in the shape of a trapezoid or quadrilateral. Along with a fifth next brighter star, they all sit within 1.5 light years of each other. The Trapezium Cluster can also be considered a sub-component of the much larger Orion Nebula Cluster which collectively has over 2,000 stars, and spans a diameter of 20 light years.

The trapezium 'pattern' and similar arrangements exist outside of the formal boundaries that have been designated to the constellations. Many of these have been created as the seasons change, such as the Summer Triangle and the Winter Triangle, both triangles being formed by linking bright seasonal stars into easily recognisable shapes. However, the use of such asterisms depends on how often they are quoted and referred to, and in the case of the Trapezium Cluster, Swiss-American astronomer Robert Julius Trumpler, who specialized in the field of star clusters studies, is considered the 'primary' source in the use of the name 'Trapezium', which became commonplace both during his life and subsequent to his passing in 1956.

The Trapezium Cluster comprises four main relatively bright stars, A, B, C, and D, with their brightness reflected by order of increasing right ascension. The brightest of the four is Theta Orionis C. Ultraviolet ionizing radiation from the Trapezium stars and, in particular, Theta-1 Orionis C forms the powerhouse behind the region's glow. The four brightest stars range from magnitude 5 to magnitude 8 and are all hot class O and B stars, recognized for their extreme temperatures and

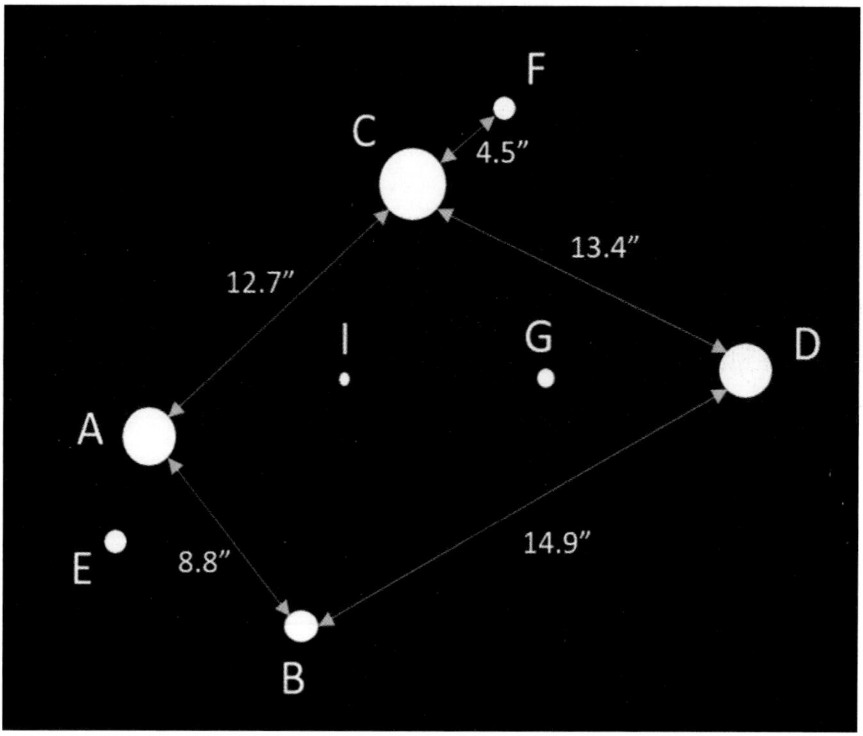

A schematic showing the Trapezium Cluster's main components: Theta-1 Orionis A (label A); Theta-1 Orionis B (label B); Theta-1 Orionis C (label C); Theta-1 Orionis D (label D). Excluded from the schematic are H1 and H2, although the fifteenth magnitude star 'I' is included. (ResearchGate/CC BY 4.0)

brightness. Under good observing conditions, a 5-inch telescope will reveal four of the cluster's stars, a 6-inch telescope or larger adding a fifth and sixth.

Following the earlier discovery of M42, it was Galileo in 1617 who proceeded to make sketches of three of the Trapezium Cluster's stars, (A, C, and D), although his drawings were not published during his lifetime.

Italian astronomer Giovanni Battista Hodierna also drew the Trapezium as a triple star before 1654, with Christiaan Huygens independently discovering the three stars and describing the cluster in 1656. French astronomer Jean Picard discovered 'B' in 1673, thereby completing the Trapezium, with Huygens also observing the fourth component in 1684. In 1826, William Struve discovered 'E', with John Herschel discovering 'F' in 1830. American astronomer and telescope maker Alvan Clark added 'G' in 1888 with Edward Emerson Barnard adding 'H' in the same year, making for eight members in total. A few of these were subsequently discovered to be binaries, with 'B' being a five-star system that can also be viewed as a mini-cluster.

The Trapezium stars display high velocities, one explanation for their behaviour possibly indicating the presence of a black hole, believed to be more than 100 times the mass of the Sun. The possible presence of a black hole was first suggested in a paper published in 2012; if indeed a black hole was present, it would be the closest one to Earth.

February

3	Mercury	Greatest Elongation East (evening)	18°
6	Moon	New: annular solar eclipse	0°
8	Moon, Mercury	0.3° apart: occultation	17°
10	2P/Encke	Perihelion	8°
11	Jupiter	Opposition	179°
14	Moon	First Quarter	90°
14	Moon, M45 (Pleiades)	1.2° apart: occultation	95°
16	6 Hebe	Opposition	176°
18	Mercury	Inferior Conjunction (evening → morning)	4°
19	Moon, M44 (Beehive)	48″ apart: occultation	157°
19	Mars	Opposition	176°
20	Moon, Jupiter	1.1° apart: occultation	170°
20	Uranus	East Quadrature	90°
20	Moon	Full: penumbral lunar eclipse	179°
25	Mars, η Leonis	0.5° apart	170°
28	Moon	Last Quarter	90°
28	Moon, α Scorpii (Antares)	0.4° apart: occultation	89°

Dates are based on Universal Time (UT). Occultations of M44 (Beehive/Praesepe) refer to sixth-magnitude star ε Cancri (Meleph). Occultations of M45 (Pleiades) refer to third-magnitude star η Tauri (Alcyone). The last column gives the approximate elongation from the Sun of the event.

Mercury is an evening sky object, better placed for viewing from northern temperate latitudes than from the southern hemisphere. The tiny planet is shining at magnitude −0.5 when it reaches greatest elongation east on 3 February. Perihelion takes place three days later. Another lunar occultation of Mercury happens this month, on 8 February, although visibility of this event will likely be confined to the uninhabited Pacific Ocean. The following day Mercury enters into retrograde motion, followed by inferior conjunction on 18 February. It reappears in the morning sky afterwards, in what will be the best morning apparition of the year for early birds in the southern hemisphere. The brightness of Mercury varies wildly this month, beginning the month around −0.7, dimming to sixth magnitude at the time of conjunction, and then recovering to +1.4 by the end of February.

Venus is *morgensteorra* or 'morning star' in Old English. It is best seen from southern latitudes where it reaches it maximum altitude above the eastern horizon this month. It is much lower for northern hemisphere observers but the darker skies and shorter twilights help with viewing this bright object. When seen through a telescope, Venus is getting smaller in diameter but rounder in appearance as it leaves Earth behind and heads towards superior conjunction in August. It is also dimming, dropping about a tenth of a magnitude over the course of the month.

The **Moon** participates in two eclipses this month, with an annular eclipse taking place on 6 February and a faint penumbral lunar eclipse occurring two weeks later. Along with the usual suspects (Antares, the Pleiades open cluster, the Beehive open cluster), waxing crescent Moon obscures Mercury on 8 February. Additionally, the Full Moon occults Jupiter this month. This last event finishes some hours before the penumbral lunar eclipse gets underway.

Mars reaches opposition mid-month. The planet is at its closest to Earth around this time and shines at magnitude −1.3. The disk of the planet is 13.81″ when seen through a telescope, with the northern hemisphere tilted towards Earth. The red planet passes half a degree south of the third-magnitude star η Leonis; the star is A-type and bluish which should make an interesting contrast to the rusty colour of Mars.

Jupiter is also visible all night as it reaches opposition eight days before Mars does. Found in the constellation of Leo, Jupiter is shining at a brilliant magnitude −2.6, presenting a disk 45.16″ across when viewed in a telescope. The Full Moon occults the gas giant on 20 February, the first of four lunar occultations of Jupiter this year.

Saturn is getting low in the west as its April conjunction approaches. Found in the constellation of Cetus until 23 February, when it moves into Pisces, the first-magnitude planet is best seen from northern latitudes.

Uranus is at east quadrature this month, meaning it is 90° away from the Sun in the evening sky. The green ice giant ends its retrogression in Taurus on 8 February and is most easily viewed from the northern hemisphere where it does not set until after midnight. By contrast, observers in the southern hemisphere lose Uranus in the west by mid-evening.

Neptune continues its slow journey across the sprawling constellation of Pisces. It is getting low in the west, especially for faint-planet enthusiasts in the southern hemisphere, where it sets early in the evening.

Supernova 1987A

David M. Harland

On the evening of 23 February 1987, Ian Shelton, an astronomer from the University of Toronto in Canada, took a long-exposure plate on one of the smaller telescopes of the Las Campanas Observatory in the high desert of Chile. This telescope, a 10-inch astrograph with a wide field of view that was ideal for surveying the sky, had a distinguished history, but on this occasion Shelton's purpose was to test its accuracy tracking across the sky. His target was the Large Magellanic Cloud (LMC), an irregular companion to our Milky Way galaxy located at a distance of around 160,000 light years.

After the three hour exposure he developed the plate and, being familiar with the field, observed an anomalously bright object on the outskirts of the Tarantula Nebula. Estimating it as being about fifth magnitude, he went outside and visually confirmed the newcomer.

On meeting some colleagues he asked whether anyone had noticed anything about the LMC. Oscar Duhalde, one of the technicians at the observatory, remarked that a few hours ago he had noted the Tarantula was unusually bright.

Several hours later, the object was independently discovered by the renowned amateur astronomer Albert Jones in New Zealand while making a routine visual survey of variable stars.

The LMC had been photographed by Robert H. McNaught using the Satellite Schmidt Camera at the Siding Spring Observatory in Australia the previous night (local time). On examining the plate on 24 February, he found he had caught the supernova early in its rise.

Designated SN 1987A, this was the first supernova for which it was possible to identify the progenitor star. On a plate taken on the evening of 22 February by Shelton there was no indication of what was to come.

The presumption was that the progenitor would be a red supergiant with an extended envelope, but it turned out to be the blue supergiant Sanduleak –69° 202. Of spectral type B3, it was around 20 solar masses and had a diameter 40 times our Sun's. One surprise was that it had an overabundance of helium in its outer layers.

Importantly, because the LMC is circumpolar for most observatories in the southern hemisphere, it was possible to gather an uninterrupted photometric

When the ESO Schmidt Telescope at the La Silla Observatory in the Chilean Atacama Desert took this image of the Tarantula Nebula in the Large Magellanic Cloud, SN 1987A (the very bright star in the middle right) was visible with the unaided eye. (ESO)

and spectroscopic record through the outburst's peak into May. In addition, it was observable in daylight in the near-infrared and mid-infrared for more than a year after the explosion. At its brightest it was visually about third magnitude.

As the nearest supernova since the invention of the telescope it gave astrophysicists their first opportunity to study the evolution of such an event in detail.

The light curve indicated a Type II supernova, implying that a very massive star had undergone sudden core collapse. However, the curve was different from previously observed core-collapse supernovas. The spectroscopic evolution gave further indications of asymmetries in the explosion. As a result, models developed

This striking *Hubble Space Telescope* image released on 19 May 1994 shows three rings of glowing gas encircling the site of SN 1987A. (Dr. Christopher Burrows, ESA/STScI and NASA)

of spherical explosions were revised to accommodate density inhomogeneities in a star's structure.

In 1989 a circumstellar ring was observed around SN 1987A for the first time, and in 1994 images by the repaired *Hubble Space Telescope* revealed the unusual remnant to consist of rings aligned along an axis of symmetry, giving it the shape of an hourglass. The rings were dense regions in the stellar wind ejected by Sanduleak –69° 202 over a period of many years, and they became visible when ionised by the ultraviolet radiation from the explosion, as so-called 'light echoes'. When the fast-moving ejecta of the supernova reached the inner ring, the shock wave released intense X-ray emission.

In 2007 simulations by Thomas Morris and Philipp Podsiadlowski of the University of Oxford supported the hypothesis that the triple-ring system was the result of a 15–20 solar mass red giant and a five solar mass star merging around 20,000 years ago. This merger would also explain why Sanduleak –69° 202 was a blue supergiant with a strange chemical composition.

Approximately two to three hours before the visible light from the supernova reached Earth, an unprecedented 'burst' of neutrinos was observed. The Kamiokande II detector in Japan observed a dozen, the Irvine-Michigan-Brookhaven detector in the USA observed eight, and the Baksan Neutrino Observatory in Russia observed five. Detecting 25 neutrinos in such a short interval was significantly above the background level. To the delight of astronomers it was soon realised that the neutrinos were from SN 1987A. As all three neutrino observatories were located in the northern hemisphere, the elusive particles had travelled through the Earth prior to reaching the detectors.

Because the emission of neutrinos occurs simultaneously with core collapse they precede the emission of visible light, which blooms only after the shock wave reaches the stellar surface. This was the first time neutrinos from a supernova were seen directly. It was consistent with theoretical models where 99% of the energy released by the collapse of the stellar core is radiated away in the form of neutrinos.

The collapse model suggested that in this case the result would have been a neutron star rather than a black hole, so astronomers were eager to find it. Infrared data from the *Herschel Space Observatory* in 2011 indicated that SN 1987A had released a great deal of fresh dust into the interstellar medium, and this was obscuring our view of the explosion site. In 2019 the Atacama Large Millimeter/submillimeter Array in Chile found evidence that suggested the presence of a neutron star located inside one of the brighter clumps of dust. Then in 2021 the *Chandra X-ray Observatory* and *NuSTAR (Nuclear Spectroscopic Telescope Array)*

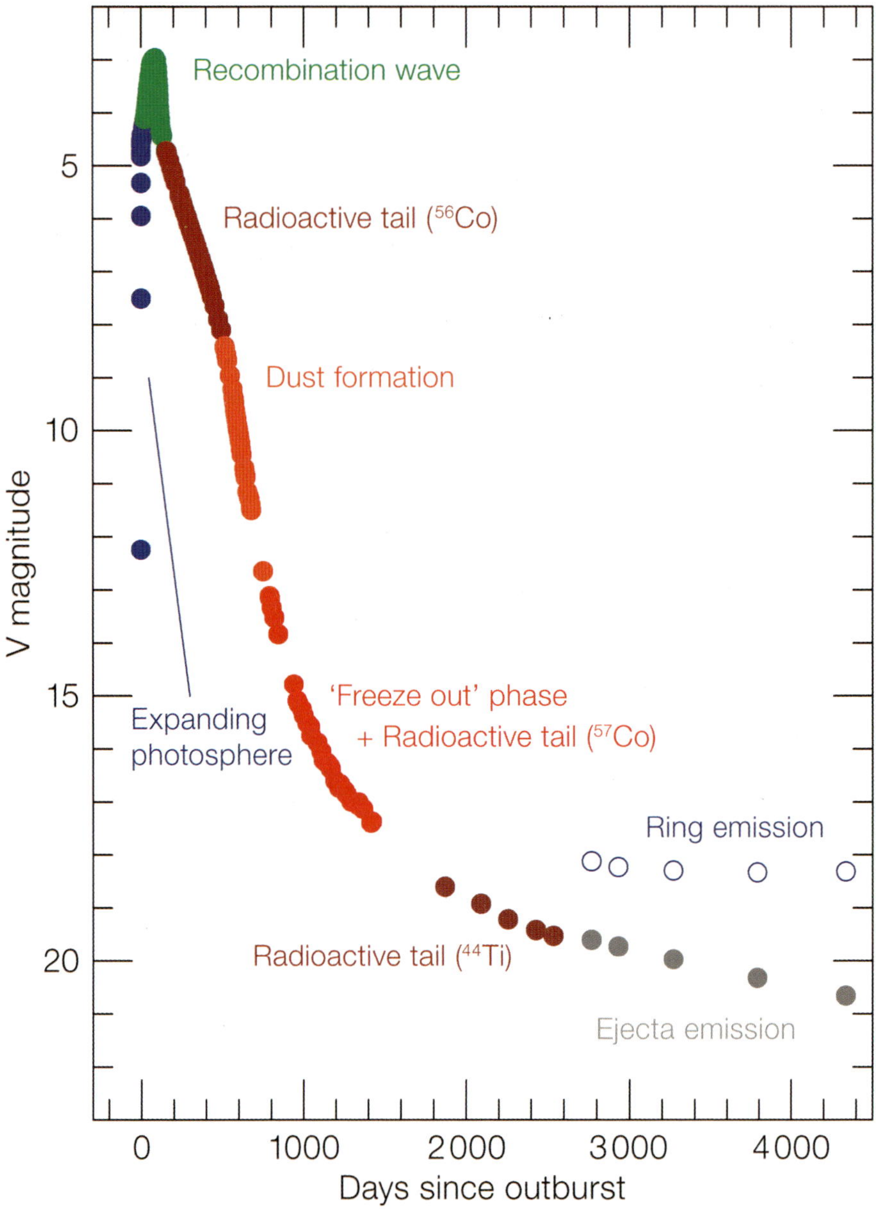

The long-term light curve of SN 1987A with characteristic phases in the evolution of the supernova indicated. (ESO)

satellites detected X-rays suggestive of a pulsar. In July 2022, one of the first targets for the newly commissioned *James Webb Space Telescope* was SN 1987A, and its spectroscopic observations confirmed the high-energy emissions of a neutron star. Astronomers now have an unprecedented opportunity to study the development of a 'baby pulsar'.

All in all, SN 1987A in the LMC provided astronomers with a good dress rehearsal for what to do when a supernova eventually occurs in our own backyard, and it would be nice if this were to be a star as well known to us as the irregularly fluctuating red supergiant Betelgeuse in the famous constellation of Orion.

Further Reading

Supernova: The Violent Death of a Star, Donald Goldsmith, St. Martin's Press, NY, 1989.

March

6	Moon, Mercury	1.1° apart: occultation	24°
8	Moon	New	2°
13	Moon, M45 (Pleiades)	1.1° apart: occultation	67°
15	Moon	First Quarter	90°
17	Mercury	Greatest Elongation West (morning)	28°
19	Moon, M44 (Beehive)	3.0′ apart: occultation	130°
19	Moon, Jupiter	0.9° apart: occultation	139°
20	Earth	Equinox	
22	Moon	Full	177°
24	Neptune	Conjunction	1°
27	Moon, α Scorpii (Antares)	0.2° apart: occultation	117°
30	Moon	Last Quarter	90°

Dates are based on Universal Time (UT). Occultations of M44 (Beehive/Praesepe) refer to sixth-magnitude star ε Cancri (Meleph). Occultations of M45 (Pleiades) refer to third-magnitude star η Tauri (Alcyone). The last column gives the approximate elongation from the Sun of the event.

Mercury continues its sprint upwards in the sky as the best morning apparition for the year for southern hemisphere early risers gets underway. It reaches its maximum altitude around the time of greatest elongation west (27.6°) on 17 March. This is a dismal morning showing for observers in northern temperate latitudes, however, with the planet lingering low to the horizon throughout. Mercury returns to direct motion on 2–3 March and is occulted by the waning crescent Moon on 6 March but this event is visible only from the Antarctic regions. The nearest planet to the Sun reaches aphelion on 22 March. It brightens steadily throughout the month, beginning at magnitude +1.4 and ending at −0.1.

Venus shines at magnitude −4.0 in the morning sky. It is higher in the sky for southern hemisphere observers than for northern astronomers but it is losing altitude above the eastern horizon as seen from all locations.

Earth arrives at an equinox on 20 March. On this day the Sun appears exactly on the celestial equator as it moves from the southern hemisphere to the northern, with

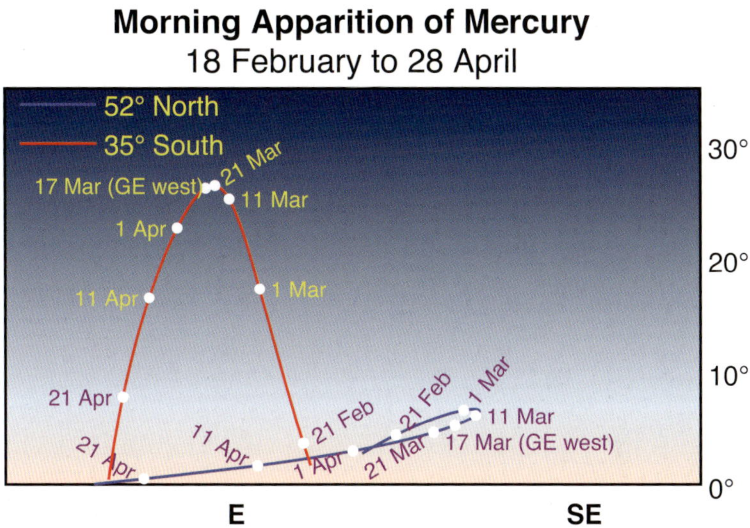

spring beginning for northern latitudes and autumn in the south. Astronomers refer to this instant as the Vernal Equinox.

The **Moon** celebrates the month of March with two planetary obscurations in addition to the current occultation series of Antares, and the Pleiades and Beehive open clusters. The third Mercury occultation on 6 March is visible only from the Antarctic and Southern Ocean whereas the occultation of Jupiter by the waxing gibbous Moon on 19 March may be observed from parts of Australia and New Zealand.

Mars is at aphelion on 4 March when it is 1.6661 au away from the Sun. It is retrograding through Leo and is best seen from the northern hemisphere where it sets well after midnight. After Jupiter, Mars is the brightest starlike object in the constellation of the lion, but it is starting to fade slightly, beginning the month at magnitude −1.1 and ending it at −0.4.

Jupiter is still in retrograde as it moves from the constellation Leo back into Cancer on 17 March. The planet shines at magnitude −2.5 and is easily the brightest point source of light in this part of the sky. However, it is out-shown by the waxing gibbous Moon two days later when our satellite occults the gas giant. Like its neighbour Mars, Jupiter is visible until at least midnight from the southern hemisphere and is best seen from northern latitudes where it sets much later.

Saturn is vanishing in the evening twilight, low in the west as the constellation of Pisces dips below the horizon. Conjunction beckons early next month.

Uranus is getting increasingly difficult for southern hemisphere astronomers, setting not long after evening twilight ends. In contrast, those inhabiting northern latitudes hang on to the sixth-magnitude planet until around midnight. Look for Uranus in the constellation of Taurus.

Neptune is lost to view this month as it undergoes conjunction with the Sun in late March.

High Speed Communication from Deep Space
The Interplanetary Broadband

Peter Rea

Introduction

I still remember that awful screeching sound. It was the sound of a 32kb modem dialling up a connection to the relatively new Internet. It was the early 1990s and I had acquired my first PC. The first website I ever visited was that of the Jet Propulsion Laboratory (JPL) in California (**jpl.nasa.gov**). Owned and operated by the California Institute of Technology it played a major role in early planetary exploration and the communication experiment that is central to this article.

Download speeds in those days were agonisingly slow. A text document came through fairly quickly; pictures took what seemed like an eternity. The ability to stream TV or a movie would not happen for many years. Whilst sitting in front of my laptop typing these words, I used an app on my phone to test my internet

The Deep Space Optical Communications flight transceiver can be identified by its large tube-like sunshade (with gold and black at one end) on the *Psyche* spacecraft, as seen here inside a clean room at JPL. (NASA)

connection. I was getting 132 megabits per second download and 20 megabits per second upload – more than fast enough for Zoom meetings and streaming video. Fibre optics had replaced old copper telephone lines and Broadband greatly increased download speeds compared to early internet connections. Now if only we could get data back from interplanetary spacecraft that fast.

First Pictures of Mars

My earliest recollection of planetary missions was *Mariner 4*, an American spacecraft launched in 1964 and which flew past Mars in the summer of 1965. I was thirteen years of age and my interest in planetary exploration was about to be given a boost that I never lost. *Mariner 4* would return just 22 images as it flew past Mars. Each image was 200 × 200 pixels (picture elements) in size for a total of 40,000 pixels. Each pixel needed six noughts and ones to define the level of grey in each 'square'. When all noughts and ones were downloaded you could reconstruct the image, albeit a low resolution one by today's standards. At a data rate of just over eight bits per second it took many hours to return each picture. They could not be returned in real time but were stored on a spool of magnetic tape for transmission to Earth after the flyby. Current spacecraft at Mars are downloading data at around four million bits per second rather than the snail pace eight bits per second of *Mariner 4*. As an orbiting spacecraft is only visible to Earth for short periods it must 'dump' the data as fast as possible before it goes behind the planet.

The Science Bit

Early lunar and interplanetary spacecraft transmitted their data using the S-band radio frequency which falls within the microwave 2 to 4 gigahertz (GHz) frequency range. More recently the X-band, (8 to 12 GHz) frequency range has been used. The higher frequency allows faster data transmission. If optical frequencies are used, then we enter the Terahertz (THz) range (100 GHz to 10 THz) when far faster data rates can be achieved. This was demonstrated in 2022 when NASA's Terabyte InfraRed Delivery System established a link with a satellite in geostationary orbit and achieved the highest ever data rate (at that time) between a satellite and a ground station. For an interplanetary 'broadband' we need to demonstrate this at lunar and planetary distances.

Early Tests from the Moon

The very first time a laser beam was detected on the Moon was during the *Surveyor 7* mission which landed on the lunar surface on 10 January 1968. Late in the mission a laser beam was radiated up to the Moon and was detected by

Table Mountain Observatory is operated by the Jet Propulsion Laboratory. Situated in Big Pines, California it was used to transmit laser communications to the *Psyche* spacecraft. (NASA)

the *Surveyor 7* camera. This was not an early test of laser communications but merely an experiment to see if a laser beam could be detected. On 7 September 2013 NASA launched the *Lunar Atmosphere and Dust Environment Explorer* (LADEE) spacecraft with a view to studying the Moon's Exosphere and dust around the Moon. As a secondary payload LADEE carried the Lunar Laser Communication Demonstration (LLCD) experiment designed to test communication between the Moon and the Earth using a laser. The LLCD demonstration consisted of a space terminal on the LADEE spacecraft and three ground terminals on Earth. Together, they demonstrated that it was possible to transfer up to 622 Mbps of data from the Moon with a space terminal that weighs less, uses less power, and occupies less space than a comparable RF, or radio frequency, system.

Demonstrating this data rate from the Moon was a small step for Earth /Moon communication.

A giant leap would be to demonstrate this from deep space.

Tests from Deep Space

The first opportunity to test laser communication with Earth in deep space, i.e. beyond the moon was with the launch of the *Psyche* mission to the metallic asteroid of the same name. The six year cruise to the asteroid gave controllers a chance to test laser communications from the distance of Mars as the spacecraft would flyby Mars for a gravity assist manoeuvre. *Psyche* carries an experiment known as the Deep Space Optical Communications experiment. It is hoped to increase data download rates from between 10 to 100 times that of radio frequency data rates. Once well beyond the orbit of the Moon, a near infra-red laser beam was sent from JPL's Table Mountain facility near Wrightwood, California. This was received successfully by *Psyche*. On 11 December 2023, at a distance of 31 million kilometres, a video file of a cat chasing a laser pointer was sent from the DSOC experiment on *Psyche*. This laser signal was picked up by the 200-inch Hale Telescope on Palomar Mountain. Data rates of 267 megabits per second were achieved. This is way beyond anything so far achieved using S- or X-band radio frequencies. A link to this video is given at the end of this article.

Located in San Diego County, the 200-inch Hale Telescope located on Palomar Mountain was used to receive laser signals from the *Psyche* spacecraft. (Peter Rea)

Conclusion

As more and more lunar and interplanetary spacecraft are launched, and with the upcoming *Artemis* crewed missions to the Moon, serious pressures are put on the Deep Space Network run by JPL and the deep space antennae of other nations. Each mission is requesting daily communication session with their spacecraft. At current data rates it can take many hours to download images and other data. At laser communication rates the same data could be returned in under an hour leaving time to communicate with more spacecraft. Only then will we have an Interplanetary Broadband.

Useful Web Links

Deep Space Optical Communications (DSOC):
nasa.gov/mission/deep-space-optical-communications-dsoc

Psyche Mission Website: **psyche.ssl.berkeley.edu**

Video of cat sent using laser communications from the *Psyche* spacecraft:
youtube.com/watch?v=k7B8lYjAOns

April

6	Moon	New	4°
7	Saturn	Conjunction	2°
10	Moon, M45 (Pleiades)	0.9° apart: occultation	41°
11	Mercury, Neptune	1.0° apart	17°
13	Moon	First Quarter	90°
14	Moon, M44 (Beehive)	0.2° apart: occultation	103°
15	Moon, Jupiter	0.9° apart: occultation	112°
19	Mercury, Saturn	0.6° apart	10°
20	Moon	Full	175°
22/23	Earth	April Lyrids (ZHR 18)	
23/24	Earth	π-Puppids (ZHR varies)	
23	Moon, α Scorpii (Antares)	0.1° apart: occultation	143°
24	Venus, Neptune	0.2° apart	29°
28	Moon	Last Quarter	90°
28	Mercury	Superior Conjunction (morning → evening)	0°

Dates are based on Universal Time (UT). Occultations of M44 (Beehive/Praesepe) refer to sixth-magnitude star ε Cancri (Meleph). Occultations of M45 (Pleiades) refer to third-magnitude star η Tauri (Alcyone). Peak activity dates and ZHRs for meteor showers are estimates. The last column gives the approximate elongation from the Sun of the event.

Mercury dives for the eastern horizon, ending the best morning apparition this year for southern hemisphere observers on 28 April. It brightens considerably as it loses altitude, beginning the month at magnitude −0.1 and ending it at −2.1. Favoured southern viewers with telescopes have the best chance of seeing Mercury a degree south of faint Neptune on 11 April. Its close approach to the brighter Saturn on 19 April will likely be a more challenging event, given that the two planets will only be 10° away from the Sun at the time. Mercury is lost to view shortly before superior conjunction on 28 April.

Venus arrives at aphelion on 17 April and pairs up with Neptune seven days later when the morning star appears 0.2° south of the eighth-magnitude world. This event is probably best viewed (telescopically, of course!) from the southern hemisphere where Venus is still quite high in the sky before dawn.

Earth experiences two meteor showers at their peak this month. However, light from the waning gibbous Moon will likely wash out all but the brightest meteors. However, the radiant of the π-Puppid shower is very low after midnight so some useful observations may be made earlier in the evening, before moonrise.

The **Moon** passes through its most distant perigee this month, on 14 April, as well as its nearest apogee 13 days later. In addition to passing over Antares, the Pleiades, and the Beehive, the Moon also occults Jupiter this month (15 April) but visibility is largely confined to the Southern Ocean and Antarctica.

Mars returns to direct motion on 1–2 April in the constellation of Leo. It sets around midnight for observers in the southern hemisphere but stays aloft until the onset of morning twilight for those in northern temperate latitudes. Mars reaches a solstice on 16 April with summer beginning in the northern hemisphere and winter in the south.

Jupiter and the waxing gibbous Moon come together again on 15 April when a lunar occultation takes place. Two days before that, Jupiter reaches a stationary point and returns to direct motion across the constellation of Cancer. The largest planet in the solar system is most easily observed from the northern hemisphere where it does not set until after midnight.

Saturn arrives at conjunction in early April and is too close to the Sun to observe, but the first-magnitude planet reappears in the east at dawn by the end of the month. Its meeting with Mercury on 19 April takes place only 10° away from the Sun.

Uranus is an evening sky object in the constellation of Taurus, already in the west as twilight ends.

Neptune is a morning sky object, found in the constellation of Pisces and best seen later in the month from the southern hemisphere. It is a degree away from Mercury on 11 April but the two planets are just 17° away from the Sun. Early risers should have a better chance of seeing the closer pairing with Venus late in April, with the two worlds appear just 0.2° apart. However, a telescope will be necessary to spot eighth-magnitude Neptune near bright Venus.

The Future of Astronomical Societies
In the Hands of Our Youth

Jonathan Powell

The astronomical society has been the very backbone of amateur astronomy for decades, embracing the core value of being able to connect and bring together likeminded people on a local, regional, and national scale. Without the presence of such a society, many of those displaying the initial spark of interest in astronomy and space would perhaps have seen that curiosity flicker, fade, and die. However, because of their existence and their longevity in serving the community, many have seen that spark ignite into a flame, turning a passion into a lifelong hobby or for a proportion of people, even a career within the science.

Founded in 1820, the Royal Astronomical Society remains a bastion for the science of astronomy, embracing amateurs and professionals alike, and encouraging the involvement of others through its outreach and educational programs. (Wikimedia Commons/Mike Peel/CC-BY-SA-4.0)

The absence of the astronomical society is unthinkable not just because of the void it would leave in the whole structure of amateur astronomy, but because of the heritage that has made such societies ingrain themselves in the community as institutions and bastions that uphold the subject. As testimony to the values held, and catering for both amateur and professional astronomers alike, the Royal Astronomical Society continues to champion the study of astronomy and related sciences.

However, as time passes, their very survival may be in question. This challenge is not faced alone as other societies covering a multitude of hobbies and pastimes contemplate their own future existence, challenged with the task of retaining the engagement of the up-and-coming generation. Over recent decades numerous attempts – some successful, some not – have been made to attract youngsters into astronomy, a good example being the 'Young Astronomers' group, set up at West Yorkshire Astronomical Society in Pontefract, West Riding of Yorkshire in Spring 2019 by society members Les Field and Pamela Ward. This involves monthly (often well attended) meetings for kids aged between four and 12, as well as activity classes and some stargazing. Happily, this venture is still going strong.

Usually sporting a stalwart core of members who are present come rain or shine, the typical astronomical society unites a common desire to meet and share acquired knowledge in an interest that can be delved into on whichever level the individual chooses; many societies sporting a wide spectrum of ages and experience, let alone diverse backgrounds in education and work. The core of members, often elected to their positions, are the heart of the society, from opening up the community hall for a meeting to organizing speakers and trips to places of interest. They represent the beating heart of affairs within the society with the blood supply interpreted as the membership. However, as time passes and more critically, focusing on the youth element within that supply, the heart may be starved and if not careful eventually flounder. The maintenance of a strong and healthy heart is underpinned by the youth who can sustain and nurture the society to its full potential through new thinking and adapting to change.

Those of a certain age, 'Generation X' for example, would have found that aside knowing someone else with an interest in astronomy, perhaps a friend or neighbour, in school or college, or perhaps a work colleague, the local library would have been one of just a handful of reference points. However, and thankfully, even as far back as 1957 and the 'Baby Boomers', the influence of Sir Patrick Moore's 'The Sky at Night' was not only groundbreaking for television, but served as a crucial link and steer, that was to influence an untold amount of viewers. The Sky at Night, which could be considered ahead of its time in the scheduling of such programmes, coupled with the role of the astronomical society, fitted as snugly as an eyepiece into the telescope; a perfect marriage of involvement in person with others in a society face to face, and that of Patrick speaking almost as if directly to you in one's own living room.

Before the advent of social media and the ability via on-line video link platforms to be present at a meeting but not actually physically there, the astronomical society offered much beside the prospect of stargazing alone. With frequent meetings, observing sessions, and outings, the technical side of affairs was always well catered for. For a number of those attending it was also a great way to socialize and engage

Members of the Bridgend Astronomical Society are seen here setting up ahead of a local event in South Wales with the aim of bringing astronomy and space to the general public. Left to right are Stephen Radford, Jason Mead and Stephen Webber. (Martyn Howells)

with others who one may never have had the chance to meet outside of the society. It also allowed for topics both astronomical and otherwise to be discussed in both formal and informal settings, at times in more convivial settings, affording some to 'come out of their shell' and feel more relaxed. Whilst many of these activities still exist, such meetings in recent times are perhaps not sporting the sort of age demographic required to sustain such endeavours for the future.

A fair proportion of amateur astronomers conduct their observing alone, so the society set-up is not for everyone. However, for some, aside the aspect of being with those who share a passion for astronomy, there is the ability to bring the science to others who perhaps without a society in the community would miss out. Public engagement at open observing sessions allows the curious and inquisitive a chance to view the heavens as never before, through a pair of binoculars or a telescope that they may not readily have access to. To inform, educate, and bring the cosmos down to Earth with a view of craters on the Moon, Jupiter's moons, or the rings of Saturn.

For the health of the 'heart', the concern is that as the long-serving veterans and in some cases founder members look to find successors, the worry exists that unless there is a steady influx of youth, societies could find that dwindling numbers force closure. Schools, colleges, and other academic outlets, along with television and the media can only inform and entertain the amateur astronomer on certain levels, but not all levels. It does not wholly quench the thirst of those who wish to become more involved; the astronomical society can quench that thirst, but unless there is sufficient youthful retention, the 'heart' will fail. Change is inevitable but without reaching out and then retaining the youth within a society, the outlook is bleak. Survival of amateur astronomy as we know it depends on it.

May

5/6	Earth	η-Aquariids (ZHR 50)	
6	Moon	New	5°
7	Moon, M45 (Pleiades)	0.8° apart: occultation	15°
7	Venus, Saturn	0.6° apart	26°
8	Jupiter	East Quadrature	90°
11	3548 Eurybates	Opposition	179°
12	Moon, M44 (Beehive)	0.5° apart: occultation	77°
12	Moon, Jupiter	1.2° apart: occultation	86°
13	Moon	First Quarter	90°
20	Moon	Full	175°
21	Moon, α Scorpii (Antares)	2.8′ apart: occultation	169°
25	Mars	East Quadrature	90°
27	Saturn, μ Piscium	0.8° apart	42°
27	Uranus	Conjunction	0°
28	Mercury	Greatest Elongation East (evening)	23°

Dates are based on Universal Time (UT). Occultations of M44 (Beehive/Praesepe) refer to sixth-magnitude star ε Cancri (Meleph). Occultations of M45 (Pleiades) refer to third-magnitude star η Tauri (Alcyone). Peak activity dates and ZHRs for meteor showers are estimates. The last column gives the approximate elongation from the Sun of the event.

Mercury returns to the evening sky for the second time this year, in what is the best 2027 western apparition of the planet for observers in northern temperate latitudes. It races upwards from the western horizon until late in the month, and then just as abruptly heads back towards the Sun. Greatest elongation east occurs on 28 May when the planet is at a solar elongation of 22.9°. Before that, however, Mercury encounters Uranus on 11 May although the planets are around 2° apart in the sky. As is customary during an evening apparition, the planet begins bright and fades throughout. In this instance, Mercury is at magnitude −2.1 on 1 May but ends at the month at +0.9. Its second perihelion of the (terrestrial) year takes place on the fifth.

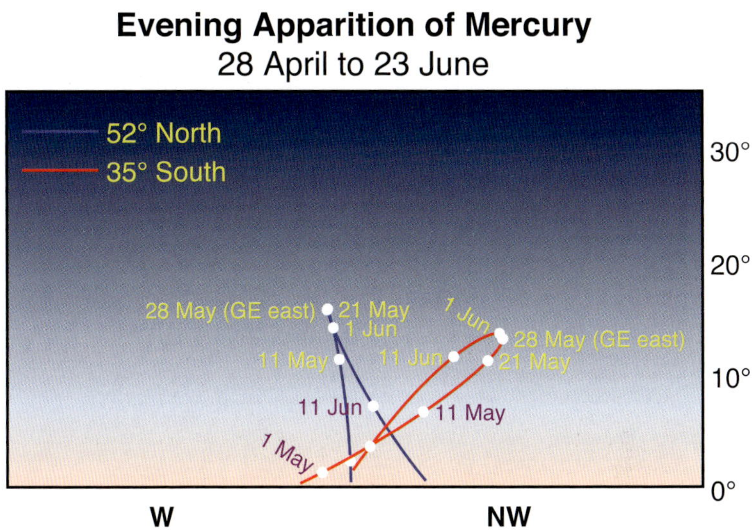

Venus, having flown past Neptune last month, now encounters nearby Saturn, with the two worlds coming within 0.6° on 7 May. The morning star is quite low in the east as seen from northern temperate latitudes, rising during twilight. The bright planet is best viewed from the southern hemisphere where it appears higher in the sky.

Earth experiences perfect observing conditions during the early-May peak of the fast-moving η-Aquariids. This shower is best viewed from the southern hemisphere and up to around 40° north latitude. Look for them in the hours before dawn.

The **Moon**, just a day short of First Quarter, occults Jupiter for the final time this year but the event is not visible outside a small region in the Southern Ocean. The other three occultation series – Antares, the Pleiades, the Beehive – are still going strong although the Sun is quite close to the Pleiades at this time of year. The Full Moon on 20 May is the third full moon in a season which makes it a 'Blue Moon' by the original almanac definition.

Mars continues its journey across the constellation of Leo. It passes a distant 1.2° north of the first-magnitude star Regulus on 15 May and reaches east quadrature ten days later. Telescopic observations of Mars around the time of quadrature reveal a distinctly gibbous phase to the planet. The red planet sets around midnight for northern viewers and a little earlier for those in the southern hemisphere. It loses half a magnitude in brightness over the course of the month, going from +0.3 down to +0.8.

Jupiter arrives at east quadrature on 8 May. This is a great time to observe telescopically the interplay of shadows between the planet and its Galilean satellites. Two days after quadrature, Jupiter departs the constellation of Cancer for Leo. Finally, on 12 May, the last lunar occultation of Jupiter this year takes place. The largest planet in the solar system is an evening sky object. It sets during mid- to early evening hours for southern hemisphere observers but those in northern latitudes hang on to the magnitude −2.0 object until midnight.

Saturn is located in the morning sky in the constellation of Pisces. Observers in the southern hemisphere will have the best opportunity to catch the ringed planet and the vastly brighter morning star 0.6° apart on 7 May. Saturn also skims past the fifth-magnitude star μ Piscium for the first time this year on 27 May but the two objects will have a closer appulse in October.

Uranus vanishes behind the disk of the Sun late this month as it undergoes conjunction. Mercury is less than 2° away from the planet on 11 May but the two bodies are at just 14° solar elongation, rendering sixth-magnitude Uranus an impossible observing target in the evening twilight.

Neptune is found in Pisces before sunrise. The faint planet is likely lost in the persistent twilight of northern latitudes at this time of year but should be accessible from southern vantage points where it rises just after midnight.

Lithium

Rod Hine

A couple of generations ago, lithium was viewed as either an obscure element in chemistry or a powerful tranquiliser used only in cases of severe mental illness. In car maintenance you might use 'lithium grease' in some places. Nowadays, lithium is synonymous with all kinds of batteries from mobile phones to electric cars. It is often mentioned in the news as a strategic resource, and the world market for lithium is insatiable. But just what makes lithium so important, and what has this got to do with astronomy?

I am always intrigued by the fact lithium is the only solid element that was produced in any quantity during the Big Bang, and to understand this we must consider its atomic structure. As the elements began to form during the first few minutes, the great majority of the first protons became hydrogen, and then deuterium was formed with one proton and one neutron. Initially the temperature

Nova Centauri 2013 seen from La Silla. (Y. Beletsky (LCO)/ESO)

was too high for the deuterium to persist, but at a certain point the temperature fell enough to allow deuterium to survive, which then triggered another burst of further fusion. In this brief phase, lithium formed with three protons and four neutrons, but within about twenty minutes, most nucleosynthesis ceased, and further changes only came about as various unstable intermediate products decayed.

In these initial moments, the ratio was set at around 75% hydrogen to 25% helium by mass. Only a tiny amount of lithium came from the Big Bang itself, but understanding the process has led to an apparent discrepancy in the expected amount of lithium. Early older stars seem to contain about a third as much lithium as might have been produced in the Big Bang. Conversely, some younger stars seem to have more lithium, and there is, as yet, no convincing explanation but plenty of theories.

Leaving aside the lithium from the Big Bang, it can also be produced in stars, particularly carbon-rich stars and is also produced as the end result of some radioactive decay processes. Recent work has found that lithium-7 is produced in nova explosions. Nova Centauri 2013, otherwise known as V1369 Centauri, was a particularly bright nova that peaked at magnitude 3.3 on 14 December 2013. Easily visible to the naked eye at maximum, it was first spotted by amateur astronomer John Seach in Australia when at magnitude 5.5 on 2 December. Bursts of gamma rays were also detected soon afterwards. On 29 July 2015, ESO's La Silla Observatory announced that lithium had been discovered in the material ejected by the nova explosion at around two million kilometres per hour. Even though the amount of lithium is small, perhaps just one billionth of a solar mass, it is enough to account for the extra lithium we can detect in our galaxy.[1]

On Earth, most of the lithium in the crust is lithium-7, plus around 5% lithium-6. Both isotopes are stable and widely distributed both in the crust and the oceans. Sea water contains about one part lithium to five million parts water, and it is found in much higher concentrations around hydrothermal vents. In the Earth's crust, it constitutes somewhere between 20 to 70 parts per million by weight, and is the 31st most abundant element. To put that in perspective, one kilogram of the crust contains around 20 milligrams of lithium – about as much as four or five caraway seeds.

Commercial sources of lithium occur as brine deposits from salt lakes, as in South America, particularly Bolivia, Argentina and Chile. In Australia, lithium is extracted from hard rock mining. Overall, Chile, Australia and China are the major

1. For further details see **eso.org/public/news/eso1531**

A freshly cut lithium sample with oxide already formed. (Wikimedia Commons / Dnn87 / GNU Free Documentation License)

producers of lithium, but other locations may have reserves too. The dry salt lakes in western Afghanistan are being investigated and Cornwall, UK, has some lithium-rich brines.

Lithium was first discovered by the Swedish chemist Johan August Arfwedson in 1817 whilst analysing samples of petalite, a crystalline mineral first found in a mine on a Swedish island in 1800. Pure lithium was not isolated until 1821 by William Thomas Brande who passed an electric current through molten lithium oxide. That same process of electrolysis had recently been used by Sir Humphrey Davy to isolate the other alkali metals potassium and sodium. During electrolysis, the metals are produced at the cathode or negative electrode.

So what does lithium actually look like? In its pure form it is a very light, silvery metal that is so soft it can easily be cut with a knife. Exposed to air, it quickly tarnishes to a grey as it reacts to form lithium oxide. It will eventually turn black as it forms lithium nitride. It has such a low density, about half that of water, it will float on water but will also react violently and usually catch fire as it reacts with the water to release hydrogen. It is this extreme reactivity that makes it so useful for modern battery technology, and about three quarters of all lithium production now goes to this application.

June

3	Moon, M45 (Pleiades)	0.8° apart: occultation	14°
4	Moon	New	4°
7	Uranus, ω^2 Tauri	0.6° apart	11°
8	Moon, M44 (Beehive)	0.7° apart: occultation	50°
11	Moon	First Quarter	90°
14	Venus, Uranus	0.7° apart	16°
15	Uranus, 51 Tauri	0.3° apart	17°
17	Moon, α Scorpii (Antares)	1.9′ apart: occultation	163°
18	Uranus, 53 Tauri	0.2° apart	20°
19	Moon	Full	176°
19–26	Earth	June ε-Ophiuchids (ZHR low)	
21	Uranus, 56 Tauri	0.4° apart	22°
21	Earth	Solstice	
23	Mercury	Inferior Conjunction (evening → morning)	4°
25–28	Earth	β-Taurids (ZHR 25)	
27	Moon	Last Quarter	90°
28	Neptune	West Quadrature	90°

Dates are based on Universal Time (UT). Occultations of M44 (Beehive/Praesepe) refer to sixth-magnitude star ε Cancri (Meleph). Occultations of M45 (Pleiades) refer to third-magnitude star η Tauri (Alcyone). Peak activity dates and ZHRs for meteor showers are estimates. The last column gives the approximate elongation from the Sun of the event.

Mercury completes its latest evening apparition on 23 June when it undergoes inferior conjunction. This is the best western appearance this year for observers in northern temperate latitudes but catch it early in the month when the tiny planet is still first magnitude. On 10 June, Mercury begins retrograde motion with aphelion taking place eight days later.

Venus encounters Uranus on 14 June, when the morning star is found 0.7° south of the sixth-magnitude outer planet. Venus is quite low as viewed from northern temperate latitudes but even favoured southern-hemisphere observers may have difficulty imaging this event as the bright planet descends into twilight. When seen through a telescope, Venus presents as a small, almost perfectly round disk about 10″ across and shining at its minimum magnitude of −3.9.

Earth passes through several meteor streams in June, among them the June ε-Ophiuchids and the β-Taurids. The June ε-Ophiuchids comprise a minor shower which will likely be obliterated by the Full Moon but the more active β-Taurids later in the month are unaffected by moonlight. The reason for this is simple: the radiant of the β-Taurid meteor shower rises during daylight hours and the meteors are detectable in radio wavelengths. Earth also reaches a solstice on 21 June, when the Sun reaches its most northerly declination. Summer officially begins in the northern hemisphere and winter takes hold in the south.

The **Moon** occults the Pleiades in the morning sky on the third day of the month but this may take place too close to the Sun to observe. The crescent Moon passes through the faint Beehive cluster five days later and then occults Antares on 17 June.

Mars is an evening sky object, found in the constellation of Leo. It shines at first magnitude and looks somewhat gibbous when viewed through a telescope. The red planet sets before midnight for all locations this month.

Jupiter is moving through Leo. It is an evening sky object for everyone but sets later for northern-hemisphere observers than for southern-hemisphere-based astronomers. Although Leo boasts a first-magnitude star, Jupiter easily outshines it at magnitude −1.9.

Saturn shines at magnitude +0.7 in the constellation of Pisces. It is confined to the morning sky at the beginning of the month but is rising around midnight by the end of June.

Uranus emerges from conjunction last month into the dawn sky. Still located in Taurus, the sixth-magnitude planet passes less than a degree away from Venus (14 June) and several fifth- and sixth-magnitude stars in that constellation: ω^2 Tauri on 7 June, 51 Tauri on 15 June, 53 Tauri on 18 June, and 56 Tauri on 21 June. However, all of these appulses take place less than 25° from the Sun, with the extended dawn twilight of northern latitudes washing out these faint objects. Observers in the southern hemisphere have a better chance of seeing these close encounters in their dark winter skies.

Neptune attains west quadrature near the end of the month which places it 90° away from the Sun in the sky. It is found in Pisces and is best seen from the southern hemisphere where it rises before midnight. Observers in northern temperate latitudes must contend with astronomical twilight and post-midnight rising times.

Gone But Not Forgotten
Corona Meridiana

Lynne Marie Stockman

Just south of the zodiacal constellation of Sagittarius lies a faint, yet distinct, semicircle of stars. The arc is distorted but recognisable in dark skies, and is known by many names in many cultures. In Australia, the Boorong call this asterism *Won* and identify it as a boomerang thrown by the mythological being Totyarguil (represented by the star Altair), the son of Neilloan (the constellation Lyra). The Arrernte tell the story of a group of women who danced along the Milky Way. One of them placed her baby in a coolamon, a wooden basket, and the basket slipped from the sky whilst she danced. The coolamon is seen today as an arch of faint stars, and the parents of the baby, the morning and evening stars, still search for their missing child. The Wiradjuri, on the other hand, see the small curve of stars as a kookaburra.

The |xam Bushmen of southern Africa call this group of stars *≠nabbe ta !nu* which means 'house of branches', and relate the tale of an Early Race girl who, upon entering puberty, was in ritual seclusion in her hut. During this time, she looked out and saw a group of men eating together in a nearby rock-rabbit's house of branches. She fixed her gaze upon them, and transformed them and the house into stars. In South America, the Barasana people of present-day Colombia see the half-circle of faint stars as a poisonous snake. However, the indigenous Peruvians regard it as a much more benign figure, an enclosure for celestial llamas.

The ancient Chinese saw this asterism as a turtle lazing on the banks of the celestial river, the Milky Way. The river turtle 鱉 (Biē) was in the 斗 (Dǒu which means 'dipper') mansion of the Symbol of the Black Tortoise (玄武 or Xuánwǔ), one of four Symbols and Three Enclosures that defined the Chinese sky of antiquity.

One tradition in Arabic astronomy depicted the circlet as a domed tent, *al-qubba*, which is oriented upright upon setting in the southwest. Another described this asterism as *udḥī al-naʿām*, an ostrich nest, with the brighter stars of Sagittarius portraying the ostriches coming and going from the area.

The European tradition of grouping these stars into a southern crown or garland was codified by Greek astronomer and mathematician Claudius Ptolemy

The southern wreath or crown lies at the very bottom of plate XX from Alexander Jamieson's 1822 *Celestial Atlas*. (Wikipedia Commons / Alexander Jamieson / United States Naval Observatory Library)

in his second-century treatise *Almagest*. He named the constellation *Στεφάνος νοτιος* or *Stephanos notios*, the southern wreath, and assigned 13 stars to it, none brighter than fourth magnitude. The 13 included the star which is now known as α Telescopii. Since then, the constellation has sported a variety of similar names in Western astronomy: Corona Australis, Corona Austrina, Corona Meridiana, Corona Meridionalis, and, for those who like to mix and match their Latin and Greek, even Corona Notis.[1]

1. *Corona* is Latin for 'wreath' or 'crown'. 'Meridiana' and 'Meridionalis' are derived from *meridiem*, the Latin word for 'south'. Notus (Νότος) was the Greek god of the south wind. His Roman equivalent was Auster.

The first General Assembly of the International Astronomical Union (IAU) took place in Rome in 1922. At this meeting, the Commission on Notations agreed on a list of 88 constellations and their three-letter abbreviations. (Actually, there were 89, with the obsolete constellation Argo Navis included as well.) The many-titled southern wreath was officially crowned *Corona Australis* and assigned the abbreviation CrA. The actual boundaries of all of the constellations were published in 1930. Then the IAU confused the issue in 1932 when they issued a new list of constellations, complete with optional four-letter abbreviations. Argo Navis had disappeared in 1930 when the constellation boundaries were established so it was not included, but Corona Australis inexplicably became *Corona Austrina*. However, in 1955 the IAU rescinded the four-letter abbreviations and 'Corona Australis' has been the preferred nomenclature ever since. Interestingly, a search of the current astronomical literature shows that **both** names are in use, with 'Australis' appearing about twice as often as 'Austrina'.

Corona Australis is a small constellation, ranked 80 out of 88 in area, and its three brightest stars are only fourth-magnitude. On Earth it is fully visible only from below 44° north latitude. However, despite its diminutive size and lack of bright stars, it has been repeatedly spotted from space. On 3 October 1962, at 02:37:38 into the third orbital Mercury flight, U.S. astronaut Walter Schirra reported, "There's the old Corona Australis, shows in beautifully; and we've got Nunki [σ Sagittarii]." Astronaut Gordon Cooper also mentioned seeing the constellation during his Mercury flight of 15–16 May 1963. The second crewed flight of Gemini, *Gemini IV*, launched on 3 June 1965. Astronauts James McDivitt and Ed White spotted the distinctive constellation 13 hours into their journey. White commented, "There's the little Corona Australis, that we decided to use as a calibration. It's out in front of me right now, and they go down to fifth order stars and I can see stars that are distinctly of a less magnitude than the stars in the Corona Australis."

Because of its proximity to the Milky Way, galaxies are difficult to see in the constellation but it does contain the Corona Australis Molecular Cloud within its borders. Molecular clouds are the birthplaces of stars, and typically contain both dark and reflection nebulae, along with Herbig-Haro objects (bright nebulae associated with new stars). The Corona Australis molecular cloud complex is thought to be one of the nearest stellar nurseries to the solar system and contains several dozen optically visible young stars, as well as hundreds more hidden in the dust.

The alpha star of the constellation was assigned the name *Meridiana* by the IAU in September 2019. This name recalls both a traditional name for this star, *Alphecca*

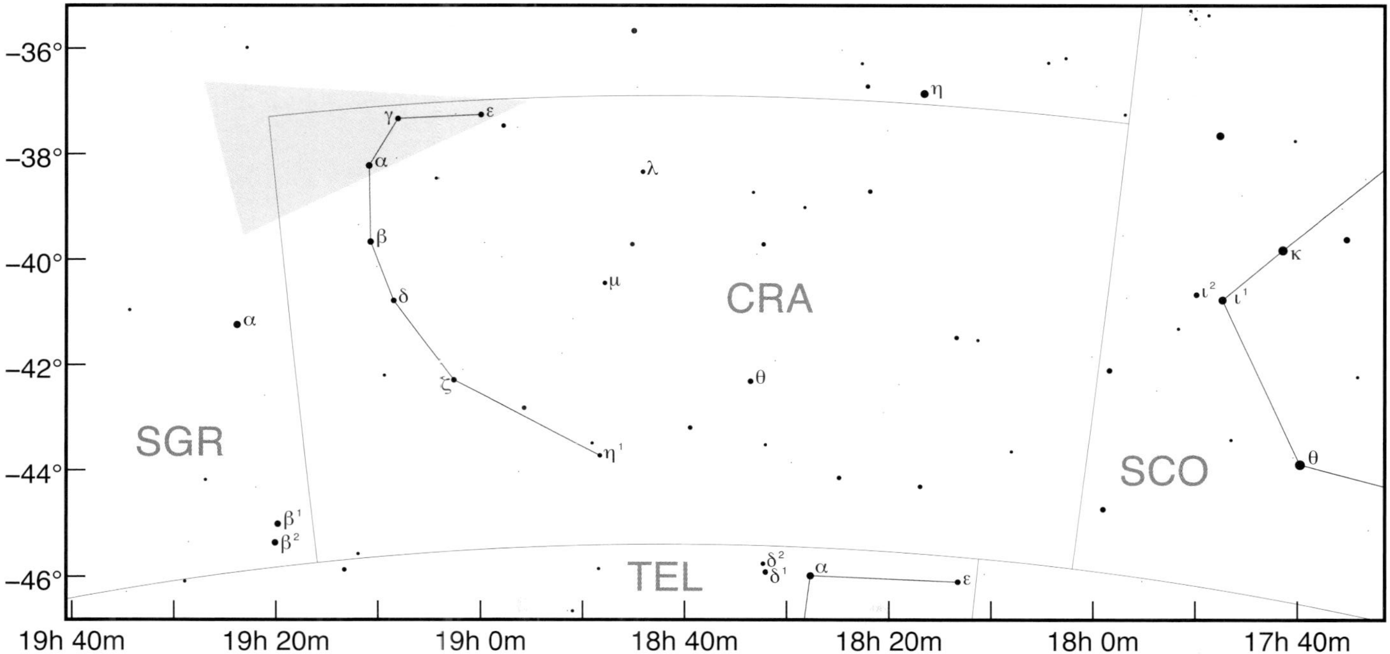

The small half-circle of stars south of Sagittarius is noticeable by whatever name you call it. The Corona Australis Molecular Cloud is located in the north eastern corner of the constellation; its position and shape is approximated by the grey area. Stars are shown to magnitude +6.5. (David Harper)

Meridiana, and the obsolete constellation name *Corona Meridiana*. Meridiana has an apparent visual magnitude of +4.09 and is an A-type main sequence star whose spectrum shows an infrared excess. This excess can be the signature of a dusty circumstellar disk but in this case, may instead be contamination from the nearby molecular cloud. A-type stars on the main sequence are typically young and rotate rapidly. Meridiana is no exception, being around 254 million years old and possessing a projected rotational velocity of nearly 200 km/s.

At the time of writing, Meridiana is the only star in Corona Australis to possess a formal name. However, given the rich history of this stellar arch from cultures around the world, the IAU will be spoilt for choice in choosing names for the other members of this constellation.

Further Reading

Allen, Richard Hinkley. (1963). *Star Names: Their Lore and Meaning*. Dover Publications, Inc.

Barentine, John C. (2016). *The Lost Constellations: A History of Obsolete, Extinct, or Forgotten Star Lore*. Springer Praxis Books.

Ridpath, Ian. (2018). *Star Tales: Revised and Expanded Edition*. The Lutterworth Press.

July

1	Moon, M45 (Pleiades)	0.9° apart: occultation	39°
4	Moon	New	3°
5	Earth	Aphelion (1.1067 au)	
5	Moon, M44 (Beehive)	0.8° apart: occultation	24°
10	Moon	First Quarter	90°
10	Mars, β Virginis (Zavijava)	0.2° apart	69°
14	Jupiter, ν Leonis	0.9° apart	36°
14	Moon, α Scorpii (Antares)	2.7′ apart: occultation	138°
15	Mercury	Greatest Elongation West (morning)	21°
18	Moon	Full: penumbral lunar eclipse	179°
19	Uranus, κ^2 Tauri	0.6° apart	48°
19	Uranus, κ1 Tauri	0.7° apart	48°
20	Saturn	West Quadrature	90°
26	Moon	Last Quarter	90°
27	Jupiter, α Leonis (Regulus)	0.4° apart	26°
28	Moon, M45 (Pleiades)	1.0° apart: occultation	65°
29	134340 Pluto	Opposition	175°
30/31	Earth	Southern δ-Aquariids (ZHR 25)	

Dates are based on Universal Time (UT). Occultations of M44 (Beehive/Praesepe) refer to sixth-magnitude star ε Cancri (Meleph). Occultations of M45 (Pleiades) refer to third-magnitude star η Tauri (Alcyone). Peak activity dates and ZHRs for meteor showers are estimates. The last column gives the approximate elongation from the Sun of the event.

Mercury is visible in the morning sky, better seen from the southern hemisphere than from the north. Starting the month at third magnitude, the closest planet to the Sun brightens throughout the apparition and by the time it attains greatest elongation west mid-month, it is magnitude +0.4. Mercury makes a distant pass by Venus on 1 July and then returns to direct motion three days later.

Venus must be getting close to the Sun if it is passing by Mercury on the first day of the month. The two planets are 4.6° apart but this will be a tricky event to watch as the morning star is getting quite low to the horizon as seen from all locations.

Morning Apparition of Mercury
23 June to 11 August

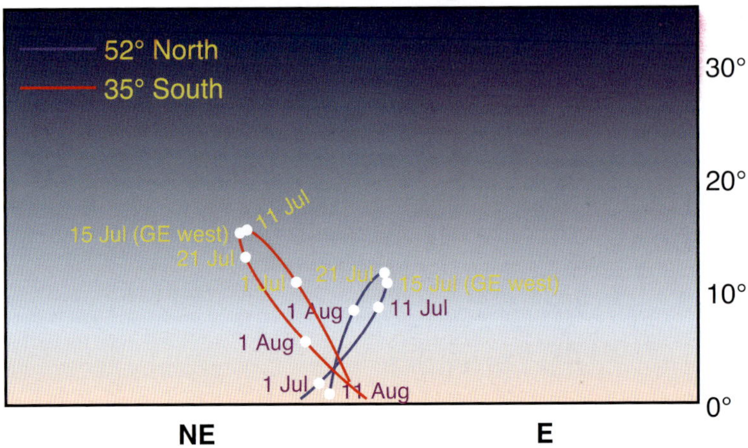

Earth is at its most distant from the Sun on 5 July when our planet passes through aphelion. The Sun's disk is 1888″ across, its smallest value of the year. (Compare this number with the Sun's apparent size at perihelion in January). *As always, never observe the Sun directly without the proper equipment as irreparable eye damage can result.* In meteor shower news, this is a good year to observe the Southern δ-Aquariids at the end of the month since the Moon is only a day or two from its New phase, leaving dark skies for this annual shower. This is one of the better showers for southern hemisphere astronomers, with the radiant rising before midnight.

The **Moon** continues its occultation series of Antares, the Pleiades cluster, and the Beehive cluster, although the faint Beehive may be lost in evening twilight. An extremely short and shallow penumbral eclipse takes place on 18 July. Lasting less than 20 minutes, only 0.3% of the Moon's disk will enter Earth's penumbra. This effect is no doubt exacerbated by the fact that this Full Moon is the 'smallest' of the year, with an angular diameter of 1766″.

Mars departs the Leo and enters the large constellation of Virgo where it will remain until mid-September. It is an evening sky object, found in the west after sunset, and disappears from view before midnight. On 10 July, the first-magnitude planet makes a close pass south of the fourth-magnitude star Zavijava which, although the beta star in Virgo, is not the second brightest star in the constellation.

Jupiter continues its journey through the constellation of Leo. It makes a distant pass by the fifth-magnitude star ν Leonis on 14 July but comes closer, 0.4°, to Regulus (α Leonis) itself on 27 July. An evening sky object now low in the west, Jupiter sets during twilight for astronomers in northern latitudes.

Saturn reaches west quadrature on 20 July. It is at quadrature, east or west, that shadows cast by the superior planets, their satellites, and their ring systems are at their most offset, leading to some interesting visual effects. The ringed planet is found in Pisces, rising in the late evening hours.

Uranus is still a morning sky object, rising after midnight in the constellation of Taurus. It passes less than a degree south of the two fifth-magnitude stars κ^2 Tauri and κ^1 Tauri on 19 July.

Neptune reverses course across the constellation of Pisces, entering into retrograde motion on 10 July. It rises before midnight for observers at all locations but is best seen from the southern hemisphere.

Gravitational Lensing

David M. Harland

One prediction of Albert Einstein's Theory of General Relativity was that the path of light from stars would be 'bent' in passing close to the Sun on its way to us, and this was confirmed by a solar eclipse in 1919. Yet, interesting though this phenomenon of an intense gravitational field deflecting light was, it did not seem likely ever to play a significant role in observational astronomy. However, it was later realised that if the light from a very remote galaxy passed close by an intervening galaxy on its way towards us, it, too, would be bent, and that, with a favourable alignment, such a 'gravitational lens' would produce a distorted image of the remote galaxy. Furthermore, if the lensing were perfect, the light would be smeared into a complete circle, a phenomenon now known as an Einstein ring.

The first gravitational lens was identified in 1979, when astronomers at the Kitt Peak National Observatory in Arizona realised that a pair of quasars were in fact two views of the same object.

In 1994, the recently repaired *Hubble Space Telescope* took a picture of Abell 2218, a rich cluster of galaxies some three billion light years away; there were a number of thin arcs of light on the line of a circle, each of which was a distorted view of the same background galaxy, located three or four times farther away and hence too faint to be seen directly. By precisely plotting the lensed artefacts and working backward, it was possible to calculate the gravitating mass that was lensing. On the basis that most of the light emitted by the galaxies was produced by stars, the masses of the galaxies were calculated. In addition, the motions of the galaxies within the cluster provided a measure of the mass of the gas, mostly hydrogen, in between the members of the cluster. The fact that the mass inferred from the lensing greatly exceeded that directly measured was a clear indication that galaxy clusters are dominated by dark matter.

The first complete ring was found in 1998, when the Hubble Space Telescope followed up radio studies by the MERLIN network operated by the University of Manchester. A giant elliptical galaxy was lensing a distant dwarf galaxy which we would not otherwise be able to see.

A productive optical partnership for finding gravitational lenses was a collaboration of the *Hubble Space Telescope* and the Sloan Digital Sky Survey at

When it imaged the galaxy cluster SDSS J1038+4849, the Hubble Space Telescope caught a 'Happy Face' where the eyes are very bright galaxies within the cluster and the arcuate 'smile lines' the result of gravitational lensing of a more distant galaxy. (NASA / ESA / Acknowledgement: Judy Schmidt)

Apache Point Observatory in New Mexico. One notable case was an almost perfect alignment of a foreground galaxy about three billion light years away with galaxies at 6 and 11 billion light years, giving rise to an inner ring and an outer ring, each comprised of multiple views of one of the more remote galaxies.

Returning to the scale of individual stars, if the light from a distant star is bent by a massive star that lies on our line of sight, we are able to infer something about the stars involved. Because the alignment for this 'microlensing' is rare, in order to

In this Hubble Space Telescope image, the precise alignment of two galaxies has produced a gravitational lens that smears out the light from the remote galaxy into an almost complete ring. Listed as LRG 3-757, it has also acquired the moniker of the Horseshoe. (ESA/Hubble & NASA)

have a reasonable prospect of success dense star fields are surveyed to measure the brightness of hundreds of millions of stars repeatedly.

A pioneer was the Optical Gravitational Lensing Experiment (OGLE) created by the University of Warsaw in 1992 at Las Campanas Observatory in the Atacama Desert of Chile.

If two stars are aligned *precisely* on our line of sight, the light from the more distant one, very likely much fainter, is magnified by the lens and makes the foreground star brighter for a period of perhaps several weeks with a symmetrical light curve. If the lensing star has a planet, the lensing by the planet will add a subsidiary peak to the light curve.

In 2006, OGLE, in collaboration with the Microlensing Observations in Astrophysics project and the Probing Lensing Anomalies Network, found a planet orbiting the star OGLE-2005-BLG-390L, located 21,500 light years away in Scorpius,

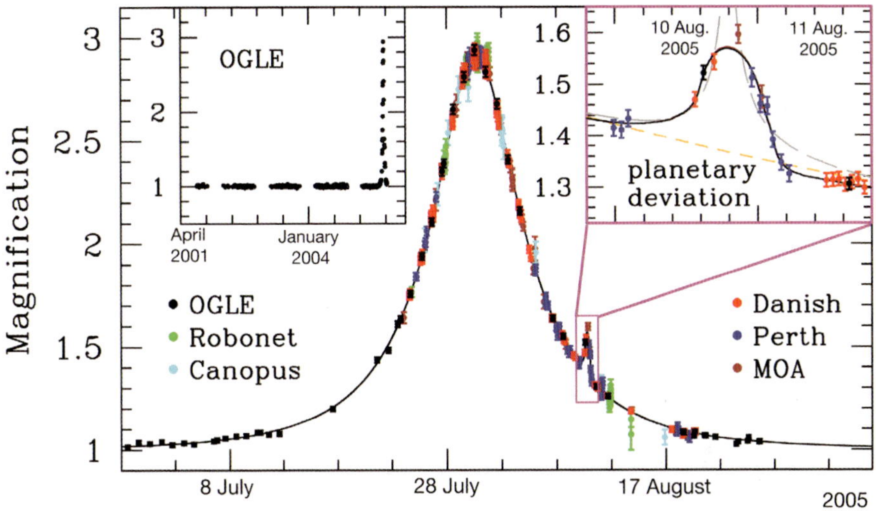

The microlensing light curve of star OGLE-2005-BLG-390L. The event peaked on 31 July 2005, and then, as it was diminishing, lensing by a planet accompanying the star imposed a secondary peak around 10 August. (ESO)

very close to the centre of the Milky Way. The star is most likely a red dwarf, and the planet, with a mass of about five times that of Earth, has an orbital period of about ten years.

Significantly, the microlensing method of detecting planets does not require the orbit of the system to be oriented for the planet to transit its star's disk. Furthermore, it can detect planets that orbit so far from their primary that any wobble that they might induce in the star is far below the threshold of the radial velocity technique.

In addition to magnifying our view of a distant object, microlensing displaces its apparent position. The change in displacement enables the mass of the lens to be calculated. This opens up the intriguing prospect of using a clearly visible remote star to detect an intervening object that is not visible. In 2022 it was reported that such observations had made the first unambiguous discovery of a stellar-mass black hole which was just drifting in space, without a companion star to feed material into an accretion ring and thereby make its presence known. The *Hubble Space Telescope* observations were obtained over a six year period, starting in 2011, shortly after the onset of a microlensing event was noticed. With a mass of about seven times that of the Sun this black hole was probably created when the core of a massive star, no longer able to support itself, underwent runaway gravitational collapse.

As there must be millions of stealthy black holes, they offer excellent prospects for discoveries during surveys conducted by facilities such as the Simonyi Survey Telescope of the Vera Rubin Observatory in Chile, whose task is to image the entire available sky every couple of nights in search of transient phenomena.

-oOo-

Thus we see that a prediction by an exotic theory has made possible observational advances that range from the discovery of planets around other stars, to stealthy black holes, to the dominance of dark matter in the largest structures in the universe.

August

1	Moon, Mercury	0.3° apart: occultation	11°
2	Moon, Venus	0.6° apart: occultation	3°
2	Moon, M44 (Beehive)	0.8° apart: occultation	2°
2	Moon	New: total solar eclipse	0°
2	Venus, M44 (Beehive)	0.1° apart	3°
3	Saturn	Maximum geocentric ring tilt: −14.4°	103°
7	(137108) 1999 AN10	Closest approach to Earth	104°
9	Moon	First Quarter	90°
11	Moon, α Scorpii (Antares)	1.2′ apart: occultation	112°
11	Mercury, Venus	0.5° apart	2°
11	Mercury	Superior Conjunction (morning → evening)	2°
12	Venus	Superior Conjunction (morning → evening)	1°
12/13	Earth	Perseids (ZHR 100)	
16/17	Earth	August δ-Aquariids (ZHR low)	
16/17	Earth	κ-Cygnids (ZHR 3)	
17	Moon	Full: penumbral lunar eclipse	179°
19	Mercury, Jupiter	0.6° apart	9°
20	Venus, α Leonis (Regulus)	0.9° apart	3°
25	Moon, M45 (Pleiades)	0.9° apart	91°
25	Moon	Last Quarter	90°
26	Venus, Jupiter	0.5° apart	4°
27	Jupiter, ρ Leonis	0.8° apart	3°
29	Moon, M44 (Beehive)	0.8° apart: occultation	29°
31	45P/Honda–Mrkos–Pajdušáková	Perihelion	31°
31	Jupiter	Conjunction	1°
31	Moon	New	2°

Dates are based on Universal Time (UT). Occultations of M44 (Beehive/Praesepe) refer to sixth-magnitude star ε Cancri (Meleph). Occultations of M45 (Pleiades) refer to third-magnitude star η Tauri (Alcyone). Peak activity dates and ZHRs for meteor showers are estimates. The last column gives the approximate elongation from the Sun of the event.

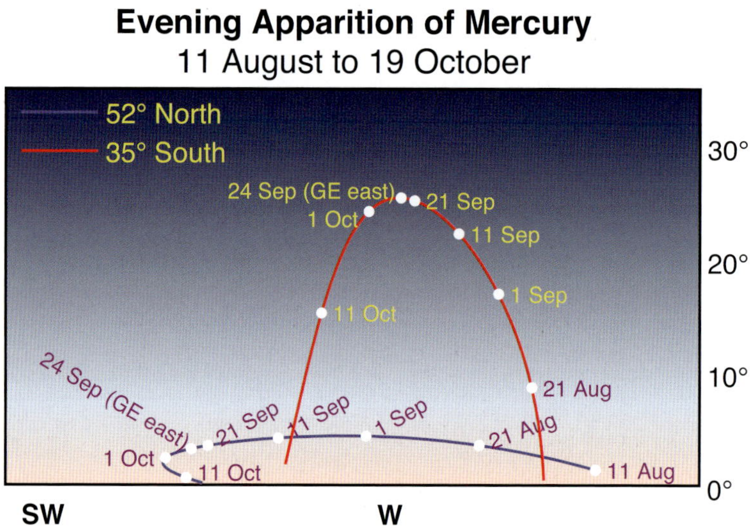

Mercury is at perihelion when the Moon occults it on the first day of the month. The Moon is less than a day from its New phase so this will be a very difficult event to capture. Mercury's close approach to Venus takes place just hours before the nearest planet to the Sun undergoes superior conjunction. Mercury and Jupiter are found together very low in the west after sunset on 19 August but both bright planets will likely be lost in the glow of civil twilight. This is the penultimate evening apparition of the year and is the best one of all for observers in southern latitudes whereas planet chasers in northern temperate latitudes will be lucky to spot Mercury at all as it never gains much altitude.

Venus exchanges skies this month. Still found in the east at dawn at the beginning of August, it goes through superior conjunction on 12 August and reappears in the west after sunset. Its occultation by the Moon on 2 August will likely go unnoticed, as will its rendezvous with the Beehive cluster on the same day. Perihelion occurs on 7 August and Venus makes a second pass by Mercury the day before superior conjunction. Additional close passes by bright objects – first-magnitude star Regulus on 20 August and bright Jupiter on 26 August – take place less than 5° from the Sun, but southern hemisphere observers have the best chance of viewing them.

Earth is blessed with a number of meteor showers this month but only the Perseids are likely to be visible in the early morning hours after the waxing gibbous Moon

Evening Apparition of Venus
12 August 2027 to 1 June 2028

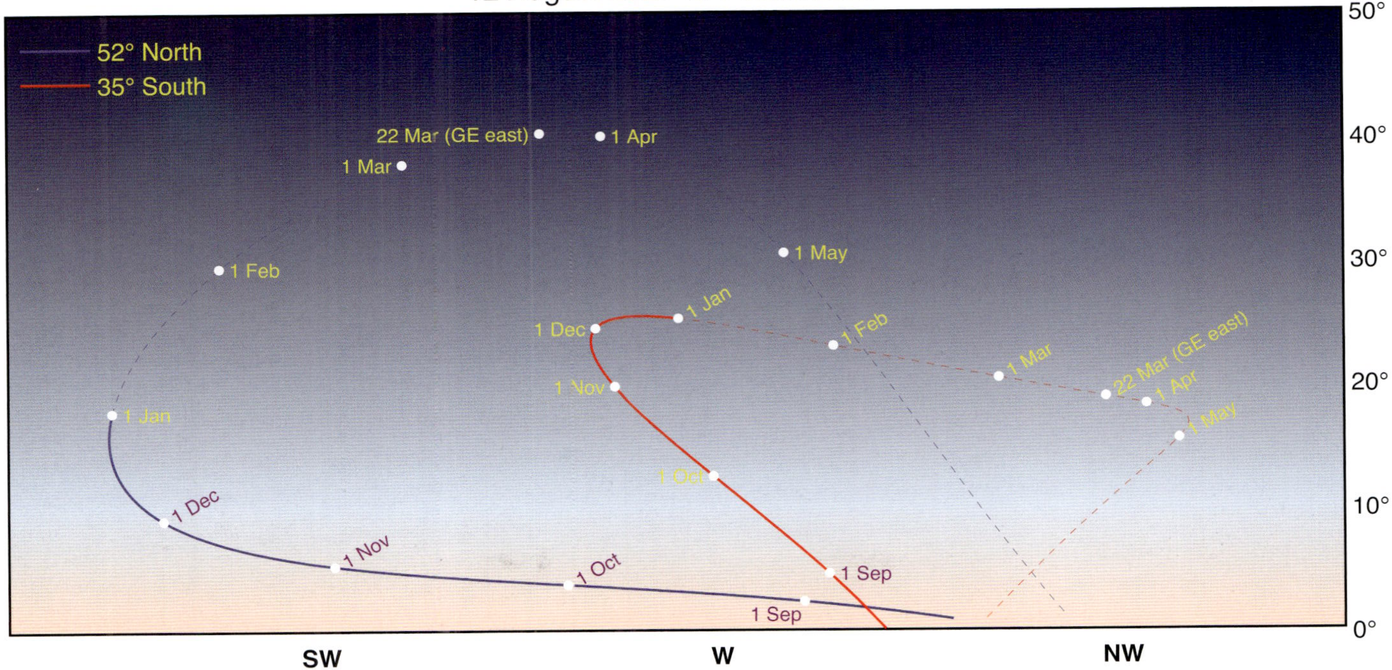

sets. The later minor showers – August δ-Aquariids and the episodic κ-Cygnids – are completely washed out by moonlight.

The **Moon** blots out the Sun on the second day of the month, leading to an annular eclipse. The fifth and final eclipse of the year, a penumbral lunar eclipse, occurs on 17 August. The New (or nearly New) Moon's occultations of Mercury, Venus, and the Beehive open star cluster takes place just before the solar eclipse and will be lost to view. However, a second occultation of M44 takes place later in the month, when the Beehive cluster is located in the morning sky. The monthly occultations of Antares and the Pleiades take place well away from the Sun and are readily visible if you are in the right place at the right time. A second New Moon happens on the last day of August. This is sometimes called a 'Black Moon'.

Mars is found in the west as skies blacken and is more easily observed from the dark skies of southern latitudes than from northern temperate regions where twilight can last all night. The first-magnitude planet is slightly fainter than Spica, the brightest star in the constellation of Virgo, when the two bodies meet, albeit at a distance of 1.9°, on 10 August.

Jupiter has close encounters with both Mercury and Venus this month, as well as the fourth-magnitude star ρ Leonis but all of these events take place too close to the Sun to see. Conjunction itself finally takes place on the last day of the month.

Saturn begins retrograde motion on 9–10 August, reversing course across the constellation of Pisces. Shining at magnitude +0.5 at the beginning of August and brightening throughout the month, the rings are at their most open on the third day, tilted at 14.4° with respect to Earth. Saturn rises during the middle of the evening and is best viewed from the southern hemisphere where evening twilight is not an issue at this time of the year.

Uranus is found in the constellation of Taurus where it rises around midnight for all observers. It is only sixth-magnitude, on the cusp of naked-eye visibility, so chose a dark location on a moonless night when you attempt to see it.

Neptune continues it retrograde motion across Pisces. It rises mid-evening for all observers but is probably easier to see in the dark winter skies of the southern hemisphere.

Astronomers and Their Cats

Katrin Raynor

There is something about astronomers and cats. They go hand in hand – or hand in paw. I could easily list at least ten friends on my social media accounts that are astronomers and cat lovers including Stuart Atkinson who loves cats so much he wrote a book called *A Cat's Guide to the Night Sky*. There is no scientific evidence behind why so many astronomers seem to have or had a cat, so the short article that follows is just for fun but makes for an informative yet light-hearted read.

Wherever I would go my cat Jasmine would follow. We would often sit together in the darkness gazing up at the stars, his soft purr the only noise breaking the silence. Occasionally he would plonk his bottom on an astronomy book I was trying to read, making it known that he was more important than silly constellations or the boring old Moon.

Like their human owners, cats are naturally inquisitive creatures and I could not start without first mentioning Sir Patrick Moore (1923–2012). Moore inherited his love of cats from his mother, Gertrude and if you search for images of Moore on the internet, you often find photos of him holding a cat. When Moore was very young, he had two cats named Ginger and Ptolemy. Later in life, when Ginger and Ptolemy had gone to the great scratching post in the sky, he had a large Norwegian forest cat named Jeannie, and another cat named Ptolemy by his side. All the black cats in Patrick's family were given the name Ptolemy.

In 2012, shortly before he passed away, Moore published his first book about cats called *Miaow! Cats really are nicer than people!* in which he gives a personal account of all the cats who have been part of his family including Bonnie, who lived to the ripe old age of 20. Cats own their humans, not the other way around, and Moore knew this better than anyone. Hanging at the front door of his home in Selsey, Sussex was a sign declaring, 'This house is run entirely for the benefit of our cats'. Even astrophysicist Brian May, a dear friend of Moore's and guitarist in the rock band Queen, wrote a song about his childhood cat Pixie passing away. 'All Dead, All Dead' featured on Queen's 1977 album 'News of the World' – a melancholic song guaranteed to bring a tear to your eye if you have ever lost a pet.

Whilst Patrick is perhaps the best-known astronomer who had a passion for cats, there are other big names in the astronomy world that were owned and inspired

Stuart Atkinson lives in Kendal, Westmorland and has been writing and editing books for nearly 30 years. He is a huge cat lover; his book *A Cat's Guide to the Night Sky* was published in 2018 and was undoubtedly inspired by his beloved cat Peggy. Stuart often takes part in astronomy outreach and 'meet the author' events, and is pictured here at a book signing event in Harrogate. (Stella Coxon/Stuart Atkinson)

by their felines. In the twilight years of his extraordinary life, famed astronomer Edwin Hubble (1889–1953) shared his quiet California estate with a regal black part-Persian cat named Nicolas Copernicus, named after the pioneering sixteenth-century astronomer.[1] Copernicus was more than a pet – he was a cosmic companion to the man who proved the universe was expanding. Hubble affectionately referred to their home as "Nicolas' estate," a nod to the cat's undeniable authority over the household. When Hubble passed away in 1953, Nicolas kept vigil, curling beside him and later waiting by the window, as if expecting his stargazing friend to return home from the heavens.

Nikola Tesla's (1856–1943) fascination with electricity was sparked – quite literally – by a cat. As a child, Tesla's closest companion was a black feline named Macak. One dry winter's evening, the cat's fur crackled with static as young Tesla stroked its back. The sight of sparks leaping from Macak's body left Tesla awestruck, igniting a lifelong obsession with the mysteries of energy. In later writings, Tesla recalled that moment as a "miracle" that made him wonder if nature itself was a giant cat, and if so, who was stroking its back. That single, electrifying encounter with Macak didn't just shape Tesla's imagination – it may have helped launch the modern age of electricity.

Theoretical physicist Albert Einstein (1879–1955) once mused, "I love cats because I enjoy my home, and little by little, they become its visible soul," a sentiment that reveals how deeply he valued their quiet presence. Einstein is famous for reshaping our understanding of space and time and clearly had a soft spot for cats. Among his pets was a cat named Tiger, who reportedly shared Einstein's contemplative moods. While there's no historical record of Tiger directly influencing his scientific breakthroughs, the quiet companionship of a cat like him may well have provided the perfect backdrop for reflection and creativity.

One cat we have all heard of is Schrödinger's cat. But this wasn't a real cat to pick up and sit under the stars with. It is a famous thought experiment introduced in 1935 by Austrian physicist Erwin Schrödinger (1887–1961) to illustrate the strange nature of quantum mechanics. In it, a cat is placed in a sealed box with a vial of poison that may be released depending on the decay of a radioactive atom. According to quantum theory, until the box is opened and observed, the cat exists in a superposition of both alive and dead states simultaneously. The paradox was never meant to describe reality, but rather to highlight the absurdity of applying quantum principles – usually reserved for subatomic particles – to everyday objects. It remains one of the most iconic mind-benders in physics.

1. For more on Hubble and Copernicus see: **huntington.org/verso/hubble-and-copernicus**

Mary McIntyre is an astronomy author, speaker and illustrator, and her nine-year-old tortoiseshell cat Eevee is her faithful companion in all things astronomy-related. Eevee follows Mary everywhere and likes to sit on her lap whenever Mary is observing the skies. Her loyal feline companion even joins Mary in the rain when she is photographing rainbows! (Mary McIntyre)

Sir Isaac Newton (1643–1727), renowned for formulating the laws of motion and universal gravitation, is also famously – if apocryphally – linked to feline history as the supposed inventor of the cat flap. According to urban legend, Newton grew frustrated when his cat repeatedly disrupted his darkroom experiments, prompting him to cut a hole in the door so it could come and go without interfering and even made a smaller hole for the kittens without realising of course, the kittens could have simply used the first hole! There appears however to be no evidence that Newton even owned a cat, and evidence suggests that cat flaps actually existed before his time.

I must end this article with a lesser-known cat, Félicette. In 1963, a small black-and-white stray from the streets of Paris made history as the first – and so far, only – cat launched into space by the French space program. She was fitted with electrodes to monitor brain activity and sent on a suborbital flight aboard a Véronique rocket. Félicette soared to a height of over 150 kilometres above Earth, experiencing a few minutes of weightlessness before safely returning. Though her mission was brief, it provided valuable data for space medicine. Long overlooked, Félicette was honoured in 2019 with a bronze statue in Strasbourg, France, a belated tribute to the quiet feline pioneer who once touched the stars.

Cats have quietly padded through the lives of some of the greatest minds in astronomy. Whether inspiring thought experiments, venturing into space, or simply curling up beside a telescope, these cats remind us that even the most brilliant astronomers found comfort and kinship in the quiet, knowing gaze of a cat.

September

2	Uranus	West Quadrature	90°
7	Moon, α Scorpii (Antares)	0.7′ apart: occultation	86°
7	Moon	First Quarter	90°
13	300P/Catalina	Perihelion	55°
15	Moon	Full	177°
21	Mercury, α Virginis (Spica)	0.3° apart	26°
21	Moon, M45 (Pleiades)	0.8° apart: occultation	118°
23	Earth	Equinox	
23	Moon	Last Quarter	90°
24	Mercury	Greatest Elongation East (evening)	26°
25	Mars, α^2 Librae (Zubenelgenubi)	0.9° apart	44°
26	Moon, M44 (Beehive)	1.0° apart: occultation	55°
28	Neptune	Opposition	179°
30	Moon	New	4°

Dates are based on Universal Time (UT). Occultations of M44 (Beehive/Praesepe) refer to sixth-magnitude star ε Cancri (Meleph). Occultations of M45 (Pleiades) refer to third-magnitude star η Tauri (Alcyone). The last column gives the approximate elongation from the Sun of the event.

Mercury is undergoing its last full evening apparition, the best one of the year for those in the southern hemisphere. These favoured observers will have the best chance of catching Mercury 0.3° south of Spica, the brightest star in the constellation of Virgo, on 21 September. Mercury spends the month getting dimmer, from −0.3 magnitude to +0.3 magnitude, and is highest in the sky around the time of greatest elongation east on 24 September. Mercury reaches aphelion on 14 September.

Venus now takes up residence in the west after sunset as *æfensteorra* (Old English for 'evening star'). It rises most rapidly above the horizon for planet watchers in the southern hemisphere although ultimately, those in north temperate latitudes will get the best views next year.

Evening Apparition of Mars
1 September 2027 to 1 February 2028

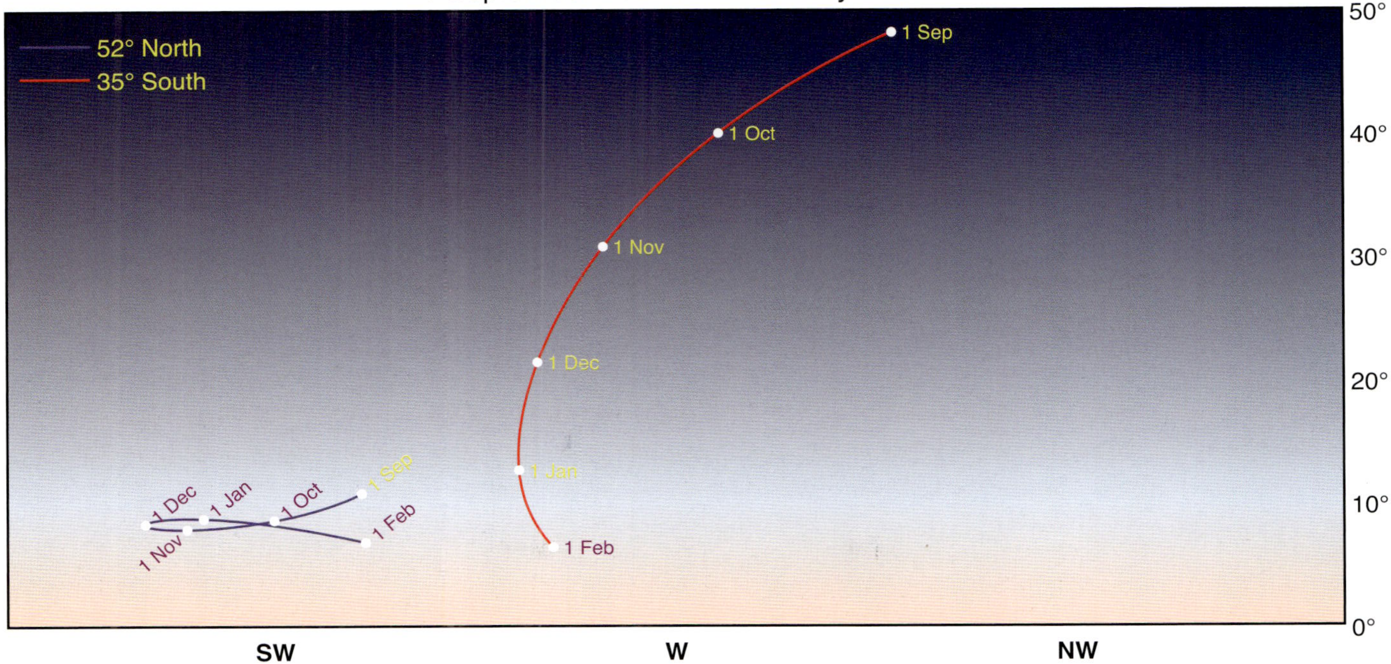

Earth is at equinox on 23 September. The Sun passes through the celestial equator on this day, leaving the northern celestial hemisphere and entering the south. Spring begins south of the equator on Earth whilst autumn gets underway in the north.

The **Moon** occults the first-magnitude star Antares on 7 September, followed by obscurations of the open clusters Pleiades (M45) and Beehive (M44) later in the month. Although the Moon will continue to pass close by M44, this will be the last time this year that our satellite occults the brightest member of the cluster.

Mars finally leaves Virgo for the constellation of Libra on 14 September and passes 0.9° south of the third-magnitude Zubenelgenubi, the brightest star in Libra 11 days later. Look for Mars ever lower in the west after sunset. It sets during twilight from northern temperate latitudes but in dark skies from seen from the southern hemisphere.

Jupiter was at conjunction the last day of August and only slowly emerges from dawn twilight when it reappears in the morning sky. Look for it low in the east later in the month.

Saturn rises during evening twilight and sets after sunrise. It is shining at magnitude +0.3 in Pisces and although the rings are closing up slightly, they are still open to over 13° as viewed from Earth.

Uranus reaches west quadrature at the beginning of the month and then enters into retrograde motion on 15 September, reversing course across the constellation of Taurus. It rises during late to mid-evening for all observers.

Neptune arrives at opposition just before the end of September which means it is visible in the sky for most of the night. It shines at magnitude +7.8 in the constellation of Pisces and a telescope will reveal a pale blue-green disk just 2.5″ across.

H. G. Wells
Time Traveller

John McCue

Herbert George Wells (1866–1946). (John McCue)

Herbert George Wells was born on 21 September 1866. I share his birthday, dear reader, which has been a lifelong source of amusement for me!

Bertie, as his mum called him, lived a humble beginning in Bromley, Kent. His mother, Sarah, was a lady's maid at Uppark, the big country house above Harting on the northern reaches of the South Downs in Sussex. It was here that she met Joseph, the gardener there, and a professional cricketer, though it must be said he was outstanding at this sport – playing for Kent he was the first county bowler to take four wickets in four consecutive balls.[1] Following their marriage in 1853, Sarah and Joe left Uppark and ran a crockery and sports shop, somewhat unsuccessfully, which probably led to Bertie's first job as an apprentice in a draper's shop; disillusioned, he was dismissed. By this time Sarah had been invited back to Uppark as head of the servants, and Bertie became absorbed in the magnificent library's books helping him, at the age of 18, to win a scholarship to Normal School of Science (now known as the Royal College of Science) at South Kensington, Middlesex to study biology with Professor Thomas H. Huxley.

His 1895 novel, *The Time Machine*, was an instant success and explored the world as it was, or will be, in the far distant past or the future. But can time travel, in reality, be done? Journey to the future is physically possible and, indeed, we do it

1. *The Young H.G.Wells: Changing the World*, Claire Tomalin, Penguin Books (2021)

all the time. On a flight across the Atlantic to New York at 800 kilometres per hour, Einstein's Special Theory of Relativity shows that time passes more slowly for the passengers against those on the ground. When the aircraft lands, therefore, those on the plane have not aged as much as those confined to Earth's surface. The emerging trippers are seeing people and their environment as being older than expected. Although only by a few billionths of a second, the travellers are in the future! The record for travel to the future, by 0.02 seconds, was held for many years by the Russian cosmonaut Sergei Avdeyev, who orbited the Earth for a cumulative time of 748 days during three missions aboard space station *Mir*, until he was superseded by fellow Russian cosmonaut Sergei Krikalev with a total time in space of 803 days.

Wells' central character, known in the book only as the Time Traveller, was not satisfied with such trivial excursions into future realms. He set the controls to the year 802,701 on his time machine, and found himself in a world of weak, gentle and beautiful people, living a life of luxurious leisure. After a few days, in an alarming moment, he encountered a white ape-like creature in the shadows and discovered that this future was a frightening place. In the Time Traveller's words, "Gradually the truth dawned on me: that Man had not remained one species but had differentiated into two distinct animals: that my graceful children of the Upper World were not the sole descendants of our generation, but that this bleached obscene nocturnal Thing, which had flashed before me, was also heir to all the ages".[2] These underground dwellers, the *Morlocks*, worked tirelessly to provide the gentle surface inhabitants, the *Eloi*, with all their needs, but they emerged to the surface at night to feed on them; Wells saw this as the inevitable outcome of the class system that he saw around him as he grew up.

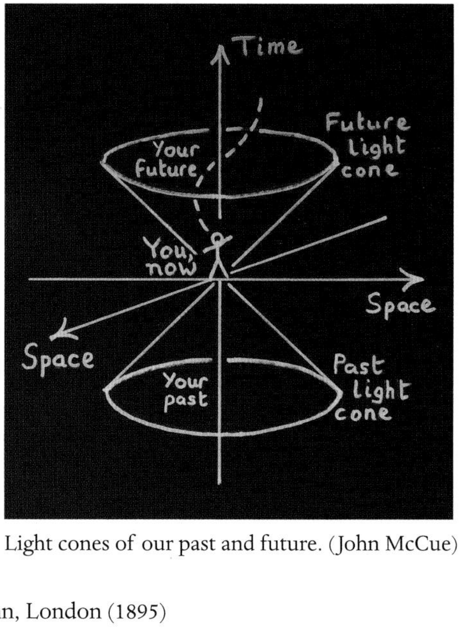

Light cones of our past and future. (John McCue)

Can we travel to our past? In theory, it can be done, but the probability is vanishingly small in comparison to journeys to the future, which happen

2. *The Time Machine*, H.G.Wells, Heinemann, London (1895)

all the time! As our life progresses, our journey through space-time can be imagined snaking through a light cone – but always remaining within the cone, at less than the speed of light. Our past life can likewise be pictured as the mirror image. If our future light cone can somehow access our past light cone, then we can visit our previous existence. This could happen if we can travel through a black hole, into its wormhole, and emerge from the white hole, arriving at our past. The light cones of the future and the past have been bent 'sideways' by the strong gravitational fields of both black and white holes respectively, thereby allowing the link.

There may well be another law of physics, though, that prevents travel to the past because of the obvious cause-and-effect paradoxes. For example, you travel to your past to see yourself as a baby in your mother's arms and try to shoot yourself! But suppose, as you aimed the gun, you felt a painful twinge in your arm causing you to let go of your weapon. Your mother, holding you as a baby, dropped you in fright, causing a recurrent injury to your arm, thereby preventing you from firing the gun in the past! (Some theorists speculate that this shows a circular passage of time, in some circumstances, instead of the usual linear time travel). Stephen Hawking proposed this yet-to-be-discovered law of physics – the 'chronology protection conjecture'[3] – which would make time travel to the past, and hence these puzzles, impossible. And yet Hawking has reconsidered; so maybe someone, somewhere, sometime may visit himself or herself in the past.

H. G. Wells did not send his Time Traveller to the past: maybe he foresaw the emergence of illogicality, but after the Traveller's disappearance, the narrator mused that the hero may have seen life long past and, "… fell among the blood-drinking, hairy savages of the Age of Unpolished Stone; into the abysses of the Cretaceous Sea; or among the gross saurians, the huge reptilian brutes of the Jurassic times." [2]

After escape from the *Morlocks* the Traveller went further forward, to a time more than 30,000,000 years in the future, and watched strange creatures on a beach – huge crabs, one of them, "… as big as yonder table, with its many legs moving slowly and uncertainly, its big claws swaying, its long antennae, like carters' whips, waving and feeling, and its stalked eyes gleaming at you on either side of its metallic front." – as the sun grew large and red. It wasn't until around the mid-twentieth century that red giants were established as old stars, so Wells' description was remarkably prescient.

The First Men In The Moon was published in 1901 as Wells continued his cycle of science fiction novels and, appropriately, he has a crater named after him on the

3. *Physics of the Impossible*, Michio Kaku, Doubleday Publishing (2008)

The crater H. G. Wells on the far side of the Moon. (NASA/Arizona State University/Apollo Browse Gallery/Apollo 16 Mapping Camera/AS16-M-3021)

far side of the Moon. Over 100 kilometres in diameter, it presents a sadly battered appearance; although the floor of Bertie's crater has suffered only insignificant pock marks, the walls have been destroyed by later impacts to the extent that it could almost be viewed as deliberate targeting! Yet the crater remains as a visible sign of Wells' literary achievements, though his books will be forever read, regardless.

October

4	Moon, α Scorpii (Antares)	0.2° apart: occultation	59°
7	Moon	First Quarter	90°
7/8	Earth	Draconids (ZHR 5)	
12	104P/Kowal 2	Perihelion	66°
15	Moon	Full	175°
18	Saturn	Opposition	177°
18	Moon, M45 (Pleiades)	0.6° apart: occultation	144°
19	Mercury	Inferior Conjunction (evening → morning)	1°
20/21	Earth	Orionids (ZHR 20)	
22	Moon	Last Quarter	90°
25	Mars, M80	0.8° apart	35°
25	Jupiter, σ Leonis	0.7° apart	43°
29	Saturn, μ Piscium	0.3° apart	168°
29	Moon	New	5°

Dates are based on Universal Time (UT). Occultations of M45 (Pleiades) refer to third-magnitude star η Tauri (Alcyone). Peak activity dates and ZHRs for meteor showers are estimates. The last column gives the approximate elongation from the Sun of the event.

Mercury begins retrograde motion on 7 October, a sure sign that inferior conjunction is not far off and indeed, the tiny planet goes past the Sun on 19 October. Retrograde motion ceases on 28 October. This is Mercury's second-to-last evening apparition, best seen from the southern hemisphere where the rapidly-dimming Mercury and bright Venus may be glimpsed together once again on 11 October. Mercury passes through perihelion on 28 October. Those arising early in northern temperate latitudes will get the best views of Mercury after inferior conjunction, with the brightening planet rising quickly above the eastern horizon at dawn.

Venus is racing upwards from the western horizon for observers in the southern hemisphere but is taking a far more leisurely path through evening twilight for those watching out for the bright planet from northern temperate latitudes. The evening star passes nearly 3° north of the first-magnitude star Spica (α Virginis) on the third day of the month and makes an even more distant swing by Mercury on 11 October.

Morning Apparition of Mercury
19 October to 11 December

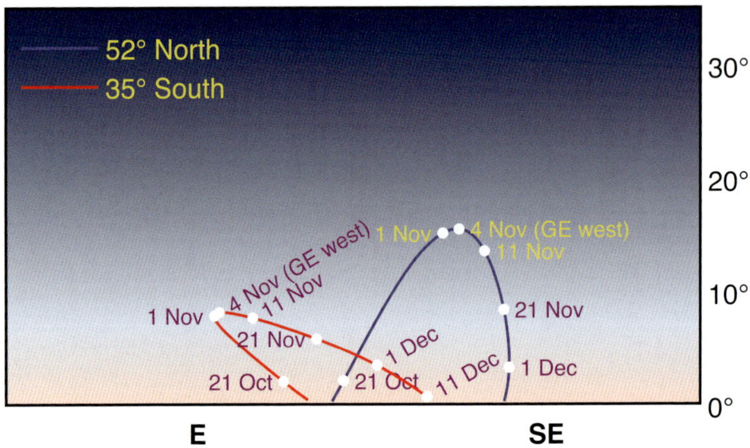

Earth intercepts a number of meteor streams in October, including the beginning of the Taurids which peak next month. The Draconids should not be too badly affected by moonlight if you seek them out after midnight, but this is not a good year for the Orionids as the waning gibbous Moon will provide plenty of light interference.

The **Moon** is down to only two occultation targets this month, with Antares disappearing on the fourth and the Pleiades on 18 October. The waning crescent Moon passes 1.2° south of M44, the Beehive Cluster, on 23 October, occulting some of the fainter members of the group but probably just missing the brightest star, Meleph.

Mars switches constellations twice this month, moving from Libra to Scorpius on 15 October and then from Scorpius to Ophiuchus on 27 October. Found 0.8° north of the globular cluster M80 on 23 October, the red planet is best viewed from the southern hemisphere where skies are still dark by the time it sets. Mars is low in the west for northern temperate latitudes and sets during twilight. There is a seasonal change on the rocky planet this month, with an equinox occurring on 17 October. Autumn begins in the northern hemisphere and spring starts in the south.

Jupiter is finally clear of the pre-sunrise twilight. Located in the constellation of Leo, Jupiter is strictly a morning sky object, not rising until after midnight. It passes less than a degree north of the fourth-magnitude star σ Leonis on 25 October.

Saturn reaches opposition this month. Found in Pisces, the gas giant shines at magnitude +0.1. A telescope will reveal its glorious rings tilted at 12.9° with respect to Earth. Saturn makes a second pass by fifth-magnitude μ Piscium near the end of the month, coming to within 0.3° of the K-type star.

Uranus is in retrograde in the constellation of Taurus. It appears in the east early in the evening and is aloft for most of the night, not setting until after sunrise.

Neptune is visible, albeit with a telescope, in the evening skies, not setting until morning twilight. Its position in the constellation of Pisces means that the eighth-magnitude ice giant is best observed from the northern hemisphere this month.

Fritz Zwicky and the Coma Cluster

David M. Harland

Ninety years ago, in October 1937, Fritz Zwicky published a detailed analysis which indicated that hundreds of galaxies in the Coma Cluster are gravitationally bound by some form of 'non-luminous' matter.

-o0o-

Fritz Zwicky. (John McCue)

Born on 14 February 1898 in Varna, Bulgaria, Fritz Zwicky gained Swiss nationality from his industrialist father. At the age of six he was sent to live with his grandparents in Switzerland. At school, he developed an interest in mathematics and physics, following up with a Ph.D. in experimental physics at the Federal Institute of Technology (nowadays known as ETH) in Zurich in 1922 for work on ionic crystals, essentially solid-state physics. Three year later, he went to the California Institute of Technology in Pasadena on a Rockefeller fellowship. By 1930 he was a member of staff with an interest in astrophysics.

At that time, astronomy was in a state of flux. In 1920 there was an inconclusive debate about whether the Milky Way system formed the entire universe, or whether the many spiral nebulae were independent systems of comparable size. This issue was settled in 1924 when Edwin Hubble proved beyond doubt that the Andromeda Spiral (M31) was an independent system. In 1929, after examining two dozen examples, he announced a linear relationship between the redshift of a spiral and its distance. The universe was not only vastly bigger than was reckoned at the start of the decade, it was expanding.

It was presumed that galaxies (of which the spirals would turn out to be only one type) were uniformly distributed in space and that chance alignments were responsible for their seeming to be populous in certain parts of the sky. Zwicky,

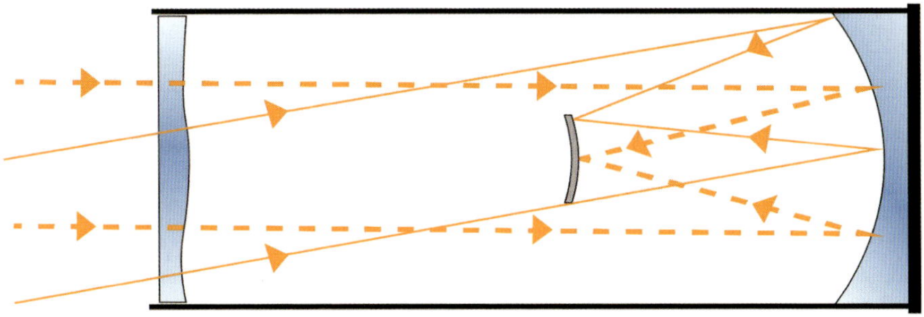

The Schmidt telescope designed by Bernhard Schmidt in 1930 had a 'corrector' at the aperture that enabled an easy-to-make spherical mirror to be employed. The photographic plate had to match the curved focal plane. The advantage for celestial photography was a wide field of view. It was for this reason that Zwicky chose the newly commissioned 18-inch Schmidt atop Mount Palomar for his study of the Coma Cluster. (Wikimedia Commons/Jamesmcardle/CC BY-SA 4.0)

however, suspected these concentrations were real. The way to prove this was to measure the redshifts of a number of putative cluster members in order to determine the average redshift for the cluster, and then subtract the average to obtain the individual motion of each galaxy in turn. The spread of velocities relative to the mean was a measure of the overall kinetic energy of the system. The greater it was, the greater would be the gravitational attraction required to prevent the members of the cluster from escaping.

After a preliminary investigation of several members of a cluster in the direction of the constellation Coma Berenices, Zwicky reported in a Swiss journal in 1933 that the cluster was not in a state of rotation; the galaxies were moving randomly around a common centre. But for the cluster to be stable there required to be more gravitating mass than was visible, leading him to speculate about the existence of 'non-luminous matter'.[1]

Intrigued, Zwicky measured the motions of 600 galaxies (by far the majority of the known membership of the cluster at that time) to refine his figure for the gravitating mass. Then he analysed the light that was captured by his photographic plates. The mass-luminosity relationship says that the more massive a star, the greater is its luminosity. The light from a galaxy provides an estimate of the overall mass of its stars. After measuring the mass of each galaxy in this way, he added them

1. Zwicky, F. (1933), The Redshift of Extragalactic Nebulae, *Helvetica Physica Acta*, **6**, 110–127.

The dome of the 18-inch Schmidt telescope on Mount Palomar in 1937, with that of the 200-inch under construction (Earl W. Gray / Palomar Observatory)

up. The results, reported in 1937, confirmed the luminous mass was insufficient to explain the gravitational attraction indicated by the kinetics.[2]

Despite its profound implications, Zwicky's discovery was ignored by the astronomical community for many years.

Aftermath

In 1968 Vera Rubin and Kent Ford of the Department of Terrestrial Magnetism of Carnegie Institute of Washington announced that they had accurately measured the radial velocity 'rotation curve' of M31 and found it to be 'flat' instead of the 'Keplerian decline' that was expected. This meant that there must be considerably more gravitating mass than was implied by the light. Radio astronomers then reported that the 21-centimetre emission from neutral hydrogen showed that this rotational state extended far beyond the visible edge of the spiral. With improved

2. Zwicky, F. (1937), On the Masses of Nebulae and of Clusters of Nebulae, *Astrophysical Journal*, **86**(3), 217–246

The Advanced Camera for Surveys of the Hubble Space Telescope imaged a portion of the central region of the Coma Cluster. Most of the galaxies there are ellipticals, composed of old red stars. Spirals appear blue because of their young stars. (NASA, ESA, the Hubble Heritage Team (STScI/AURA), D. Carter (Liverpool John Moores University), and the Coma HST ACS Treasury Team)

equipment, Rubin determined that for spirals, flat rotation curves were the rule rather than the exception. The motions implied that the fall off in luminous matter with distance from the centre of a galaxy was more than compensated by an increase in non-luminous matter.

Later work established that this is not confined to spirals; other types of galaxy are also surrounded by such material. Indeed, it is pervasive throughout clusters.

The merit of Zwicky pioneering study was belatedly recognised. The matter that we perceive by its electromagnetic radiation is now known to be only a fraction of the matter in the universe, and the search is on to determine the nature of what has come to be known as 'dark matter'.

November

1	Moon, α Scorpii (Antares)	0.4° apart: occultation	32°
4	Mercury	Greatest Elongation West (morning)	19°
5	Earth	Southern Taurids (ZHR 7)	
6	Moon	First Quarter	90°
12	Earth	Northern Taurids (ZHR 5)	
14	Moon	Full	175°
14	Uranus, κ^1 Tauri	0.7° apart	163°
14	Uranus, κ^2 Tauri	0.6° apart	163°
14	Moon, M45 (Pleiades)	0.5° apart: occultation	171°
14	Mars, θ Ophiuchi	0.9° apart	30°
16	Mars, 44 Ophiuchi	37″ apart	29°
17/18	Earth	Leonids (ZHR 10)	
21	Moon	Last Quarter	90°
25	Venus, Mars	0.3° apart	27°
27	Mars, M8 (Lagoon Nebula)	0.1° apart	26°
28	Moon	New	4°
28	Moon, α Scorpii (Antares)	0.5° apart: occultation	6°
30	Uranus	Opposition	180°
30	Moon, Venus	0.7° apart: occultation	28°

Dates are based on Universal Time (UT). Occultations of M45 (Pleiades) refer to third-magnitude star η Tauri (Alcyone). Peak activity dates and ZHRs for meteor showers are estimates. The last column gives the approximate elongation from the Sun of the event.

Mercury is in the morning sky one last time this year. This morning apparition is the best one of 2027 for observers in northern temperate latitudes. The planet brightens by over half a magnitude this month, beginning at −0.2 and ending at −0.9, and is already at −0.4 when it reaches greatest elongation west on the fourth day of the month. Afterwards Mercury heads back towards the rising Sun.

Venus is still low in the west for northern temperate latitudes but has reached a useful altitude for those in the southern hemisphere should they wish to observe the evening star's close encounter with Mars on 25 November when the two worlds are 0.3° apart. Venus reaches aphelion two days later and is then occulted by the crescent Moon on the last day of the month.

Earth has been in the middle of the Taurid meteor complex for over a month. The Southern Taurids peak first, when the Moon is a waxing crescent, so this is a good year for observations. The Northern Taurids, on the other hand, peak much closer to the time of Full Moon so viewing will be more problematical. The Leonids are also badly affected by moonlight, reaching maximum activity during the waning gibbous phase of the Moon.

The **Moon** passes over Antares twice this month, on the first day of the month and again on 28 November, but the second event takes place too close to the Sun to observe. The Pleiades are located in the opposite part of the sky from Antares so that occultation will be easily visible. The very young crescent Moon occults Venus for the second time this year but you will need to be in the Antarctic to view it.

Mars begins the month in the constellation of Ophiuchus, passing near the third-magnitude star θ Ophiuchi on 14 November and very close by the fourth-magnitude star 44 Ophiuchi two days later. This latter star is a bluish A-type giant; a telescope may bring out the colour contrast between the two objects. Mars moves into the constellation of Sagittarius on 21 November and is found 0.3° north of the evening star three days later. An astrophotography opportunity occurs on 27 November when the red planet is 0.1° south of the famous Lagoon Nebula (M8), an interstellar cloud and emission nebula. However, Mars is low in the west after sunset; the best chance to see Mars in dark skies is in the southern hemisphere.

Jupiter rises around or just after midnight for most locations on Earth. Shining at magnitude −1.8, it moves from Leo to the constellation of Virgo on 22 November.

Saturn is already above the horizon as the sun sets, having undergone opposition late month. It sets in the early morning hours, and, at magnitude +0.1, is easily the brightest starlike object in the constellation of Pisces.

Uranus trundles past the two fifth-magnitude stars κ^1 Tauri and κ^2 Tauri for the second time this year thanks to retrograde motion. Opposition takes place on the last day of the month, when the green ice giant is at its biggest (apparent disk diameter of 3.7″) and brightest (magnitude +5.6). The lack of light from a very young crescent Moon that day will make observations just that little bit easier.

Neptune continues its retrogression through Pisces. The northern hemisphere is the favoured location for observing the planet this month, with Neptune not setting until an hour or more past midnight.

The Cat's Whiskers
On the Radio Astronomy

Jonathan Powell

The advent of radio broadcasting introduced the public to a new medium of communication where the printed word no longer monopolized a potential audience. In a world where numerous inventors worked tirelessly toward the resulting revolution, the dawn of a new era would see the culmination of not only the first radio broadcast but less than two decades later, the first television broadcast. In terms of astronomy and space, radio was to never really fulfil its potential with regard to sustained and regular output, and aside the obvious about visuals, that doesn't translate as the whole reason.

The history and background in the evolution of radio has a great number of persons of note attached to the credits. Names in that list include Hans Christian Oersted, Michael Faraday, James Clerk Maxwell, Nikola Tesla, and Guglielmo Marconi, the latter two of which very much at odds with each other over the actual invention of radio itself. That honour was eventually attributed to Marconi who conquered both sea and air waves with his transmission of the first radio message across open water in 1897. Beamed across the Bristol Channel from Brean Down Fort, just south of Weston-Super-Mare, to Lavernock in Wales, the landmark moment fulfilled Marconi's desire to produce 'telegraphy without wires'.

However, what could be considered the first entertainment in radio broadcasting is attributed to Reginald Fessenden, a Canadian-American inventor. Fessenden, an electrician by trade and employed by Thomas Edison, is credited with being the first person to broadcast a voice message over the radio in 1900. Then on Christmas Eve 1906, and transmitted from Brant Rock in the county of Plymouth, Massachusetts, Fessenden broadcast what can only be interpreted as a variety show, with the sound of 'O Holy Night' being played on the violin, plus the reading of a passage from the bible.

In later years, radio broadcasting was to find itself playing the part of key role of messenger for news and information, changing the face of how reports would reach the masses, and with two world wars ravaging much of Europe, a vital link to the outside world beyond the newspaper which despite the reach of radio still

Cat's Whisker Crystal Radio: Utilizing a thin wire akin to that of cat's whisker, the 'whisker' itself made contact with a crystalline material which in turn created a semiconductor junction; the completed set-up subsequently allowed the production of audible sounds to be derived from radio frequency signals. (Science Museum / Gift of Mr. M.R. Bailey / Y2008.56 / CC0 1.0 Universal / CC BY 4.0 / Attribution 4.0 International)

held a noteworthy position. The 'Cat's Whisker' radio device, which involved fine 'cat-like whiskers' wiring, was used in occupied Guernsey during World War II to keep citizens informed.

The extensive growth of radio as the primary communication medium saw private individuals and large corporations alike display a keen interest in what could be broadcast and the demographics of who might be listening. For the many parties following the rapid evolution of radio broadcasting, even terrestrial television or the introduction of cable and satellite television followed by online formats, still didn't detract from radio's capability to maintain its audience share.

With such a vast and varied output since radio broadcasting began, there would seem sufficient room for astronomy and space to feature more readily and consistently than it has, which begs the question as to whether or not such a format has been overlooked? There were a few astronomy-themed broadcasts, notable amongst which were those of astronomy writer and populariser Joseph Henry Elgie (1864–1937); the publishing of Elgie's books and appearances in newspapers and journals were supplemented by numerous broadcasts on the recently-formed BBC during the 1920s. These broadcasts, usually of around quarter of an hour or so in duration, were transmitted from studios in London and Nottingham, the topics covered including both astronomy and the weather. Also worthy of note is author and science educator Rupert Thomas Gould (1890–1948) who was asked by the BBC to give a series of talks for Children's Hour (the BBC's main recreational service for children) during the 1930s, in the course of which he went under the name 'The Stargazer'. The talks covered a number of topics (including astronomy) and were subsequently collected together and published in his book *The Stargazer Talks* (1943).

Sir Patrick Moore: Pictured here at the family home in Sussex with one of his many telescopes, Patrick brought the world of astronomy to the general public in a yet unparalleled manner and style. His medium was predominantly television with the long-running monthly programme 'The Sky at Night', but also through numerous and notable radio broadcasts. (Brian Jones)

Yet in spite of the efforts of Elgie, Gould et al., radio would have appeared to have missed a trick, especially in the United Kingdom, television stamping its authority with Patrick Moore's 'The Sky at Night'. For radio listeners, astronomy found itself only worthy of perhaps a short series, banished for the most part to be used as just news bulletin items or feature-length programmes, but not as a running ad-hoc programme or a short series of broadcasts like the monthly 'Sky at Night'.

One could contest that at least radio output wasn't completely void of astronomy and space topics, but given the diversity of radio stations that as time passed saw the introduction of non-music cerebral broadcasting, it remains a mystery as to why on a national scale an astronomy and space radio programme, with news and features, plus a basic look at the night sky, didn't materialize? In a format where the axe could be readily wielded 'The Sky at Night', with a broadcast time of twenty minutes, has survived for decades despite the possessive nature of television executives eager to secure the best possible output for their scheduling.

With the growing amount of sports coverage output appealing to the masses, astronomy found itself even more out of contention, despite sporadic heightening of interest such as the *Apollo* landings or the return of Halley's Comet in the 1980s, which brought much media attention, but spawned little in the way of new radio shows after the events had passed.

In recent years the advent of the podcast has been well received by many quarters of the listenership as yet another outlet to receive astronomy-based news and features, serving as a complimentary source to more sizable output from other sources. However, one could argue that the podcast, (especially in the case of the UK), further dilutes the pool that could have sprung forth a monthly radio broadcast. An example of such a podcast is 'StarTalk', hosted by astrophysicist Neil deGrasse Tyson, although the content does draw on a number of other topics in its format, such as the world of entertainment and especially comedy, with the show co-hosted by celebrities such as comedian Chuck Nice and multisport television and radio broadcaster Gary O'Reilly.

For some reason astronomy on the radio in the United Kingdom has never been taken up on the same scale as some other countries such as the USA where 15,445 stations exist (Statista – September 2024), compared to 600 licensed stations in the UK, which includes the BBC. A prime example that proves that an astronomy programme on the radio is sustainable is 'StarDate' which is produced by the University of Texas and has been running since 1978. However, given the number of stations, it is a welcome but ultimately solitary programme.

Despite British shores being home to such bastions of astronomy as the Royal Observatory at Greenwich, and with all its great contributors to astronomy – Sir Edmond Halley, Stephen Hawking, and Jocelyn Bell Burnell to name but three – it would appear that astronomy just hasn't been recognized in broadcast radio as something that could possible be the 'cat's whiskers'.

December

3	Mars, M28	0.5° apart	25°
6	Moon	First Quarter	90°
7	Earth	Puppid-Velids (ZHR 10)	
7	Mars, M22	0.3° apart	24°
11	Mercury	Superior Conjunction (morning → evening)	1°
12	Moon, M45 (Pleiades)	0.5° apart: occultation	160°
13	Moon	Full	177°
14	Earth	Geminids (ZHR 150)	
18	Uranus, 56 Tauri	0.4° apart	161°
18	Jupiter	West Quadrature	90°
20	Moon	Last Quarter	90°
21	Uranus, 53 Tauri	0.2° apart	157°
22	Earth	Solstice	
22/23	Earth	Ursids (ZHR 10)	
25	Moon, α Scorpii (Antares)	0.4° apart: occultation	24°
25	Neptune	East Quadrature	90°
27	Moon	New	2°
28	Moon, Mercury	0.7° apart: occultation	10°
29	Moon, Mars	0.5° apart: occultation	19°
29	Uranus, 51 Tauri	0.3° apart	149°

Dates are based on Universal Time (UT). Occultations of M45 (Pleiades) refer to third-magnitude star η Tauri (Alcyone). Peak activity dates and ZHRs for meteor showers are estimates. The last column gives the approximate elongation from the Sun of the event.

Mercury completes its final morning apparition, reaching superior conjunction on 11 December (aphelion takes place almost simultaneously) and returning to the evening sky one last time for the year. It is best viewed from the southern hemisphere late in the month as it gains in altitude above the western horizon. The Moon occults the tiny planet on 28 December but the Moon is less than a day old, making observations of this event improbable.

Venus finally climbs out of the evening twilight for astronomers in northern temperate latitudes but is still best seen from the southern hemisphere. However, the view from the north only gets better as the month progresses whereas the planet's increase in altitude ceases for those in the southern hemisphere and it actually begins to fall back towards the western horizon before the end of the year. Venus also begins to brighten slightly as it slowly approaches Earth; through a telescope it appears in its waning gibbous phase but the increase in overall angular diameter more than compensates for the decrease in the illuminated fraction. Indeed, Venus is always at its brightest when it is a fairly thin crescent close to Earth.

Earth reaches solstice on 22 December. The Sun attains its most southerly declination on this day before heading north again. This marks the longest day in the southern hemisphere and the beginning of summer. For the north, it's the longest night and start of winter. The finest meteor shower of the year peaks this month. Sadly, it does so right around the time of Full Moon, so this is a very poor year to go hunting for the Geminids. However, the Puppid-Velids should not be too inconvenienced by the First Quarter Moon early in the month, and it is worth trying to catch a glimpse of the Ursids which peak just days before New Moon nearer the end of the month.

The **Moon** has three major occultation targets this month, the Pleiades on 12 December (the open cluster is aloft for much of the night), the first-magnitude star Antares on 25 December (a morning sky object), and finally Mercury on 28 December. This last event takes place very close to the Sun, however, and is unlikely to be observable.

Mars continues its passage through the constellation of Sagittarius this month, passing close by two globular clusters, M28 and M22, on 3 and 7 December respectively. These clusters are relatively faint so a telescope will be necessary to observe these events. Mars ends the year with a lunar occultation on 29 December when the young crescent Moon passes in front of it. Mars is low in the west as skies begin to darken and is setting during twilight for all locations by the end of the month.

Jupiter arrives at west quadrature on 18 December. It is largely a morning sky object, rising just before midnight for southern hemisphere astronomers but not appearing until after midnight for those further north. Look for it in the constellation of Virgo.

Saturn is an evening sky object in the constellation of Pisces, setting on or after midnight depending upon the observer's latitude, but the planet is best viewed from the northern hemisphere. Its rings have been closing up since August but reach a shallow minimum angle of 11.7° this month and begin opening up again. Saturn returns to direct motion on 24–25 December.

Uranus is visible for most of the night as it is just past opposition. Retrograde motions takes it past a trio of naked-eye stars bearing Flamsteed designations: 56 Tauri (0.4° on 18 December), 53 Tauri (0.2° on 21 December), and 51 Tauri (0.3° on 29 December). Uranus first glided past these stars in June but it was too near to the Sun to be observed.

Neptune returns to direct motion on 15 December and reaches east quadrature ten days later. The telescopic object may be found in Pisces and is best observed from northern vantage points as the planet does not set until after midnight.

The Mythical Vulcan
A Planet That Never Was

John McCue

The famous confrontation between Le Verrier and Lescarbault at Orgères in December 1859 is imagined here. (John McCue)

It was December 1859, and the nationally-celebrated French astronomer Urbain Jean Joseph Le Verrier (1811–1877) was impatiently knocking on the farmhouse door of the French medical doctor and amateur astronomer Edmond Modeste Lescarbault (1814–1894) in Orgères, about 150 kilometres west of Paris. Le Verrier was becoming increasing anxious to prove the existence of a new planet, predicted by himself, and located between Mercury and the sun. He had baptized it Vulcan, and had received by letter Lescarbault's claim to have seen it in March of that year. Le Verrier had dashed from his desk at the *École Polytechnique* in Paris, eager to interrogate the good doctor. Seeing the door creaking open, Le Verrier began, "It is then you, Sir, who pretend to have observed the intra-Mercurial planet, and who have committed the grave offence of keeping the observation secret for nine months. I warn you that I have come here with the intention of doing justice to your pretensions and of demonstrating either that you have been dishonest or deceived."

Born in 1811 at Saint-Lô, Normandy, Le Verrier's lasting fame rests on his discovery of Neptune in 1846 by purely mathematical, rather than observational, means. Rumour has it that he never looked through a telescope,[1] though this is to diminish his Herculean efforts in applying Newton's Law of Gravitation to predict, with high

1. Eggen, O. J. (1953), Vulcan, *Astronomical Society of the Pacific Leaflets*, **6**,(287), 291–298.

precision, the positions of the planets across the whole solar system, forwards and backwards over aeons of time. Le Verrier knew that because of the gravitational pulls from other planets (which are in elliptical orbits as well, and also slightly inclined to each other), Mercury's orbit gradually, and very slowly, swivels around the sun – an effect known as its perihelion precession and which is illustrated in the diagram seen below. However, it swivels a little faster than it should, by about 43 arc seconds every century, eight percent more than expected. Le Verrier was convinced that there must be another planet closer to the sun, tugging Mercury along.

One wonders what the meticulous Le Verrier would have made of French mathematician Henri Poincaré's later game-changing discovery, in the 1890s, of the existence of chaos? In a dynamic system such as the planets possess, a small uncertainty in the original position of a planet can lead to a dramatic difference in the final outcome. Computers have demonstrated that an initial discrepancy of ten metres in the position of the earth will result in an error of a million kilometres after 100 million years![2]

Returning to the encounter at Orgères, Lescarbault was taken aback by Le Verrier's famously abrupt manner. Nonetheless, he showed the latter his observation of March 1859. Le Verrier was doubtless sceptical when he saw that Lescarbault wrote his notes on a wooden plank, shaving them off with a plane afterwards, and followed the seconds by counting the pendulum swings of an ivory ball hanging by a silk thread from a nail. Eventually though, rather scrappy and soiled, the original observation was found, and Le Verrier left for his return to Paris, convinced that the rural doctor had seen the new planet.

Back at his desk, Le Verrier then calculated, based on Lescarbault's figures, that Vulcan orbited the sun in nearly 20 days, at a distance of a little under 21 million kilometres, just over a third of the way out to Mercury. He was sure it was there.

As mentioned, Le Verrier's discovery of Neptune 13 years earlier, following his study of Uranus' deviations from its predicted positions, had sealed his reputation as a master of celestial mechanics. It was simply sensational. He figured out where the unseen planetary culprit must be, sent his right ascension and declination figures to the Berlin observatory, whence the director Johann Galle and his assistant Heinrich d'Arrest found the mysterious planet on the first night of trying, very close to Le Verrier's forecast. Likewise, Vulcan must be there, the Frenchman thought, and the world continued to search.

2. Hoskin, M. (Ed.) (1997), *Cambridge Illustrated History of Astronomy*, Cambridge University Press.

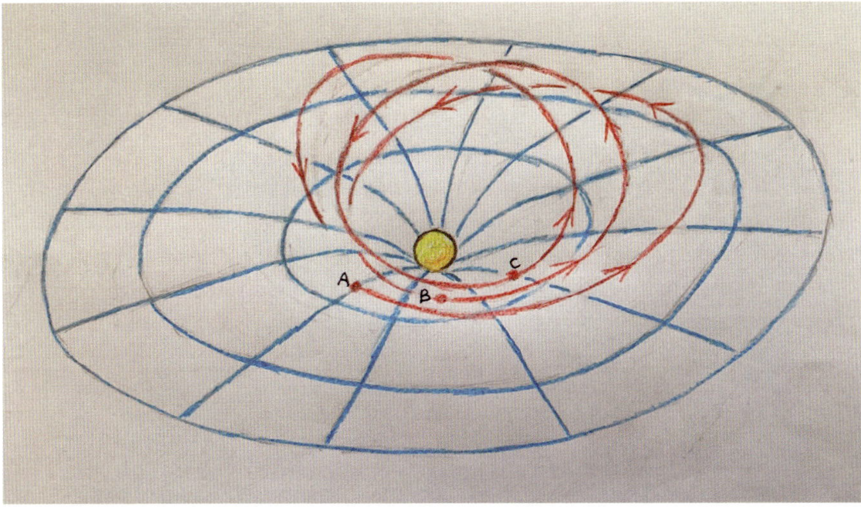

Precession of an elliptical orbit. The perihelion advances from A to B then C in three successive orbits, showing the open nature of the planetary ellipses. (John McCue)

Three years later, an amateur astronomer, Mr. Lummis, observing from Manchester in 1862, saw a spot on the sun which was moving too rapidly to be a sunspot. The orbit of this 'Vulcan, mark II', agreed quite well with Vulcan's original orbit, and so transits across the face of the sun were predicted, but still Vulcan was elusive. A total eclipse of the sun on 29 July 1878, across the western United States, was perfect for Vulcan-searching in the sun's immediate vicinity, and two observers thought they'd found it! One was the celebrated comet hunter Lewis Swift of New York. The other was Professor James Watson, director of the University of Michigan observatory. The eclipse lasted two minutes and forty seconds, and both found two interlopers near the sun. Despite criticisms, both astronomers were convinced that intra-Mercurial planets existed. The hunt continued …

With a deft touch of cosmic irony, it was the total eclipse of 1919, and an expedition led by Arthur Stanley Eddington to observe it, that inadvertently solved the problem of Vulcan – it never existed at all! Eddington was testing Einstein's new theory of gravitation, namely general relativity, which described gravity not as a force but as a result of the curvature of space-time caused by the presence of mass. Light from stars observed near the sun at the time of the solar eclipse would bend, following these curved lines of space-time. On his photographic plates of the eclipse, Eddington therefore expected to find star images located slightly further from the Sun than they would normally be – and he did. Einstein was vindicated and the world gasped in admiration. However, another consequence of a curved continuum was more swivelling of planetary orbits. In Einstein's own words, "On the basis of general relativity, it is found that the ellipse of every planet round the sun must necessarily rotate exceedingly slowly in the plane of the orbit and in the sense of the orbital motion, and that for all the planets, with the exception of Mercury, this rotation is too small to be detected ... but that in the case of Mercury it must amount to 43 seconds of arc per century".[3] There was no need for the existence of Vulcan; general relativity had topped up Mercury's orbital precession to the observed amount.

Curved space-time is difficult for humans to comprehend. The story is told of a question asked of Sir Arthur at a dinner party: "Professor Eddington, you must be one of only three persons in the world who understands general relativity." Eddington hesitated, and then replied, "I'm trying to think who the third person is".[4]

3. Einstein, A. (1920), *Relativity, the Special and General Theory*, Methuen & Co. Ltd.

4. Aczel, A. D. (1999), *God's Equation*, Piatkus.

Article Section

Recent Advances in Astronomy

Rod Hine

The New Kid on the Block

In the *Yearbook of Astronomy 2022*, we described the new Vera C Rubin Observatory, then being built in Chile. In June 2025 the first images were released by University College London (UCL) researchers at the very beginning of a ten-year programme to create an ultra-high definition time-lapse record of the Universe. We are used to describing research in superlatives and then having these descriptions exceeded in turn by new discoveries, and we shall not be disappointed with this latest new development Just as the *Hubble Space Telescope* amazed us more than thirty years ago, only to be upstaged somewhat by the *James Webb Space Telescope (JWST)* in 2022, the Vera C. Rubin Observatory looks set to raise the standard even higher.

What makes it unique is that it is not just a telescope but actually a complete system designed not only to capture images at ultra-high resolution but to process huge amounts of data rapidly to detect transient events and also to accumulate

This stunning image – appropriately nicknamed 'The Cosmic Treasure Chest' – shows just part of the southern region of the Virgo Cluster of Galaxies, almost 55 million light years from Earth, captured here in a small part of the entire field of view. (NSF–DOE Vera C. Rubin Observatory)

A close-up on the spiral galaxy Messier 61 in the Virgo Cluster (flanked by NGC 4301 to its left) as captured by NSF–DOE Vera C. Rubin Observatory. The image reveals a profusion of asteroids, seen here as tri-coloured streaks scattered throughout the picture, making their way across this portion of the night sky. (NSF–DOE Vera C. Rubin Observatory)

a long-term history of observations. It does this by combining a huge 8.4-metre primary mirror with a 3.2 Gigapixel prime focus camera. The field of view is 3.5 degrees and it covers wavelengths from near ultraviolet to near infrared; it will be taking a 15-second exposure three times every minute. Each image consists of 64 Gigabytes of raw data. Any novel events or unexpected changes will be flagged for investigation within a minute.

Professor Jason McEwen, based at UCL's Mullard Space Science Laboratory, said: "Rubin is about to change the way we see the Universe. It will find millions of exploding stars – supernovae – that can help us understand dark energy, one of the biggest mysteries in science." The UCL group will focus on dark energy and the origin of the Universe. Serious scientific work will begin later in 2025 when all the initial testing and commissioning work has been completed. The Rubin Observatory is supported by funding from the US National Science Foundation

(NSF) and the US Department of Energy (DOE). The scientific work is organised as eight separate collaborations based in premier universities and research institutions around the world.

Witnessing the Birth of a New Solar System

Researchers working with ALMA telescope and the JWST have spotted a very young star in the very process of forming a new solar system. The object is called HOPS-315 and is some 1,300 light years from Earth in the constellation of Orion. For the very first time, observers have detected the presence of silicon dioxide both as vapour and in a crystalline state in the protoplanetary disk surrounding the

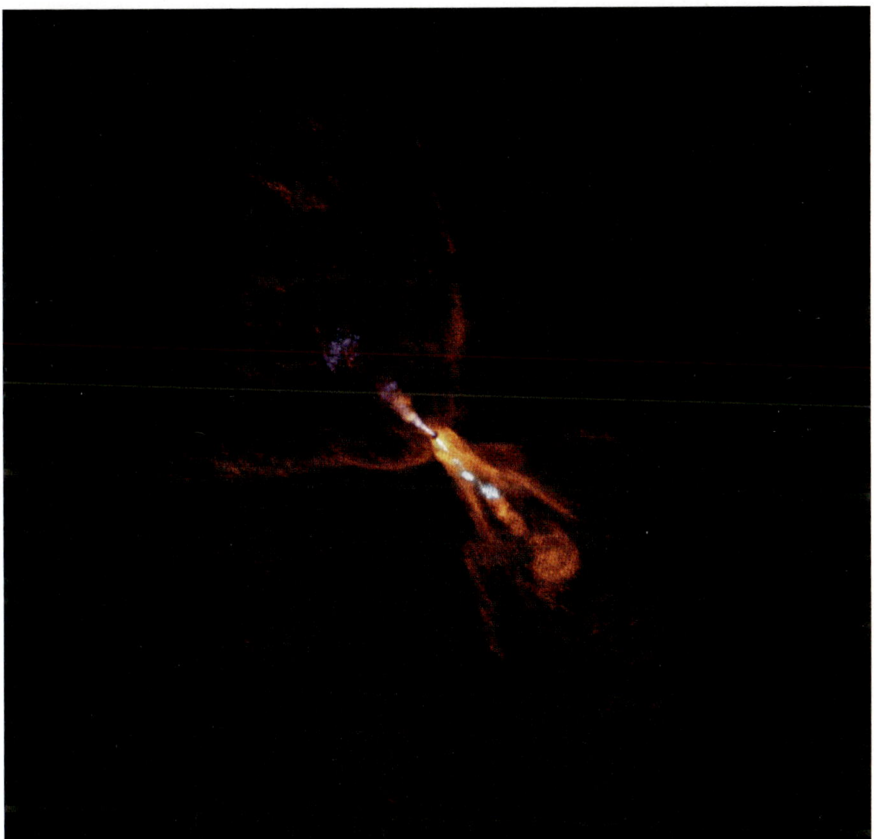

ALMA image of HOPS-315, a still-forming planetary system. (ALMA(ESO/NAOJ/NRAO)/M. McClure et al)

star. We know this mineral as silica, and it is very common on Earth as the major constituent of sand and quartz. It suggests that the silicon dioxide is in the very process of solidifying and thus providing the material for a future rocky planet, perhaps not too dissimilar to Earth.

"For the first time, we have identified the earliest moment when planet formation is initiated around a star other than our Sun," said Melissa McClure, a professor at Leiden University in the Netherlands and lead author of the new study, published in *Nature* in July 2025.

The spectrum from the minerals was first discovered by the JWST but to confirm the exact location of the origin of the signals, the team used the Atacama Large Millimeter/submillimeter Array (ALMA), located in Chile and operated by ESO. The signals were found to be coming from a small region of the disk corresponding approximately to the location of the asteroid belt in our own Solar System. This makes HOPS-315 a wonderful analogue to study our own cosmic history.

The Most Powerful Fast Radio Burst Ever?

Fast radio bursts, or FRBs, are enigmas that have puzzled scientists for decades. Brief bursts of intense energy that can outshine a whole galaxy for a few milliseconds, they have been seen all over the sky like random flashbulbs popping off in a sports stadium, but their origin is very obscure.

Perhaps the most puzzling FRB ever known was detected in June 2022 by the Australian Square Kilometre Array Pathfinder radio telescope (ASKAP) in Western Australia. Named FRB 20220610a it was particularly powerful. During follow-up observations with HST and with the European Southern Observatory's Very Large Telescope (VLT) in Chile, astronomers realised that the signal had come from an obscure and distant collection of galaxies that existed when the Universe was only five billion years old. So old and distant was it that the FRB was at least four times more energetic than closer FRBs.

Lead author Alexa Gordon of Northwestern University in Evanston, Illinois, praised the sharpness and sensitivity of the HST which was vital to pinpoint the exact location of the FRB, the first to be observed in association with a group of very old galaxies. Most FRBs come from isolated galaxies. The Hubble image suggests that up to seven of these distant galaxies may be on a path to merging, which may perhaps be related to the extreme intensity of the FRB.

There are many possible explanations for FRBs, but one very likely source could be some cataclysmic incident involving black holes or neutron stars. The most extreme kinds of neutron stars are called magnetars, and they have the most intense magnetic fields in the Universe. A disturbance of the magnetic field of a

A *Hubble Space Telescope* image of the host galaxy of the exceptionally powerful fast radio burst FRB-20220610a. (NASA, ESA, STScI, Alexa Gordon (Northwestern))

magnetar, like a solar flare but trillions of time stronger, or perhaps an interaction between a black hole and a magnetar – all these scenarios may be possible.

The signal from FRB 20220610a had travelled across the universe for about eight billion years before registering as a few milliseconds blip in the receivers at ASKAP. Maybe an even more intense one is well on its way to puzzle future astronomers.

Going, Going, Gone – The Disappearing Star

Look up 'supernova' in any astronomy textbook and you will find the explanation that massive stars, bigger than eight or so times more massive than our Sun, explode as supernovae at the end of their lives. While the star becomes a black hole or a neutron star, the outburst of the explosion often outshines the host galaxy for months. After it has died away, there is still an expanding cloud of glowing debris which is visible for millennia.

Now, however, astronomers appear to have spotted a massive star that turned directly into a black hole. To be precise, the star simply brightened for a while and then disappeared suddenly and completely.

The star in question is named M31-201-DS1, located in Messier 31, otherwise known as the Andromeda Galaxy, about 2.5 million light years from Earth. M31 is our nearest major galactic neighbour and bears a close resemblance to our own Milky Way. The star was known to be a massive, hydrogen depleted supergiant star. Observations in the mid-infrared during 2014 noticed a brightening, after which it remained at constant brightness for about a thousand days. Then, between 2016 and 2019 it faded dramatically and was totally undetectable by optical and near infrared imaging.

Research posted on the arXiv open-access archive investigates this unusual 'failed supernova' phenomenon (see **arxiv.org/abs/2410.14778**). The lead author is Kishalay De, a postdoctoral scholar at the Kavli Institute for Astrophysics and Space Research at MIT. They estimate that the star initially weighed about 20 solar masses and had reached its terminal phase with about 6.7 solar mass remaining. It was surrounded by a recently ejected dense cloud of dust. Normally such stars reach a point where the build-up of iron on the core results in abrupt cessation of nuclear fusion processes, and the sudden reduction in pressure in the core causes the collapse of the star and a rebound as all the stored gravitational energy is released.

So why should this process fail? According to the researchers, an important part of the collapse process depends on the extreme density forces electrons and protons to combine to form neutrons and neutrinos. Some 10% of the star's rest mass energy is carried in this ferocious blast of neutrinos. Despite the fact that neutrinos are reluctant to interact with regular matter, some of this energy goes to heat up the surrounding stellar material thus generating a shock wave. The neutrinos themselves escape from the core very rapidly but usually dump enough energy to trigger the supernova explosion and expel the outer layers of the star.

However, if the initial burst of neutrons fails to dump enough energy before the neutrinos escape, the whole of the star collapses into the inevitable black hole leaving no trace at all. In this case about 98% of the mass of the star collapsed leaving a black hole of 6.5 solar masses. In more common supernovae, a much higher proportion is ejected and the black hole is correspondingly much smaller relative to the initial star mass.

M31-2014-DS1 isn't the only 'failed supernova' known, because in 2009 a supergiant red star in NGC6946 was also spotted as it faded from view. By 2015 that too had disappeared from optical observations. Clearly there is still a lot to learn about the precise way supernovae occur, and a recent survey suggested that as many as 20% to 30% of massive stars do just fade out as failed supernovae. However, only these two examples have been studied and confirmed. Supernovae are just not sufficiently common – only one every hundred years in the Milky

This image of the well observed Supernova SN1987a in the Large Magellanic Cloud was taken eight years later by Hubble's Wide Field and Planetary Camera 2. The picture shows the orange-red rings surrounding Supernova 1987A, the glowing debris of which can be seen at the centre of the inner ring. (George Sonneborn (Goddard Space Flight Center)/Jason Pun (NOAO)/STIS Instrument Definition Team/NASA/ESA)

Way – to be sure of this projection. The most spectacular supernova to occur in our neighbourhood recently was SN1987A in the Large Magellanic Cloud, a mere 168,000 light years from Earth. It was first observed in February 1987 and remained bright for several months before fading from view. The glowing debris and the glowing, expanding rings are still visible to telescopes, as seen in this image. A burst of neutrinos was detected a few hours before the visible event, confirming that neutrinos play an important role in supernovae events.

Recent Advances in Solar System Exploration

Peter Rea

This article was written during the spring of 2025. As the missions mentioned are either active or due for launch imminently, the status of some missions may change after the print deadline. The mission websites are shown in bold in each section and these can be visited for the very latest information.

Star Performers

The Sun rose today. Not surprising really. It rose yesterday as well and the day before that. In fact it has risen every day since the Earth came in existence. We often don't give it a second thought except perhaps when there is a glorious reddish sunrise. We should give it a second thought though. It doesn't just give us great sunrises and sunsets; it provides the light and warmth and energy without which life on Earth could not exist.

Around 4.6 billion years ago (give or take a week) a bright light appeared at the centre of a slowly rotating cloud of hydrogen and dust. A star was born and after a few million years a retinue of planets condensed out of that cloud. Some were rocky and some were mostly hydrogen gas. Within a billion years primitive life arose on the third planet from that star. A few billion years later a sentient race of hominids referred to that star as their 'Sun' and one particular group of these hominids called themselves 'astronomers' and built domed buildings on tops of mountains to study the Cosmos. They also built other instruments to study the Sun and sent machines to orbit the sun for a closer look. One such machine carried the name *Parker Solar Probe (PSP)* in honour of the solar

Eugene Parker was an American solar physicist. After gaining his PhD from Caltech he spent some time at the University of Utah, before joining the University of Chicago in 1955. The *Parker Solar Probe* is named in his honour. (NASA/Kim Shiflett)

physicist Eugene Parker (1927–2022) who carried out most of his research from the University of Chicago and excelled in solar physics. Launched on 12 August 2018, the *Parker Solar Probe* was destined to approach the Sun at an incredibly close distance of 6.9 million kilometres. This distance was five solar diameters from the surface of the Sun; by comparison the Earth is 107 solar diameters or 149 million kilometres away. *PSP* did not reach this close distance straight away. The first closest approach to the Sun on 6 November 2018 was much further out at 24.8 million kilometres. The spacecraft repeatedly used the gravity of Venus to slightly change the shape of *PSP*'s orbit and slowly reduce the perihelion or closest point to the Sun. It was the seventh and last Venus flyby on 6 November 2024 that set up the 22nd flyby of the Sun and the first very close flyby on 24 December 2024 at a distance of 6.9 million kilometres. The intense gravity of the Sun accelerated *PSP* to a staggering velocity of 688,000 kilometres per hour. This flyby meant that *PSP* became the closest any object from Earth has come to our Sun. Five more flybys at the same distance are scheduled through 2025. This would conclude the primary mission. Readers are invited to visit the mission website shown below for a detailed description of the mission and the dates and flyby distances of each of the solar passes.

Parker Solar Probe (NASA): **parkersolarprobe.jhuapl.edu**

The European Space Agency also has an active mission in orbit around the Sun. *Solar Orbiter* was launched on 10 February 2020 and is currently in a highly elliptical orbit around our parent star, with a perihelion or closest point to the Sun of 42 million kilometres and a aphelion or furthest point from the Sun of 136 million kilometres. Observations from *Solar Orbiter* complement those of the Parker Solar Probe. Interested readers can find extensive information on the mission website.

Solar Orbiter (ESA): **esa.int/Science_Exploration/Space_Science/Solar_Orbiter**

Readers may also be interested in the Living With a Star Program, which is a fascinating look at why we need to understand our Sun and what missions are under development to further study the Sun.

Living With a Star Program (LWS): **lws.gsfc.nasa.gov**

Mercury Orbiters

One of those planets that condensed out of the solar nebula travels around the Sun in an elliptical orbit which ranges from 69.82 million kilometres at aphelion and 46 million kilometres at perihelion. This 4,879 kilometre diameter lump of rock moves very quickly, completing one orbit in just 88 days. Named Mercury after the Roman god of messages and communication, the planet is difficult to observe due to its proximity to the Sun but can be glimpsed in the morning and evening skies at certain times. Mercury has been visited twice before by spacecraft from the USA. *Mariner 10* made three close flybys of the planet on 29 March 1974, 21 September 1974 and 16 March 1975. The USA returned to Mercury with their *MESSENGER* spacecraft (MErcury Surface, Space ENvironment, GEochemistry, and Ranging.) *MESSENGER* was launched on 3 August 2004, arriving in orbit around Mercury on 18 March 2011. The mission came to an end on 30 April 2015.

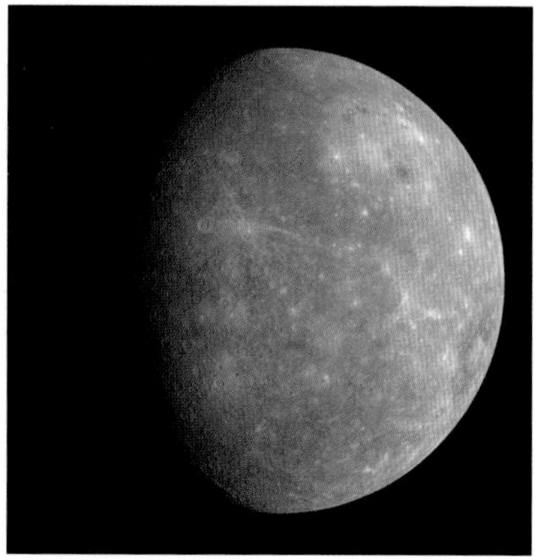

View of Mercury in colour taken with the *MESSENGER* spacecraft's Wide Angle Camera on 14 January 2008. The colour image shown here was generated by combining the mosaics taken through the WAC filters that transmit light at wavelengths of 1000 nanometers (infrared), 700 nanometers (far red), and 430 nanometers (violet). These three images were placed in the red, green and blue channels, respectively, to create the visualization presented here. (NASA/JPL)

The European Space Agency, in partnership with the Japanese Aerospace Exploration Agency JAXA, have launched another Mercury orbiter called *BepiColombo* after Italian scientist Dr Giuseppe 'Bepi' Colombo (1920–1984) who played a pivotal role in designing the *Mariner 10* flyby trajectories of Venus and Mercury. *BepiColombo* is two spacecraft in one. The *Mercury Planetary Orbiter (MPO)* is built by ESA and is scheduled to be placed into an elliptical orbit around Mercury with a period of 2.3 hours. The *Mercury Magnetospheric Orbiter (MMO)* is built by JAXA and is scheduled to be placed into a much more elliptical orbit with a period of 9.3 hours.

Getting to Mercury is a long process due to the intense gravity of the Sun. The highly elliptical orbit of *BepiColombo* needs to be slowly circularised at Mercury. To do this by onboard thrusters alone would require so much fuel that the spacecraft would be too heavy to launch. *BepiColombo* needed to use the gravitational energy of three planets to assist with this and reduce the amount of onboard fuel required.

The spacecraft also used ion thrusters rather than conventional chemical fuelled thrusters. These provide very low thrust but can be operated for very long periods of time. Ion thrusters are also very fuel efficient and, for a given mass of fuel, can produce a far greater change in velocity.

Mercury is receding into the distance in this image taken by the ESA/JAXA BepiColombo mission in the early hours of 5 September 2024 as the spacecraft sped by for its fourth of six gravity assist manoeuvres at the planet. (ESA)

BepiColombo flew past the Earth once, Venus twice and Mercury six times. In combination with ion thrusters the spacecraft gradually circularised the orbit so that the relative velocity of Mercury and the spacecraft was gradually reduced. The slower the relative velocity between the two the less energy – and hence fuel – is required to achieve an orbit around the planet.

The original flight plan called for *BepiColombo* to enter orbit around Mercury in December 2025. However, in September 2024 ESA announced that the ion thrusters were not operating at full capacity. There was an issue with the power distribution and the thrusters were not receiving enough electricity from the solar panels. ESA managed to get the thrusters up to 90% of nominal thrust but a new trajectory had to be calculated to account for this reduction. The December 2025 orbital insertion would be missed, the adjusted trajectory now being scheduled to put the spacecraft into orbit around Mercury in November 2026.

BepiColombo carries a multitude of instruments to study Mercury and interested readers can read about these from the mission website.

BepiColombo (ESA/JAXA): **esa.int/Science_Exploration/Space_Science/BepiColombo**

Ion thruster: **en.wikipedia.org/wiki/Ion_thruster**

To a Rocky Moon

It is never easy landing on the Moon. It requires many things to work perfectly in the right order and at the right time. The early part of 2025, when this article was being written was not a good time for lunar exploration and was proof of how difficult lunar exploration can be. The first lunar missions of 2025 were launched on 15 January with a *Falcon 9* rocket carrying two lunar bound payloads. These were *Blue Ghost 1* built and operated by Firefly Aerospace and the Japanese *Hakuto-R Mission 2* which comprises the *Resilience* lander and *Tenacious rover*. On 2 March 2025 *Blue Ghost 1* landed successfully in Mare Crisium, carrying a suite of NASA payloads under the *Commercial Lunar Payload Services (CLPS)* program. The *Hakuto-R Mission 2* attempted a landing within the Mare Frigoris on 5 June 2025, although the landing was not successful due to an anomaly in the Laser Range Finder which gives height from the Moon information. On 27 February 2025 a second *Falcon 9* launch despatched the Intuitive Machines' *IM-2 Athena* lander and the *Lunar Trailblazer* mission. The landing of *Athena* occurred on 7 March 2025 and ran into immediate trouble. A later image from the lander showed the vehicle had toppled over on its side, just like the first Intuitive Machines' *IM-1* lander in 2024. This effectively ended the mission. Intuitive Machines has another lander lined up for launch with the future of the company riding on it. The *Lunar Trailblazer* mission was intended to put the spacecraft into orbit around the Moon in a search for water, which may exist in areas of the lunar polar regions that never see daylight. This mission ran into trouble shortly after launch when the spacecraft went into an uncontrolled spin. Contact with the spacecraft was lost soon after. Our nearest neighbour is proving a difficult objective. Mission websites are shown below for further details, and readers may also want to visit the *Commercial Lunar Payload Services (CLPS)* website for additional details on that program and its objectives.

Model of the Japanese Hakuto-R spacecraft on public display. (Wikimedia Commons/Artvill)

Blue Ghost Mission 1 (NASA/Firefly Aerospace): **fireflyspace.com/missions/blue-ghost-mission-1**

Hakuto-R Mission 1 and Mission 2 (ispace Inc): **2022.ispace-inc.com/m2**

Lunar Trailblazer (NASA): **trailblazer.caltech.edu/index.html**

IM2 Athena (NASA/Intuitive Machines): **intuitivemachines.com/im-2**

Commercial Lunar Payload Services: **nasa.gov/commercial-lunar-payload-services**

Four en route to Asteroids

Situated between the orbits of Mars and Jupiter is a region commonly referred to as the asteroid belt, comprising a circumstellar ring of rocks, both large and small, that may well have been there since the formation of the solar system. For as long as this writer can remember, diagrams showing the location of the asteroid belt always showed the belt as if it were filled with asteroids almost touching each

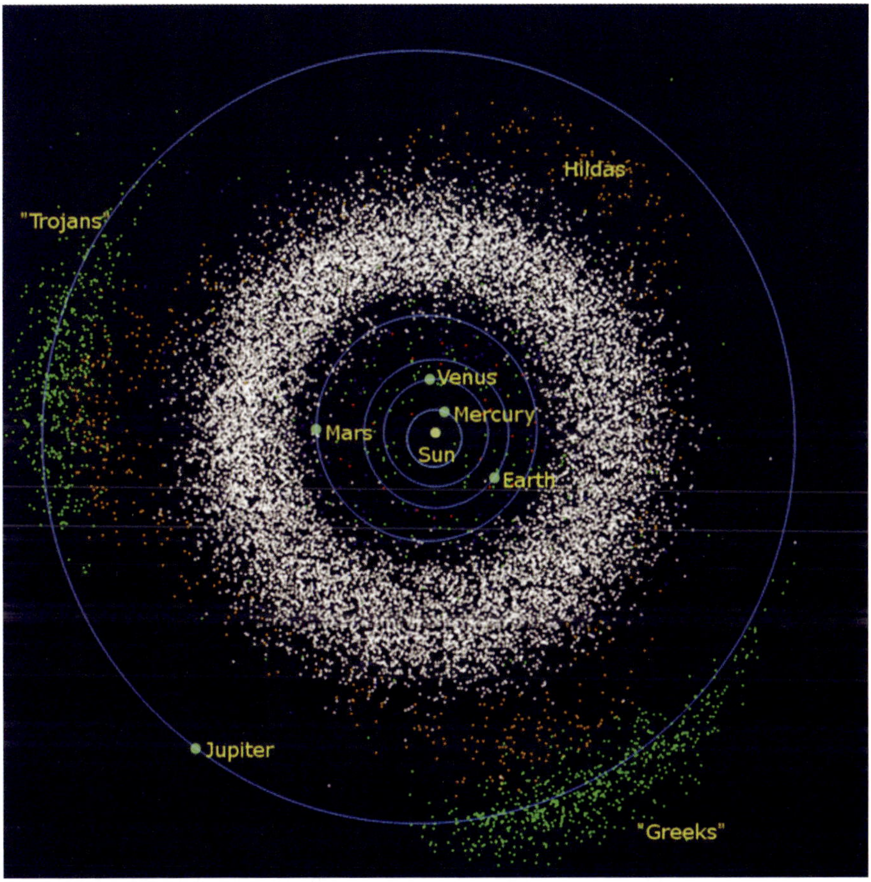

The inner Solar System, ranging from the Sun out to Jupiter. Also includes the asteroid belt (the white donut-shaped cloud), the Hildas (the orange 'triangle' just inside the orbit of Jupiter), the Jupiter Trojans (green), and the near-Earth asteroids. The asteroids that lead Jupiter are called the 'Greeks' and the trailing group the 'Trojans'. Although the asteroid belt looks quite crowded in this representation, in reality there are great distances between the asteroids. (Wikimedia Commons/Mdf)

The *Lucy* spacecraft flies past the Trojan asteroid (617) Patroclus and its binary companion Menoetius in this artist's concept. *Lucy* will be the first mission to explore Jupiter's Trojan asteroids – ancient remnants of the outer solar system trapped in the giant planet's orbit. (NASA's Goddard Space Flight Center / Conceptual Image Lab / Adriana Gutierrez)

other. Almost without exception the asteroid belt has been illustrated this way, and it is wrong! If you added up the masses of the approximately one million known asteroids their combined mass would be around three percent that of the Moon. The average distance between asteroids in this circumstellar ring is roughly a million kilometres – not exactly close to touching each other. Many spacecraft have safely traversed the asteroid belt without coming even remotely close to colliding with one of these rocky bodies. I am sympathetic with the artists who draw these diagrams; if you drew the asteroid belt as it really is it would not resemble a ring around the Sun at all.

The asteroid belt is far more complicated than I was taught 60 years or so ago. It has families of asteroids in certain orbits and the belt includes one dwarf planet, Ceres at a mean diameter of 939 kilometres. There are also two groups of asteroids sharing the same orbit as Jupiter and which are situated at the Lagrange points L_4, located 60 degrees ahead of Jupiter and L_5, situated 60 degrees behind Jupiter. It is into these two groups of asteroids that the American spacecraft *Lucy* is headed for flybys of five of these asteroids, with flybys of two asteroids en route. Full details of this mission and a list of the asteroids to be visited go to the mission website.

Lucy (NASA): **lucy.swri.edu**

Most of the asteroids are composed mainly of rocky material but some are rich in metal. One such asteroid is 16 Psyche, discovered by Italian astronomer Annibale de Gasparis on 17 March 1852 from the Naples Observatory. Psyche is an M-type or metallic asteroid and is rich in iron and nickel. No spacecraft has yet visited an M-type asteroid. The American spacecraft bearing the same name was launched toward Psyche on 13 October 2023 and is due to arrive at the asteroid in August 2029. More will be reported in future yearbooks but interested readers can read more about the mission at the mission website.

Psyche (NASA): **psyche.asu.edu**

We know the Earth has been hit by asteroids in the past, and there is ample evidence of this in the form of craters, perhaps the most famous of these being the Barringer Meteor Crater near Winslow, Arizona. The effect on the Earth of a large asteroid hit would be catastrophic. Science fiction movies have shown attempts to stop an asteroid should it be on a collision course for Earth, but what is the reality? Could we deflect an approaching asteroid? NASA launched

ESA's *Hera* mission follows up the work of NASA's *DART* mission to the 800-metre diameter asteroid Didymos, which is accompanied by a 170-metre diameter secondary body called Dimorphos. *Hera* will study the aftermath of the impact caused by *DART* on Dimorphos. (ESA – Science Office)

a mission to answer this important question. *The Double Asteroid Redirection Test (DART)* mission was launched on 24 November 2021 toward the double asteroid 65803 Didymos with its smaller companion Dimorphos, the purpose being to collide with Dimorphos to see if a change in velocity could be detected. If an asteroid were on a collision course with our planet it would only take a small change in velocity to redirect the asteroid away from the Earth as long as this was done millions of kilometres away. The mission was also a test of a targeting system. There is no use travelling millions of kilometres only to miss the target by one kilometre. The mission was successful, and the half tonne *DART* spacecraft collided with Dimorphos at a velocity of 22,000 kilometres per hour. The released kinetic energy slowed the asteroids orbital period by 32 minutes, far more than expected. Had that particular asteroid been on a collision course with Earth, a similar impact could well have deflected it towards a very close pass of Earth rather than a collision with it.

The European Space Agency launched a follow up mission, *Hera*, toward the double asteroid on 7 October 2024, with a scheduled arrival at Didymos in December 2026. *Hera* will go into orbit around Didymos to begin six months of observations of this double asteroid. Further details can be found at the *DART* and *Hera* mission websites listed below.

Dart (NASA): **dart.jhuapl.edu/Mission**

Hera (ESA): **heramission.space**

On 24 September 2023 the *OSIRIS-Rex* spacecraft returned to Earth a sample return capsule carrying samples collected from the asteroid 101955 Bennu. After capsule release the main spacecraft flew past the Earth and continued along its heliocentric (Sun-centred) orbit. The spacecraft was fully operational and was now flying under the new name *OSIRIS-APEX (APophis EXplorer)*. Its follow-on mission is to rendezvous with the potentially hazardous asteroid 99942 Apophis, at which it should arrive in April 2029.

OSIRIS-Rex (NASA/Lockheed Martin): **osirisrex.arizona.edu/osiris-apex**

Exploring the Red Planet

Mars has always been given plenty of space in my 'Recent Advances in Solar System Exploration' articles in this publication. This year I have featured other missions, so space is at a premium. The two American rovers *Curiosity* and *Perseverance* continue

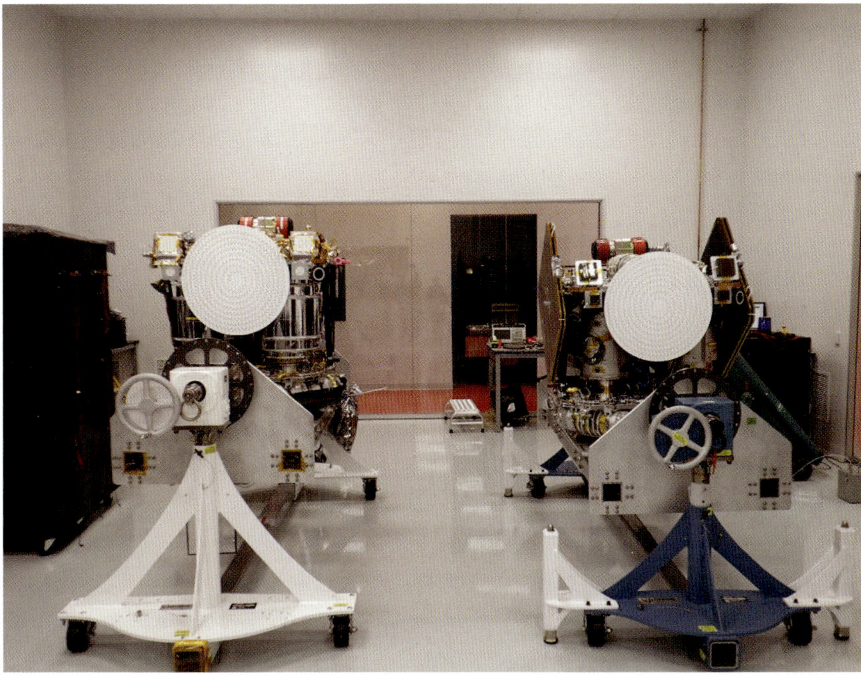

EscaPADE spacecraft development images from Rocket Lab's Spacecraft Production Complex and headquarters in Long Beach, California. The *EscaPADE* mission comprises two separate spacecraft called Blue and Gold. (Rocket Lab/NASA's Scientific Visualization Studio - KBR Wyle Services, LLC/Beth Anthony)

to explore the region around their landing sites. A retinue of orbiters continue to return valuable science. As I write these words, two missions are awaiting launch; NASA's *EscaPADE (Escape and Plasma Acceleration and Dynamics Explorers)* and JAXA's *Martian Moons eXploration*.

List of Mars Missions: **en.wikipedia.org/wiki/List_of_missions_to_Mars**

EscaPADE (NASA): **rocketlabusa.com/missions/escapade**

Martian Moons eXploration (JAXA): **mmx.jaxa.jp/en**

Two en route to the Icy Moons

The further exploration of Europa, Ganymede, and Callisto is a high priority for NASA and the European Space Agency. To that end, ESA launched the *Jupiter Icy Moons Explorer (JUICE)* mission on 14 April 2023. When it arrives at the planet in

Artists rendering depicting the NASA spacecraft *Europa Clipper* passing over Europa with Jupiter in the background. (Flickr/Kevin M. Gill)

2031 it will perform several flybys of Europa and Callisto before going into orbit around Ganymede for long term observations. Rather than going into orbit around any of the moons, NASA's *Europa Clipper* mission will make repeated close flybys of Europa. *Europa Clipper* was successfully launched on 14 October 2024 by a *Falcon Heavy* launch vehicle and is scheduled to arrive at the Jovian system in April 2030. Interested readers can follow the missions at the websites listed below.

JUICE (ESA): **esa.int/Science_Exploration/Space_Science/Juice**

Europa Clipper (NASA): **jpl.nasa.gov/missions/europa-clipper**

As always, Solar System exploration continues to excite and inspire, and next year promises to be no different.

Anniversaries in 2027

Neil Haggath

Sir Isaac Newton (1642–1727)

Perhaps the greatest mathematician and physicist who ever lived died three centuries ago.

Sir Isaac Newton; portrait by the German-born English painter Godfrey Kneller, 1689. (Wikimedia Commons / PD-US / Godfrey Kneller / Isaac Newton Institute)

Isaac Newton was born on 25 December 1642 (Old Style) in Woolsthorpe, Lincolnshire. He graduated from Trinity College, Cambridge in 1665. Soon after, the university was closed due to the Great Plague; during the next two years, working alone at his home, Newton laid much of the foundations of his great works, including developing the branch of mathematics which he called the 'Method of Fluxions' – now known as calculus. Gottfried Leibniz (1646–1716) also independently developed calculus, but Newton beat him to it by several years; however, we now use Leibniz's system of notation.

He returned to Cambridge in 1667, and was elected a Fellow of Trinity College. Two years later, aged only 26, he was appointed Lucasian Professor of Mathematics; his predecessor and mentor, Isaac Barrow, resigned to make way for him.

In 1687, at the urging of his friend Edmond Halley, Newton wrote his masterwork, *Philosophiae Naturalis Principia Mathematica* (Mathematical Principles of Natural Philosophy), which established the principles of classical mechanics, including his Three Laws of Accelerated Motion and the Universal Law of Gravitation. He used the latter to derive mathematically Kepler's Three Laws of planetary motion – which Kepler had determined empirically – thereby dispelling any remaining doubt that the planets orbit the Sun.

The *Principia* has been described as the greatest ever mental effort by a single person.

Newton's other great work was *Opticks*, published in 1704. Many years earlier, he had shown that white light can be split into colours by a prism and then recombined. Realising that refracting telescopes suffer from chromatic aberration, he made the first reflecting telescope in 1668 (James Gregory described the principle five years earlier, but never actually made one).

Newton served as President of the Royal Society from 1703 until his death. In his later life, he also served as a Member of Parliament and as Master of the Royal Mint. He was knighted by Queen Anne in 1705.

Newton was often irritable, and made enemies; in particular, he had bitter public disputes with Robert Hooke. But he was fairly modest about his own achievements, writing in 1676, "If I have seen further, it is by standing on the shoulders of giants", i.e. acknowledging that he had built upon the work of those who had gone before him.

Sir Isaac Newton died in London on 20 March 1727 (OS). He is buried in Westminster Abbey alongside royalty.

The Moons of Mars

Once every 15 years, Mars' opposition occurs close to its perihelion, and it therefore comes closer to the Earth than at any other time, offering the best opportunity to observe it. It was during one such opposition, 150 years ago, that the planet's two tiny satellites, Phobos and Deimos, were discovered.

Asaph Hall (1829–1907) was Professor of Astronomy at the US Naval Observatory (USNO) in Washington, D.C. For some reason, he believed that Mars must have satellites, and was determined to find them. During the 1877 opposition,

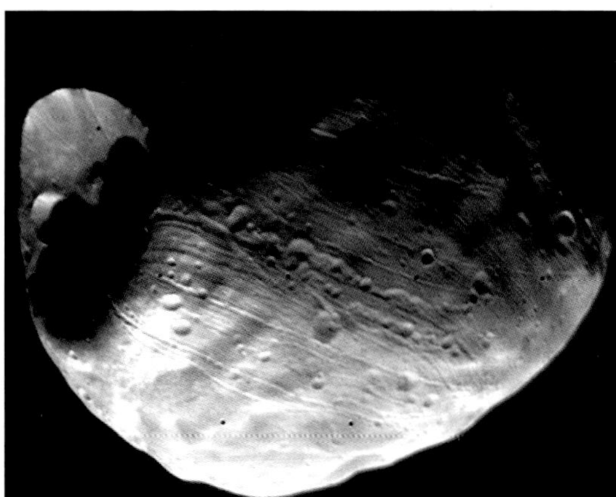

Image of Phobos, taken by the Viking 1 Orbiter in 1977, showing the 10 kilometre diameter Stickney crater at left. (NASA/Viking Orbiter 357A64)

he repeatedly observed Mars with the USNO's 26-inch refractor, which was then the world's largest. He was assisted by his wife Angeline, who was a lecturer in mathematics, and analysed his data.

He found the outer satellite on 12 August and the inner one six days later. They were later named Deimos and Phobos respectively, meaning 'terror' and 'fear' in Greek, after the two horses which were said to pull the War God's chariot.

The moons were later found to be tiny – Phobos is only 28 kilometres across and Deimos 16 kilometres – and both are probably captured asteroids, rather than genuine satellites.

In 1971, the *Mariner 9* probe returned the first close-up images of the moons, and two years later, a number of surface features on them were named. Phobos has several craters; while one is named after Hall himself, its biggest one, which dominates it, is named Stickney, which was the maiden name of his wife. Legend has it that after numerous long cold nights without success, Hall was ready to give up, but Angeline urged him to, "Get back out there and find those moons!" – and that very night, he found Deimos!

While that story may well be apocryphal, Hall did later write that, "I might have abandoned the search, had it not been for the encouragement of my wife."

The 'Canals' of Mars

The same Martian opposition of 1877 gave rise to one of astronomy's greatest controversies.

Italian astronomer Giovanni Schiaparelli (1835–1910) claimed to have seen a network of fine dark lines criss-crossing the surface of Mars. He thought they were watercourses, and called them *canali*, meaning 'channels' in Italian.

But in the English-speaking world, the word was wrongly translated as 'canals', implying an artificial origin. Over the next few decades, many other astronomers claimed to see the 'canals', though some very eminent observers, most notably Edward Emerson Barnard (1857–1923) could see no sign of them. A widespread belief arose that the 'canals' were indeed the work of intelligent Martians.

In particular, the wealthy American businessman-turned-astronomer Percival Lowell (1855–1916) established the observatory which bears his name at Flagstaff, Arizona, with a 24-inch refractor – which is still in use today – specifically to study Mars. His drawings of the planet showed an ever more complex network of 'canals', and he created an elaborate mythology of a Martian civilization building vast canals to carry water from the polar caps to irrigate the dry deserts. He realized that no canal could actually be visible from Earth, and believed that the lines were the strips of fertile land bordering the canals, like those on the Nile.

The concept led to numerous works of early science fiction, most notably H.G. Wells' *The War of the Worlds*.

By the time of Lowell's death, most astronomers had come to accept that the 'canals' were an illusion, but the myth persisted until 1971, when images from *Mariner 9* finally proved conclusively that they do not exist.

Urbain Le Verrier (1811–1877)

One of astronomy's greatest theoreticians died 150 years ago.

Urbain Jean Joseph Le Verrier was born on 11 March 1811 in Saint-Lô, France. After studying astronomy at the *École Polytechnique* in Paris, he spent most of his professional life at Paris Observatory, initially working under François Arago, then France's leading astronomer. When Arago died in 1853, Le Verrier succeeded him as Director.

While a brilliant astronomer and mathematician, his qualities as a person were less admirable. He was arrogant, opinionated and bad-tempered, and was described by a contemporary as the rudest man in all of France! In 1870, he was forced to resign as Observatory Director, due to his bullying manner – but he was reinstated three years later, after the premature death of his successor, and held the post until he died.

Le Verrier is best known for his prediction of the existence of Neptune, which eventually led to its discovery. In the decades after the discovery of Uranus in 1781, the planet was found not to be behaving quite as it should; its position in the sky differed from that predicted by calculations of its orbit. In 1846, Le Verrier postulated that its motion was being gravitationally perturbed by another as yet unknown planet, even further from the Sun, and predicted where in the sky the planet must be.

Urbain Jean Joseph Le Verrier. (John McCue)

On 23 September that year, Johann Galle and Heinrich d'Arrest at Berlin Observatory, on their first night of searching, found the planet within a degree of Le Verrier's predicted position. The discovery was hailed as a great triumph of Newtonian physics. Le Verrier was subsequently elected to the *Académie des Sciences*, the French equivalent of Britain's Royal Society.

Englishman John Couch Adams (1819–1892) independently predicted the existence of Neptune, but was strangely ignored. Much controversy, and even a diplomatic dispute, arose over priority. While Adams had done his calculations first, Le Verrier was first to publish, and his calculations were used by Galle and d'Arrest; Adams himself readily acknowledged that Le Verrier deserved the credit. Today, they are rightly given joint credit for the prediction.

Le Verrier's grave in Montparnasse Cemetery, Paris. (Neil Haggath)

Another problem baffling astronomers was an anomaly in the orbital motion of Mercury. Perhaps unsurprisingly, Le Verrier proposed that this was due to yet another unknown planet, orbiting closer to the Sun than Mercury. This was later found not to exist; we now know that Mercury's behaviour is due to effects of General Relativity.

But Le Verrier somewhat tarnished his reputation by becoming involved in the fiasco of 'the planet that wasn't.' In 1859, an obscure amateur astronomer claimed to have observed the hypothetical planet; Le Verrier convinced himself that it did indeed exist, and stubbornly refused to budge, despite mounting evidence to the contrary.

Le Verrier died at Paris Observatory on 23 September 1877, aged 66. He will be forever remembered as, in the words of François Arago: "the man who discovered a planet with the point of his pen."

Henry Norris Russell (1877–1957)

In the same year as Le Verrier's death, another great theoretician was born, one who is often referred to as 'the father of modern astrophysics'.

Henry Norris Russell was born on 25 October 1877 in Oyster Bay, New York. He graduated from Princeton in 1897, gained his Ph.D. two years later, and became a professor in 1911.

In 1913, Russell made the discovery for which he is best known. He found that there is a relationship between the luminosity of stars and their colours, which indicate their temperatures. Danish astronomer Ejnar Hertzsprung (1873–1967) had made the same discovery seven years earlier, but his work, published only in an obscure German journal, was unknown to Russell. They are jointly credited with what we now call the Hertzsprung-Russell Diagram, an essential tool of modern astrophysics.

In 1927, Russell published the two-volume *Astronomy: A Revision of Young's Manual of Astronomy*, which became a standard textbook for the next two decades.

Henry Norris Russell died in Princeton, New Jersey on 18 February 1957, aged 79.

Henry Norris Russell; portrait by an unknown photographer. (Wikimedia Commons/*The World's Work*, 1921/Princeton University)

A Historic Mission

Fifty years ago, two small space probes were launched, to begin what was arguably the most productive planetary exploration mission in history.

Project *Voyager* – a scaled-down version of NASA's 'Grand Tour' concept, which had proved too complex and expensive – took advantage of a rare planetary alignment, which occurs only once in 175 years, and made it possible to use a series of gravity assists or 'slingshots' to send a single spacecraft to Jupiter, Saturn, Uranus and Neptune. (For an excellent explanation of gravity assists, see 'Gravity Assists: Something for Nothing?' by Peter Rea, in the *Yearbook of Astronomy 2022*).

Voyager 2 was launched on 20 August 1977, with *Voyager 1* following on 5 September, by Titan IIIE-Centaur rockets. The one launched first was designated *Voyager 2*, as its twin was launched on a trajectory such that it would overtake it and reach Jupiter and Saturn first. Each had a mass of 825 kg at launch.

Both spacecraft flew by Jupiter, in March and July 1979, and Saturn, in November 1980 and August 1981. *Pioneers 10* and *11* had previously flown by Jupiter, and the latter by Saturn, but the Voyagers had far more sophisticated cameras and instruments.

Voyager 1 was targeted to make a close flyby of Saturn's largest moon, Titan; this meant altering its trajectory such that it was not able to continue to any other planet. *Voyager 2*, however, used further gravity assists to reach Uranus in January 1986 and Neptune in August 1989; it remains the only spacecraft to have visited those two planets. (The Titan flyby was considered so important that had *Voyager 1* failed, *Voyager 2* would have been diverted to repeat it, and the Uranus and Neptune part of the mission sacrificed. Thankfully, that wasn't necessary!)

The probes' trajectories had been calculated with such remarkable accuracy that when *Voyager 2* reached Neptune, after a journey of 12 years and over

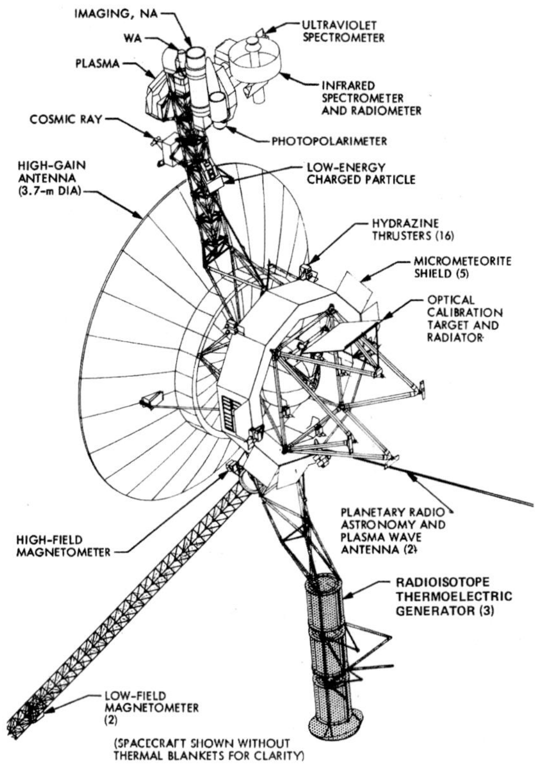

Diagram of a Voyager spacecraft, indicating important components. (NASA/JPL)

four billion kilometres, the estimated time of closest approach was in error by *three minutes*!

Among the many discoveries made by the two *Voyager* craft were: the existence of the rings of Jupiter and Neptune; active volcanoes on Jupiter's satellite Io; the mysterious 'spokes' in Saturn's rings; the dense atmosphere of Titan; the 'two-faced' nature of Iapetus, which explains its brightness variation; the bizarre surface features and towering cliffs of Uranus' small moon Miranda; and the ice geysers of Neptune's large moon Triton.

Like the Pioneers before them, the Voyagers were accelerated to a velocity greater than the escape velocity of the Sun, and would therefore leave the Solar System into interstellar space. They are the first human-made objects to do so, having overtaken both *Pioneer* spacecraft; by one definition, *Voyager 1* did so in 2012 and *Voyager 2* in 2018, when they crossed the boundary known as the heliopause. They are still transmitting data about the interstellar medium.

Other probes have since been sent to the Gas Giants – *Galileo* and *Juno* to Jupiter and *Cassini* to Saturn – which returned far higher quality images and data. But the *Voyager* missions paved the way; they discovered the features and phenomena which the later missions investigated in detail. It's fair to say that they taught us more about the outer planets than we had previously learned in over three centuries! They also inspired a generation of astronomers and planetary scientists.

There are, of course, always those who complain that spaceflight is "a scandalous waste of taxpayers' money", or words to that effect, and there were many who claimed that about *Voyager*. The total cost of the project was $850 million in 1970s and '80s money, which indeed sounds like a huge amount. But if we divide that by the number of years of development plus mission duration, and by the US population, it equates to about *25 cents* per taxpayer per year. A waste of money? I would call that the greatest bargain in history!

Jocelyn Bell Discovers Neutron Stars by Their Radio Pulsations

David M. Harland

Susan Jocelyn Bell was born on 15 July 1943 in Belfast, Northern Ireland. Her father was architect for the Armagh Observatory. Her first school was Lurgan College, where she failed the 11+ examination. She received most of her secondary education at The Mount School, York, a Quaker boarding school in northern England, where she developed a keen interest in physics. Bell subsequently gained a B.S. in physics from Glasgow University in 1965. The next step was Cambridge University for a doctorate under the supervision of Antony Hewish.

Hewish was born in 1924 in Fowey, Cornwall and attended King's College in Taunton. His undergraduate studies at Cambridge were interrupted by the war. He worked at the Royal Aircraft Establishment at Farnborough on radar research. On

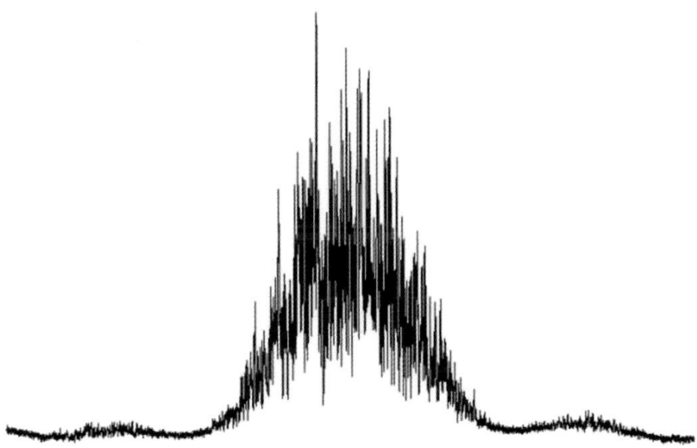

A representation of scintillation as a point source moves through a radio beam. (S. K. Glubokova, et al/*Advances in Astronomy and Space Physics*)[1]

1. Extracted from a diagram in 'Changes in the spectral index of the interplanetary plasma turbulence in the period of low solar activity from observations of strongly scintillating source 3C 298', S. K. Glubokova, et al, *Advances in Astronomy and Space Physics*, **3**, 94–97, 2013.

graduating from Cambridge in 1948 he joined the Radio Astronomy Group at the Cavendish Laboratory, from which he gained his Ph.D. in 1952.

Hewish sought to exploit the way in which radio waves travel in a transparent but inhomogeneous medium, in particular the 'solar wind'. The diffraction of the signal by the electrically charged particles causes the intensity of a source having a small angular diameter to 'scintillate'; in effect, the radio signal 'twinkles' in the same manner that bright stars flicker to visual observers. On a clear night, the brightest star in the sky, Sirius, can twinkle spectacularly. But even though the planet Venus can outshine Sirius, it does not twinkle. This is because stars are essentially point sources and planets show disks. Only a point source will twinkle. It is caused by starlight being perturbed by random air currents in the atmosphere.

In the case of celestial radio sources, the scintillation is produced by turbulence in the solar wind. This was first noticed by Margaret Clarke, one of Hewish's graduate students. It had already been seen for objects which were believed to be

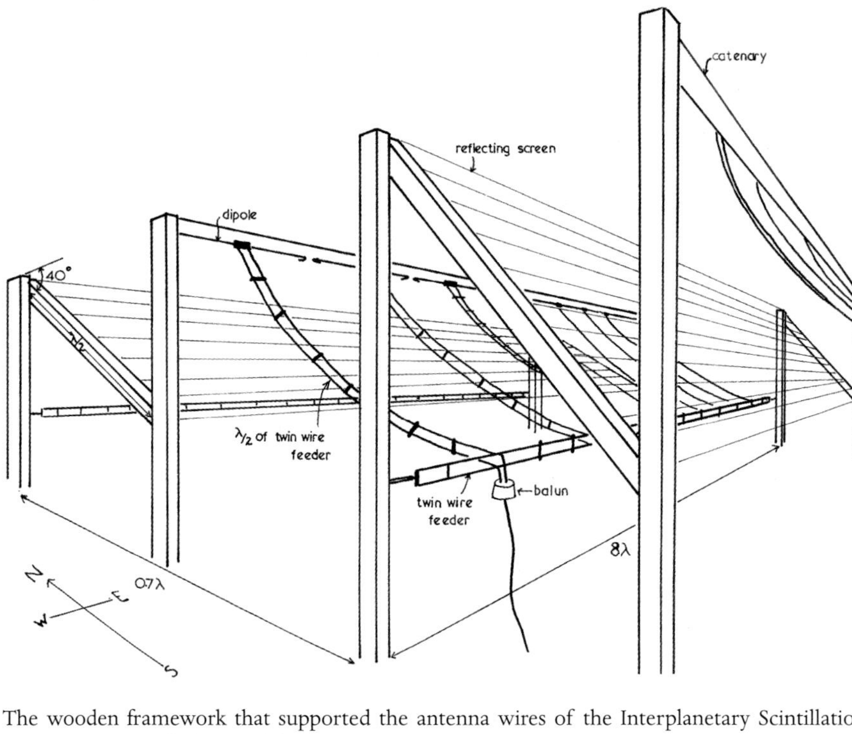

The wooden framework that supported the antenna wires of the Interplanetary Scintillation Array. (Jocelyn Bell's doctoral thesis)

compact in the 3C catalogue of radio sources observed by a telescope at Cambridge using the interferometry method and an array of four fixed elements.[2,3]

To the Interplanetary Scintillation Array (IPSA) designed by Hewish, random fluctuations of point sources would readily stand out. The goal was to measure the twinkling of the recently discovered class of quasi-stellar radio sources, or 'quasars', to find new celestial point sources, and to measure the angular diameters of such sources from the way in which their scintillation varied with solar elongation as the Sun circled around the sky annually.

In 1965 Hewish received a grant of £17,286 from the Department of Scientific and Industrial Research to construct the telescope and Jocelyn Bell joined him in October.

Unlike the large steerable antenna at Jodrell Bank in Cheshire, which could track a source across the sky, the IPSA was to observe sources as they transited the meridian using a fixed array of dipoles, with several beams taking simultaneous measurements at a range of elevations. The obliquity of Earth's axis confines the Sun to within 23° of the celestial equator, so the reflectors located behind the dipoles were to be angled to maximize the signal from south of the zenith. The telescope was an east-west array, with its eastern and western halves electrically isolated as the elements of a phase-switching interferometer. It comprised 2,048 full-wave dipoles for a frequency of 81.5 MHz (or a wavelength of 3.68 metres) in 16 rows of 128 elements. Also known as the Four Acre Array, the telescope would be able to view any particular source for at most a few minutes as it made its transit.

In 1965 the field chosen for the IPSA was surveyed by Leslie T. Little, one of the research students, with Bell's assistance. Over the next two years she, with students and technicians, sledge-hammered metal into the ground, erected woodwork, and strung the various dipoles, feeders and reflectors to make the array. Bell was responsible for all the cables, transformers, and connectors.

The phasing device designed by John D. H. Pilkington was used to slew a beam to any given angle of declination. This was done by inserting appropriate delays in the signals from the different rows of the array, prior to combining them. If the delays were correctly chosen, then only the signals from a specific declination would be in-phase upon arriving at the receiver. Each of the 16 rows provided a different beam, and the beam selector permitted a given receiver to be set to observe one of

2. Ryle, M., & Hewish, A. (1955). The Cambridge Radio Telescope. *Memoirs of the Royal Astronomical Society*, **67**, 97–105.
3. Hewish, A., Scott, P. F., & Wills, D. (1964). Interplanetary Scintillation of Small Diameter Radio Sources. *Nature*, **203**, 1214–1217.

The Interplanetary Scintillation Array installed at the Mullard Radio Astronomy Observatory, Cambridge, seen here in 1967. (University of Cambridge, Department of Physics)

four declinations. As Earth rotated, the beams scanned strips three or four degrees across. In the final configuration there would be four receivers, and the grouping of beams into four fours would be determined by that day's observing schedule. Overall, the preliminary survey would scan the declination range –08° to +44°.

The scintillation meters were developed by Leslie Little. In contrast to the usual practice of integrating and smoothing a signal to achieve a desired sensitivity, the IPSA was meant to measure fluctuations in the signal as rapid as one-tenth of a second. The percentage of scintillation was represented as an index based on the root mean square fluctuation. This would be the primary output of the telescope.

In its final form, the four receiving/recording systems would be able to simultaneously observe meridian transits at four declinations. Most of this apparatus was installed in the Laboratory of the Mullard Radio Astronomy Observatory, a quarter of a mile away, and it was Bell's task to operate it. There was a 3-track pen recorder for each receiver, usually running at 12 inches per hour. One pen displayed the sidereal time, the second showed the high-pass filtered output of the scintillation meter, and the third gave the rectified smoothed scintillation meter. The plan called for at least 15 observations of any scintillating source during the 3 months while it transited within 3 hours (45°) of the Sun, or roughly once a week, to measure how its scintillation index varied as a function of solar elongation. Observations started on 19 July 1967. For the first two months only three recorders were available.

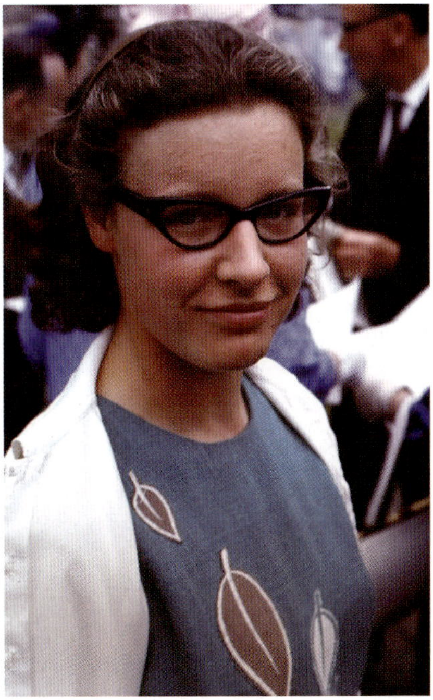

Jocelyn Bell in 1967. (Wikipedia Creative Commons/ CC BY-SA 2.0/Flickr/Roger W. Haworth)

Bell's job included visiting the IPSA on a daily basis to verify that the equipment was operating properly and that the recorders had sufficient paper and ink. At the normal (slow) setting each recorder produced 24 feet of paper per day. It was her responsible to analyse the squiggly line traces. This was done visually, and she soon learned to distinguish the scintillations of celestial sources from terrestrial interference ('noise'). She took note of the sidereal time of a scintillating source as an initial indicator of its celestial position.

In late September she noticed "a bit of scruff" (as she would later call it) on the beam for a declination of 23°. It drew her attention because it did not appear exactly like a scintillating source and yet didn't look like interference. She tagged it with a question mark, and soon realised that the peculiar signal was present in some of the records dating back to early August. With a right ascension (RA) of 19h 19m, it was crossing the meridian late in the evening. It was unlikely that the solar wind at such a large elongation would cause a celestial source to scintillate. It wasn't listed in the 3C catalogue. The fact that it was not always present was curious, but the signal was close to the telescope's limit of detectability.

After showing the anomaly to Hewish, it was decided to investigate further using a faster recorder in order to examine the signal at a higher time resolution. In late October, by which time all four recorders were operating, Bell made special visits to the observatory when the source was due to pass over, hoping to catch it with the fast recorder, but for weeks it did not show up. This was surprising,

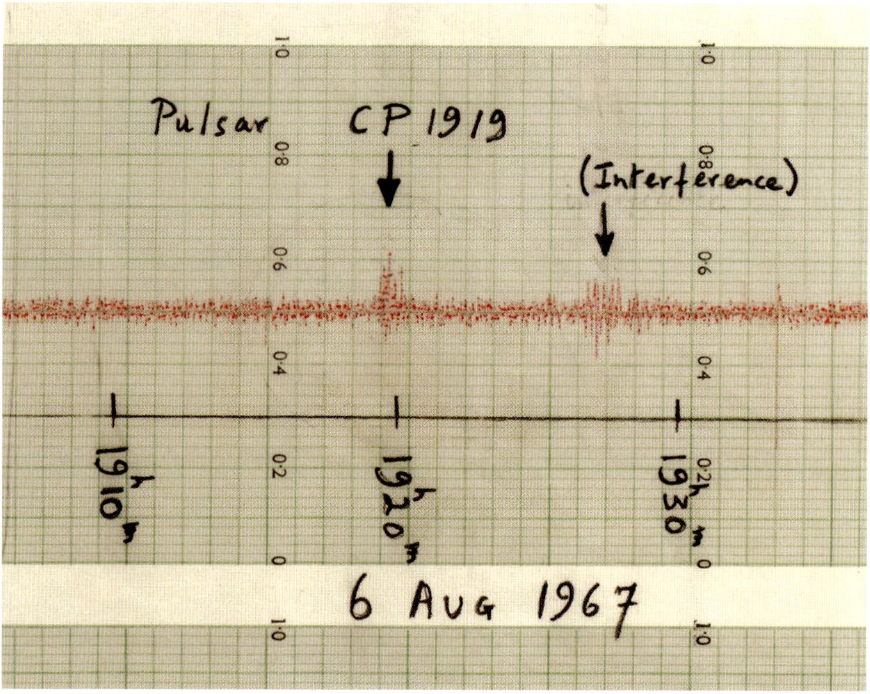

An examination of the daily charts traced the intriguing signal back to 6 August 1967. (The Papers of Professor Antony Hewish, HWSH 1/1, Churchill Archives Centre, Cambridge)

because by this time the diminished solar elongation would imply the scintillation should have increased. Then one day she skipped the scheduled transit to attend a lecture. The system continued recording at the normal slow rate. On returning the next day, 28 November, she found that the scruff had returned. Later that day, she was all set up and managed to record it at a rate of 1 centimetre per second. "As the chart flowed under the pen, I could see that the signal was a series of pulses and my suspicion that they were equally spaced was confirmed as soon as I got the chart off the recorder." The pulses were arriving every 1.33 seconds, lasted about 0.3 seconds, and were of differing amplitudes.

At this point, Bell telephoned Hewish to inform him the signal was a series of regular pulses. His natural reaction was to suspect it was man-made interference, as nothing natural was known to emit pulses in that way. Nevertheless, he attended when the source transited the next day, when it was once again seen to be pulsating with the same period.

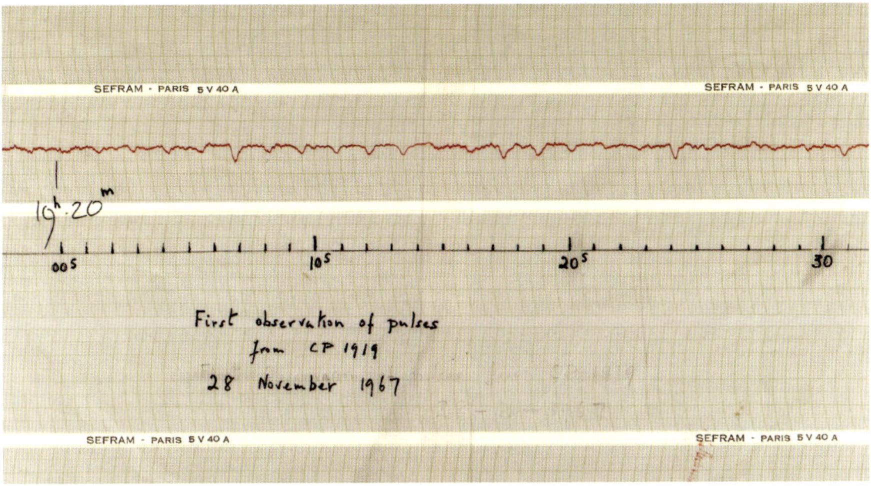

When the source was displayed using a fast recorder on 28 November 1967, it showed a series of pulses spaced one-and-one-third seconds apart. Deflections are downward. (The Papers of Professor Antony Hewish, HWSH 1/1, Churchill Archives Centre, Cambridge)

As Bell would wryly say a decade later, "This is where our problems really started."[4]

In order to rule out the signal being an artefact of the IPSA system, Hewish asked Mullard colleagues Paul F. Scott and research student Robin A. Collins to look for it using the cylindrical paraboloid telescope whose compilation of the 4C catalogue of sources had recently finished. Refitted with a receiver for the same frequency as the IPSA, it verified the existence of the pulsating source.[5]

The source was only rarely detected for the entire three minutes that it spent within the beam of the IPSA, and was typically present for only a minute or so. It was mostly seen in the 23° beam, sometimes in the 20° beam, never in the 26° beam. By 'weighting' the sightings in the beams, it was calculated to be between 21.5° and 22.5°. Another analysis found that in the four months for which data was then available, the RA had not changed by more than ten seconds (the limit of measurement). What is more, the fact that the signal showed up several minutes earlier per day meant it was keeping sidereal time and was a celestial object.

4. See the 'Petit Four' after dinner speech that Bell gave in 1977.

5. Bell Burnell, J. (2004). Presidential Address: Pliers, pulsars and extreme physics. *Astronomy & Geophysics*, **45** (1), 1.7–1.11.

Meanwhile, John Pilkington was able to estimate the distance to the source by simultaneously observing it at slightly different frequencies. He found that the lower frequency pulse arrived slightly later. This was known as 'dispersion'. The signal was emitted simultaneously across the spectrum but the speed of a radio wave depends upon the medium through which it travels, with lower frequencies being progressively retarded. Making some assumptions about the electron density in the interstellar medium, Pilkington calculated the source of the pulsating signal must be several hundred light-years distant to account for the measured dispersion.

Although the mood was that this must be some hitherto unsuspected natural phenomenon, the signal was jokingly referred to as 'The Little Green Man'. Regular observations of the source were continued and Hewish inspected the signal for evidence of a planet orbiting a star. As a planet first approaches us and then recedes again during its orbit, the Doppler effect would make the rate of pulsations vary in a cyclical way. Earth's travel around the Sun was evident, but there was nothing in the data to suggest an alien planet.

On 21 December Bell was checking the latest slow-rate paper traces and found a second anomalous signal. This one was in the 17° beam at an RA of 11h 33m. Her archive revealed that this too was appearing on an intermittent basis. Braving the overnight conditions, she returned to the telescope to monitor its next transit on the fast recorder, and caught pulses with a period of 1.19 seconds. It was inconceivable that two alien civilisations could be transmitting in this way on the same frequency. It *had* to be a natural mechanism. Bell left the trace of this new source on Hewish's desk and left to visit her family in Ireland at Christmas and inform them of her engagement to Martin Burnell, a local government officer. Hewish attended to the telescope during the holiday period and stacked the pen traces on her desk.

On 6 January 1968, back at work and now knowing what to look for in the slow traces, Bell soon found two more possible sources at an RA of 8h 34m and an RA of 9h 50m. With milder weather conditions on 16 January she verified the 08h 34m source was pulsating at 1.27 seconds. A good recording of 09h 50m on 18 January revealed a pulsation period of only 0.25 seconds, a rate so rapid that it would later rule out some of the proposed mechanisms for generating such pulses.

A search of all of the archived traces turned up no other candidates, but the discovery of four cases of an entirely new phenomenon was a good haul for a new instrument. The discovery was remarkable in three ways.

Firstly, because no one had predicted rapidly pulsating radio sources no one had created a telescope specifically to search for them. The IPSA had been designed to sense tiny, rapid fluctuations in signal strength because Hewish was interested in scintillation. It was sheer luck that it was well suited to detecting sources pulsating with periods of 1 second or so. Then as Bell related in her doctoral thesis in 1969,[6] if the aerial output had been digitised and fed directly into a computer, the programming would have sought the characteristics of scintillation and have rejected these unexpected results as noise. In contrast, her eye had been drawn to the scruff because it wasn't the usual noise. Thirdly, if Bell had not been so conscientious she could so easily have dismissed the scruff as not being what she had been told to watch out for.

At the end of January, with rumours rife that an amazing discovery had been made, Hewish led the writing of a scientific paper. This was received by the journal *Nature* on 9 February. The fact that it passed through the peer review process in only two weeks was an indication of the interest it stirred.[7] A few days ahead of publication on 24 February, Hewish gave a seminar entitled 'Pulsating Radio Sources'. The venue was the austere Maxwell Lecture Theatre of the Cavendish Laboratory. Almost all the Cambridge astronomers were present: radio, optical, and theoretical.

Once Hewish had outlined the manner in which the IPSA operated, he presented the results for the first source because it was the one with the story to tell, but at the end he pointed out that three others had been found. As to what they might be, he speculated on possible oscillations in white dwarf stars.

After the paper came out, Anthony Michaelis, science correspondent of *The Daily Telegraph*, coined the moniker 'pulsars' as a contraction of pulsating radio sources.

Astronomers elsewhere were frustrated that Hewish did not publish the coordinates of the four sources, but this was because positions were not easily measured by the IPSA. Right ascension could be taken from the transits, but determining the declination required an analysis of the signal strength in adjacent beams. There was sufficient data to place the first source at 19h 19m 38s ± 3s, 22° 00' ± 30' in terms of the 1950 epoch. Such an 'error box' was too large to allow a serious search for an optical identification, particularly in the constellation of Vulpecula, close to the galactic plane between Cygnus and Pegasus. The positions of the other sources were initially defined only to within 20 seconds in RA and

6. Bell, S. J. (1969). *The Measurement of Radio Source Diameters using a Diffraction Method* [Doctoral dissertation, University of Cambridge]. p 231.
7. Hewish, A., Bell, S. J., Pilkington, J. D. H., Scott, P. F., & Collins, R. A. (1968). Observations of a Rapidly Pulsating Radio Source. *Nature*, **217**, 709–713.

45 minutes of arc in declination. After further study, refined coordinates were published on 13 April.[8]

A conference in New York in May was dominated by papers seeking to explain the pulsating radio sources, mostly in terms of white dwarfs. The general presumption was that a white dwarf was oscillating like a ringing bell, and emitting pulses of electromagnetic energy. The rate at which a star is able to vibrate depends on its mass, gravity, radius and elasticity. In 1966, David W. Meltzer and Kip S. Thorne at Princeton had calculated that white dwarfs would naturally oscillate with periods in the range 1-10 seconds.[9] It would be difficult, however, for the model to accommodate a period as short as 0.25 seconds.

One possibility that was not widely explored at the conference was that a star might be rotating at a phenomenal rate. The maximum rate at which a star can rotate depends upon its mass and its diameter. The smaller an object of a given mass, the greater its density and hence the faster it can spin. If a white dwarf were to spin with a period as short as 1 second, that would cause its poles to flatten and its equator to start to create a disk. Any faster and it would tear itself apart. But why would it issue radio pulses?

The cores of stars that suffered supernova explosions were believed to be neutron stars. This possibility was raised in 1934 by Walter Baade and Fritz Zwicky at the California Institute of Technology after a study of the likely sources of cosmic rays.[10] Was a supernova the transition of an ordinary star into one consisting mainly of neutrons, they pondered. The general character of such stars was examined in 1939 by J. Robert Oppenheimer and his student George M. Volkoff.[11] The possibility that such objects might exist was ignored by astrophysicists. They had form, however, because they had rejected the prediction earlier in the decade by Subrahmanyan Chandrasekhar of a maximum limit to the mass of a white dwarf, and thus evaded the question of what would happen to a star larger than this limit.[12] It now seemed that a massive star might go supernova and its core become a neutron star with a

8. Pilkington, J. D. H., Hewish, A., Bell, S. J., & Cole, T. W. (1968). Observations of Some Further Pulsed Radio Sources. *Nature*, **218**, 126–129.

9. Meltzer, D. W., & Thorne, K. S. (1966). Normal Modes of Radial Pulsation of Cold Stars at the Endpoint of Thermonuclear Evolution. *Astrophysical Journal*, **145**, 514–543.

10. Baade, W., & Zwicky, F. (1934). Cosmic Rays from Super-Novae. *Proceedings of the National Academy of Sciences*, **20**, 259–263.

11. Oppenheimer, J. R., & Volkoff, G. M. (1939). On Massive Neutron Cores. *Physical Review*, **55**(4), 374–381.

12. I discussed Subrahmanyan Chandrasekhar's work in *Yearbook of Astronomy 2026*.

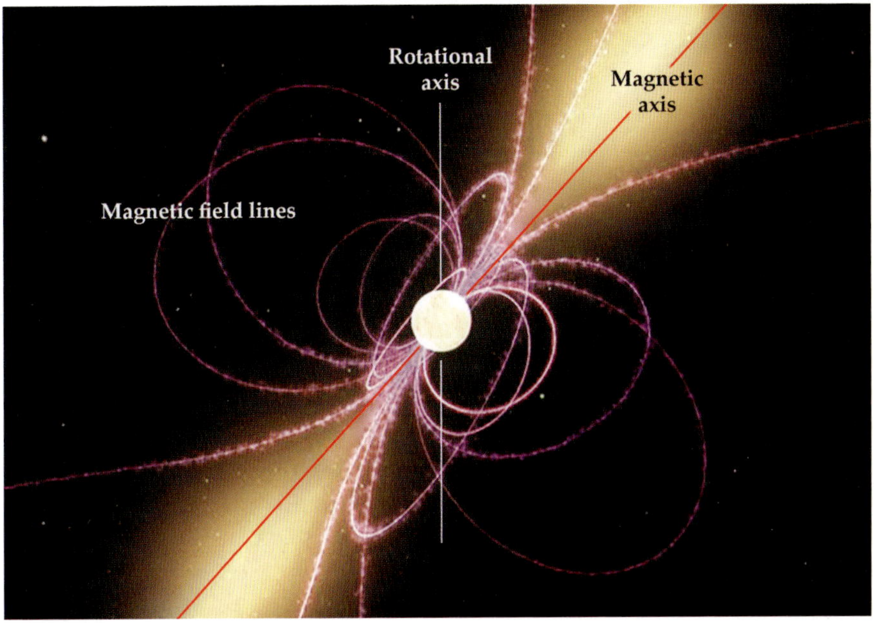

A pulsar is a spinning neutron star whose magnetic axis projects beams of electromagnetic radiation in opposite directions. Depending on the obliquity of the magnetic axis relative to our line of sight we might detect one or two pulses per revolution. (NASA illustration annotated by the author)

mass comparable to the Sun squeezed into a sphere with a radius of 12 km. The gravity would be hundreds of billions times greater than that at the surface of Earth. It would bend light to such an extent that it would be possible to see for tens of degrees beyond the horizon. In that realm the physics of Isaac Newton yields entirely to the relativity of Albert Einstein. Clocks would run at half the rate they do on Earth.

Franco Pacini graduated from the University of Rome in 1964 with a Ph.D. thesis on the properties of neutron stars. After postdoctoral work in Paris, in 1967 he became a visiting professor at Cornell University in America. In a paper that was published in *Nature* on 11 November 1967 he pointed out that the core of a star that went supernova would retain a great deal of energy as heat, vibration, magnetism, and rotation.[13] In particular, as a collapsing core was spun up by conservation of angular momentum its magnetic field would 'wind up', increasing its strength

13. Pacini, F. (1967). Energy Emission from a Neutron Star. *Nature*, **216**, 567–568.

by millions-fold. Furthermore, if the magnetic axis weren't aligned with the spin axis, it would form an 'oblique rotator'. A strongly magnetised neutron star could release its rotational energy by a flow of relativistic particles from its magnetic poles, and those particles would in turn emit electromagnetic radiation. As the star rotated, the observed intensity of that radiation would vary.

Several weeks after Pacini's paper appeared, Jocelyn Bell made the first high-speed recording of a rapidly pulsating radio source. Shortly after the Cambridge announcement in February 1968, Thomas Gold, in an office a few doors from that of Pacini, independently developed the oblique rotator model. Predictions are a welcome aspect of a scientific theory. Gold made two. The first was that with so much energy being emitted in the radio spectrum, the rotation rate should slow down. His second prediction was there must be *younger* neutron stars which were pulsating more rapidly than those already discovered.

When Gold submitted a paper making these predictions to the New York conference, it was rejected. When he sought permission to speak for five minutes, he was refused because the organiser said his idea was 'outlandish'. Gold immediately sent the paper to *Nature*, which turned it around within a week and published it on 25 May 1968.[14]

The week the Cambridge paper appeared, Alan E. Vaughan started his doctorate at Macquarie University in Sydney, Australia, by visiting the Molonglo Radio Observatory near Canberra to use the Mills Cross Telescope. This comprised a pair of mile-long narrow radio antennas in the shape of a cross, configured to aim its beam at the meridian. It was designed by Bernard Mills of Sydney University.[15] It had a wide field of view and could scan large tracts of the sky to zero in on sources whose positions were ill-defined.

Vaughan's supervisor, Michael I. Large, had assigned him the job of developing a phase-delay apparatus that would enable the telescope to undertake three independent observations of a source during a transit, as a way of greatly improving its sensitivity.[16] It transpired that the phase-delay made the pulsating radio sources stand out.

14. Gold, T. (1968). Rotating Neutron Stars as the Origin of the Pulsating Radio Sources. *Nature*, **218**, 731–732.

15. Mills, B. Y., & Little, A. G. (1953). A High-Resolution Aerial System of a New Type. *Australian Journal of Physics*, **6**, 272–278.

16. Vaughan, A. E. (1974). Pulsar Observations at Molonglo [Doctoral dissertation, University of Sydney].

The Mills Cross Telescope at the Molonglo Radio Observatory, near Canberra in Australia, employed east-west and north-south arrays of dipoles arranged in a cross configuration with the receiver hut adjacent to the centre. (Wikimedia Commons/CSIRO Radio Astronomy Image Archive, under CC-BY-4.0)

Utilising this mode, the telescope made a discovery in October 1968. The pulsation was too rapid for the pen recorder to isolate. On the next transit they ran the recorder at its fastest rate, still without success. The next time, they improvised by pulling the paper through the machine by hand, and were finally able to measure the pulsation rate as 11 times per second. This 89 millisecond period was far beyond the limit of what was explicable in terms of white dwarf stars unless the situations were contrived. When informed of the position of this source, Mills checked a catalogue of radio sources and found that it was right at the centre of a supernova remnant in the constellation of Vela.[17] This alignment was either a remarkable coincidence, or the core of the star that detonated as a supernova had become a rapidly pulsating neutron star. Their paper was published on 26 October 1968.[18]

On 10 November of that year, the 1,000-foot radio dish at Arecibo on Puerto Rico discovered a pulsar in the Crab Nebula, the remnant of the supernova that was observed in AD 1054.[19] Not only was the period an astonishingly fast 33 milliseconds, within a week the pulsation rate was shown to be slowing down by 38 nanoseconds per day.[20] This ruled out a radially vibrating white dwarf, because over time it would increase its rate and reduce its amplitude. The slowing rate distinguished between two possible neutron star models for the radio pulses. It was consistent only with the axial rotation of a neutron star. As a spinning top loses energy it naturally reduces its rate of spin, *lengthening* its period. The observed slow down eliminated the possibility of two neutron stars in a close binary system, almost in contact, because as such a system loses energy the separation between the components reduces and the period *decreases*. So the pulsating radio sources could *only* be rapidly spinning neutron stars whose magnetic fields were aligned obliquely.

With two links between pulsars and cataclysmic stellar detonations, the situation was starting to make sense. The two fastest pulsars were so young that their supernova remnants had not had time to disperse. The nebulae of the longer period pulsars, being older, had merged with the interstellar medium. Also, many pulsars were found to be travelling so rapidly that they would have escaped the bounds of their supernova remnants.

Postscript

Bell Burnell carried out research in the emerging branches of X-ray and gamma-ray astronomy, taught at a number of research institutes, and performed key leadership tasks. She remained active in the Quaker movement. In 1999 she was awarded a CBE and in 2007 received a DBE for her services to astronomy.

---

17. Bhathal, R. (2012). Bernard Mills and Australian Radio Astronomy. *Astronomy & Geophysics*, **53**, 2.19–2.21.
18. Large, M. I., Vaughan, A. E., & Mills, B. Y. (1968). A Pulsar Supernova Association. *Nature*, **220**, 340–341.
19. Lovelace, R. B. E., Sutton, J. M., & Craft, H. D. (1968). Pulsar NP 0532 Near Crab Nebula. *IAU Circular No. 2113*.
20. Richards, D. (1968). NP 0532. *IAU Circular No. 2114*.

In 2018 Bell Burnell was awarded the Special Breakthrough Prize in Fundamental Physics for her 'contributions to the discovery of pulsars, and a lifetime of inspiring leadership in the scientific community'. She donated the $3 million to the 'Bell Burnell Graduate Scholarship Fund' to be administered by the Institute of Physics to assist female, minority, and refugee students seeking to become physics researchers.

- oOo -

I am grateful to Jocelyn Bell Burnell for sharing her recollections and for improving the original manuscript.

Further Reading

'The Measurement of Radio Source Diameters using a Diffraction Method', Susan Jocelyn Bell, Ph.D. thesis, University of Cambridge, Submitted in September 1968, approved 18 February 1969 (available at **www.repository.cam.ac.uk/handle/1810/260694**).

Petit Four by S. Jocelyn Bell Burnell, an after dinner speech at the Eighth Texas Symposium on Relativistic Astrophysics, published in the *Annals of the New York Academy of Sciences*, **302**, 685–689, 1977 (available at **jila.colorado.edu/~ajsh/astr2030_12/sn/Bell.html**).

'The Discovery of Pulsars', Jocelyn Bell Burnell, *Serendipitous Discoveries in Radio Astronomy*, Proceedings of a Workshop held at the National Radio Astronomy Observatory, Green Bank, West Virginia, edited by K. Kellermann and B. Sheets, 160–170, 4–6 May 1983 (available at **library.nrao.edu/public/collection/02000000000280.pdf**).

'Pliers, Pulsars and Extreme Physics', Jocelyn Bell Burnell, *Astronomy & Geophysics*, **45**, #1, 1.7–1.11, February 2004 (available at **academic.oup.com/astrogeo/article/45/1/1.7/229920**).

The Vera Rubin Observatory

Prof. Catherine Heymans, Astronomer Royal for Scotland

The Vera Rubin Observatory on Cerro Pachón, Chile. (RubinObs/NOIRLab/SLAC/NSF/DOE/AURA)

Let me take you back to the year 2000: the *International Space Station* has just become operational, Britney Spears is topping the charts, a 'Google search' is something only really used by academics, and I am about to experience a baptism of fire on my first day as a PhD student at the University of Oxford. We still have a few more years before PowerPoint presentations become the norm, and so I'm standing at the front of the lecture theatre with my hand drawn acetate slides. I nervously place my 'The Future of Cosmology' title sheet on the overhead projector, and over the next hour I introduce my colleagues to the American plans for an upcoming observatory that would utilise a brand-new field of research to revolutionize our understanding of the invisible dark side of our Universe. During the Q&A that then followed, it became clear that very few in the audience believed this future vision could ever be realised. Not only would the Americans need to find over five-

The Simonyi Survey Telescope inside the Vera Rubin Observatory. (RubinObs/NSF/DOE/NOIRLab/SLAC/AURA)

hundred million dollars to construct this facility, but the scientific methodology underpinning the whole design was unproven. "Pie in the sky" was the unhelpful comment from one of the more doubtful academics assessing my thesis plan. But I am delighted to share that there was only one thing wrong in my presentation: the initial ten-year timeline. It took until 2025 before this new Vera Rubin Observatory started operations, and it has been well worth the wait.

High in the Andes, on the arid desert mountaintop of Cerro Pachón, you'll find lizards, scorpions and spiders scuttling around a bright white futuristic building that is the Vera Rubin Observatory. This is one of the world's best sites for astronomy. It is so remote from cities and towns that the night sky is pristine, allowing astronomers to see the sky as our ancestors once viewed it, uncontaminated from the streetlights that usually tinge our night skies orange. The climate is so dry that it's rare to lose any night time observing opportunities to the cloud cover that curses many a back garden astronomer back home. Peer inside the opening of the giant dome and you'll see one of the world's largest telescopes. The primary mirror for this beast is 8.4 metres in diameter, and its design is unique because it's actually two mirrors in one. The outer ring catches the first light from the skies, reflecting it up to a 3.4 metre secondary mirror. Normally that would be the end of the light-bouncing story, but this telescope is novel. The light bounces back down to the inner 5.0 metre core of the first mirror which acts as a tertiary mirror, finally focusing the light into the camera. It is this revolutionary and unorthodox design that allows this telescope to image a patch of sky seven full moons across (9.6 square degrees) in a single snap. And wait until you hear about the camera. This piece of technology is so advanced that it holds not one, but two Guinness World Records. Rubin is kitted out with the world's largest digital camera, physically the size of a small car and boasting 3.2 billion pixels. To see a single image from this camera at full resolution, you would need 400 high-definition televisions! The camera also contains the world's largest optical lenses, with the diameter of the primary lens as tall as an average woman.

Back in 2000, the primary objective for this new piece of kit was to create the deepest widest image of the Universe ever created and that is exactly what Rubin is now working on, imaging the entire observable Southern sky. By 2035 Rubin will have captured images for over twenty billion galaxies, some of which will be so far away that their light will have travelled through almost the entire history of the Universe to reach us. And while some of my colleagues are very excited about understanding how these galaxies formed and evolved, for me I'm more interested in the dark side, the stuff that we can't see or touch, but that we know is there from the effect that it has on the things that we can see. Our current best understanding of our Universe tells us that the Earth, Moon, Sun, you, me and all the stars and galaxies that shine brightly in the cosmos, makes up less than five percent of everything that

The team prepares the world's largest digital camera ahead of its installation on the telescope. (RubinObs/NOIRLab/SLAC/NSF/DOE/AURA/A. Pizarro D.)

is out there. The rest is dark and invisible. There are two entities on the dark side. The first is called dark matter, the strong gravitational force in the Universe that dictates where and when galaxies form. The second is called dark energy, a label we use to describe a curious phenomenon that not only is our Universe expanding, but it appears that the expansion rate is getting faster and faster each and every day. Rather embarrassingly, we don't understand either of these dark components. According to the physics we teach at University, neither should exist, but a wide range of different observations of the Universe all conclude that there has to be something else out there. This is one of the biggest questions in science today as it's widely believed that in order to understand the dark side of our universe, there will have to be a revolution in physics that forever changes our cosmic view.

When astronomers are faced with a big problem like not understanding what makes up 95% of the Universe, the standard approach is to build a bigger telescope to find a clearer answer. But how could any telescope see these hidden dark entities? Luckily for us, even the invisible leaves a mark. Albert Einstein, in his 1915 theory of General Relativity, showed that mass curves space and time. If you look at a nearby cluster of hundreds of galaxies, their immense mass causes the very fabric of spacetime to curve to such an extent that it visibly warps and arcs the images of background galaxies. This spectacle is known as strong gravitational lensing, and using Einstein's theory you can deduce exactly how much matter is inside the cluster based on the shape and position of the arcs. Suffice to say there is a lot more mass inside these cosmic conglomerations than the mass you would associate with the galaxies that can be seen.

During my PhD, we developed a method to exploit a weaker version of this gravitational lensing effect. Imagine the light from two nearby galaxies, travelling for billions of years through the cosmos. Each time that light passes a clump of matter, the local curvature of space and time will bend the path of the light, imperceptibly warping the shapes of the neighbouring galaxy images. The more structure the light passes on its journey through the cosmos, the more these weak distortions will build up. As a result, when we look up to the night sky, the galaxies appear to very weakly align with one another. But this is only a gravitational illusion. By imaging billions of galaxies, we can map this faint alignment in the heavens to deduce exactly where the dark matter lies – the more dark matter, the stronger the alignment illusion. What we've found in our pilot studies over the last two decades has revealed a giant cosmic web of invisible matter spanning the Universe, providing the gravitational scaffolding for the shining spiralling glittering galaxies that we can see. This is no longer "Pie in the sky" science – it is a fully proven technique that has driven the technological development of the Rubin Observatory to become the most precise and efficient survey telescope in the world.

A *James Webb Space Telescope* observation of the galaxy cluster SMACS 0723. The white galaxies in the foreground are housed within a massive clump of dark matter that curves space and time. This strong gravitational lens distorts our view, with distant redder galaxies appearing as giant warped arcs across the image. (NASA, ESA, CSA, STScI)

Now would be a good moment for me to introduce you to my all-time astronomical hero, Vera Rubin. In the late-1960s Rubin and her colleague Kent Ford were using a new state-of-the-art spectrograph to measure the orbital speed of stars in our nearest major galactic neighbour, the Andromeda Galaxy. This was painstaking work because at the time the telescopes were hand-steered, and these observations took all night. What Rubin discovered changed the course of science. The stars at the outer edges of the Andromeda Galaxy were moving so quickly that they should really be flying off out into space. The best way to explain this is for you to imagine spinning a lasso around your head. The faster you spin it, the tighter you

Vera Rubin operating the 0.9 metre telescope at Kitt Peak National Observatory. (Carnegie Institution for Science)

have to hold on. It's the same with stars in a galaxy, except they are not tied on with a rope: it is gravity that keeps them connected. Rubin calculated the strength of the gravity based on how many stars she could see in the galaxy, and that is where the problem arose. There was simply not enough matter there to keep the stars orbiting at this fast speed. She knew that Fritz Zwicky had proposed the existence of invisible dark matter to solve a similar problem he had encountered with 1930's observations of speeding galaxies in the Coma Cluster. But everyone thought he was mistaken, and Rubin knew that she had to be absolutely sure before backing his extraordinary claim. It took her and Ford another decade of observations, building a bank of twenty-one galaxies that all exhibited the same behaviour. She concluded that dark matter existed, and thanks to her meticulous, perfectionist and indisputable work, the community agreed with her. You would think you would win a Nobel Prize for Physics for such an exceptional career, but unfortunately for Vera, a man called John is twice as likely to win this award as any woman is. Vera Rubin passed away in 2016, but if she had to pick between a Nobel Prize or the ultimate dark matter mission being named after her, my bet is that she would have chosen the observatory any day.

We use the Rubin Observatory to collect wide, deep sky images, looking at the subtle alignment of billions of galaxies to map the invisible web of dark matter across the Universe. But how can this data also tell us about the mysterious dark energy? First, we can look at how the dark matter structures change over time. You can think of dark matter and dark energy as warring siblings. Over cosmic time, the gravitational pull of the dark matter works to create clumpy structures within the cosmic web. But with each day, the power of the dark energy builds, accelerating the expansion of the Universe, making it ever harder for gravity to try and pull it back together again. Second, we can use a cosmic candle to directly measure the change in the expansion rate over the history of the Universe. There is a special type of supernova explosion that occurs when a white dwarf star collapses. Its outburst outshines the rest of the galaxy, and no matter where this special supernova explodes it always shines at the same brightness. If we can spot these cosmic candles we can then use them to measure distances in the cosmos. Combine distance with speed measurements and you can determine the evolving impact of dark energy on our expanding Universe.

You would think that the prospect of bringing about a revolution in our understanding of fundamental physics would have been more than enough for the US Congress to approve funding for the construction of the Rubin Observatory.

Rubin spent its first seven hours imaging the Trifid Nebula (top) and the Lagoon Nebula, revealing faint clouds of gas and dust several thousand light years away from Earth. (NSF–DOE Vera C. Rubin Observatory)

But in 2011 the Nobel Prize for Physics was awarded for the discovery of Dark Energy and many a politician felt this meant 'job done'. It was time to change the narrative and sell Rubin as more than just a Dark Universe explorer. If Rubin could spot a supernova going flash-bang in the night, it could also spot nearly all the asteroids and comets whooshing through our solar system. By accurately mapping their trajectory, Rubin would be able to inform the US government if anything was on a direct collision course with planet Earth. Killer asteroids and dinosaur-like extinction events captured Congress' imagination and Rubin was finally approved for construction in 2014.

In 2025 the completed Rubin Observatory started making the first ever movie of the Universe. Using its wide-field vision, it maps out the entire visible Southern night sky every three nights. And then it repeats, visiting the same patches of sky again and again, building up a time-lapse of the millions of objects that change their position or brightness across our skies every night. In just the first few nights of observations, Rubin discovered over 2,000 new asteroids and by the end of its ten-year mission, Rubin is expected to find almost four million solar system objects, on top of the one million that are already known. Fingers crossed that none of them are hurtling directly towards planet Earth any time soon! Stack many years of these time-lapse images, and you create the wide ultra deep image of the Universe that we need for our dark universe experiment.

Rubin's 'first look' view of the Virgo Galaxy Cluster, revealing two prominent spiral galaxies, three merging galaxies, galaxy groups both near and far, and stars within our own Milky Way galaxy. (NSF–DOE Vera C. Rubin Observatory)

I am writing this article in late 2025, delighted with the amazing progress that the dedicated Rubin Commissioning team has made in Chile. We are all in awe of the stunningly beautiful 'first look' images that Rubin made and are now very busy starting to analyse the stream of data that flows every night from this fabulous new facility. Here in the UK, we are responsible for processing a quarter of all the data collected by Rubin, and we host one of the main data centres that allows the thousands of astronomers across the globe to access this ultimate sixty-petabyte data set. Let me peer into my crystal ball for you to see the 'Future of Rubin' in 2027. I would place good money on the discovery of Planet Nine, a currently hypothesised planet, bigger than Earth but smaller than Neptune, that is expected to lurk in the outer solar system. If any observatory is going to find it, it will be Rubin. I have no doubt that there will also be stunning images of a massive galaxy cluster with gravitational lensing used to reveal the invisible dark matter surroundings. You'll have to wait a little while longer for the promised revolution in physics though. Rubin will need time to build the depth we need to peer deep into the cosmos but do keep watching the news for Rubin's ground-breaking new visions of the dark side of the Universe!

The Rubin Observatory under the Milky Way, with Venus shining brightly on the left and Comet C/2023 A3 (Tsuchinshan–ATLAS) appearing just above the centre of the observatory. (H.Stockebrand/RubinObs/NOIRLab/SLAC/DOE/NSF/AURA)

Alcor and Mizar
More Than Meets the Eye

Jonathan Powell

Double stars have always proved an excellent target for the amateur astronomer with in some cases the ability to observe such a pairing without the need for binoculars or a telescope. One such double system that can be discerned with the naked alone is Alcor (80 Ursae Majoris) and Mizar (ζ Ursae Majoris) located in the well known asterism of the 'Plough' or 'Big Dipper'. The instantly recognizable shape of the 'Plough' resides within the much larger configuration of Ursa Major (the Great Bear), one of the Northern Hemisphere's circumpolar constellations.

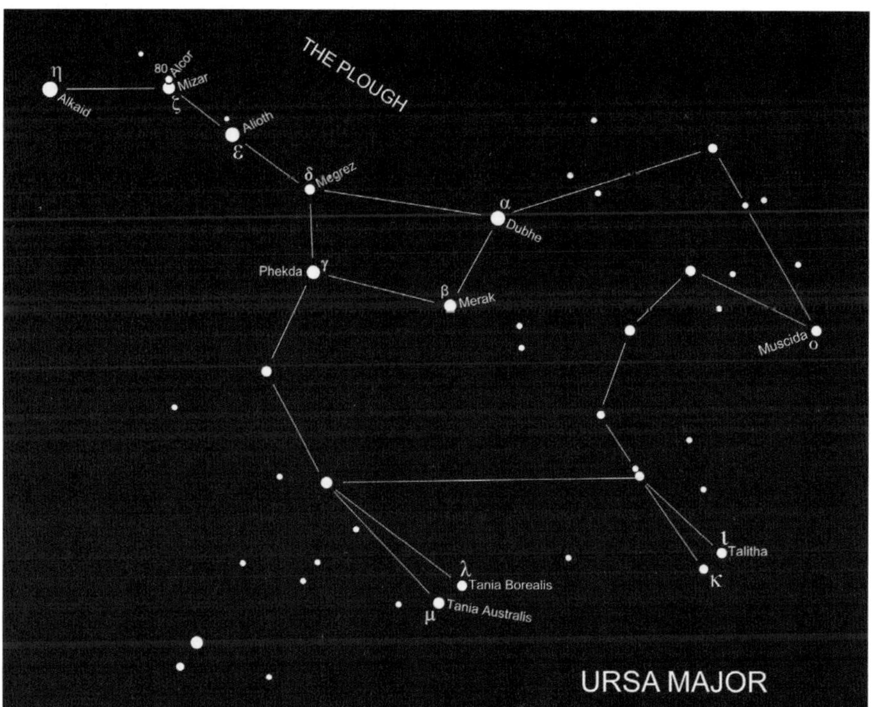

The stars of Ursa Major including Alcor and Mizar. (Brian Jones/Garfield Blackmore)

With access to a pair of binoculars or a telescope, one can really start to delve into the realms of double stars such as Alcor and Mizar, unlocking for oneself further detail that the naked eye alone cannot resolve. With even greater instrumentation, the astronomer has the ability to probe further, pushing back the boundaries to unveil the circumstances surrounding the birth and subsequent evolution of such star systems. With such enhanced capabilities, Alcor and Mizar reveal themselves to not only be an optical double star, but part of a larger system consisting of other members.

By the nature of their birth, which would usually entail conception within large star clusters, science dictates that for a proportion of stars it was perhaps always destined that they would end up with a gravitationally-bound partner for much of their lives. The existence of such pairings reveals another dimension to the life and death of stars that have companions, opening up a whole new level of questioning about the associated mechanics involved, given that no two double star systems are identical.

In recent years, double and multiple star systems may not have been at the forefront of the amateur astronomer's mind; other preferred targets offering much more visually stimulated reward. Indeed, there was a time that double stars would have definitely been the 'go-to' choice by way of learning one's way around the night sky. However, the presence of other more distant objects, coupled with cutting-edge imaging techniques, seems to offer a greater return in the satisfaction stakes. That said, double stars remain a sound starting point in the field of tracking down deep sky objects and, at least for the beginner, offer a feeling of achievement having initially viewed them through an eyepiece.

Double Star Classification

Binary stars are star systems that consist of two or more stars that are linked by gravity, orbiting around their common centre of mass. They make up approximately half the stars in the Milky Way. Optical doubles, on the other hand, may appear to be close together although this is only because they happen to lie in the same line of sight as seen from our planet but in reality can be many light years apart. In other words, they are a chance alignment and not physically related.

Further categories include non-visual binaries that cannot be separated by telescope. These are subdivided into interferometric binaries whereby electromagnetic waves are engaged to extract information on the pairing of stars that are very close together. Then there are astrometric binaries whereby the potential binary is inferred from the perturbation of its movements; spectroscopic binaries and the measuring of electromagnetic radiation emitted by stars; eclipsing binaries where one star literally eclipses the other; and contact binaries, those involved in the extraction of mass exchange.

The two brightest stars in this image are Alcor (top) and Mizar, the fainter Sidus Ludoviciana forming a flattened triangle with these two and visible here just to the left of a line joining the Alcor and Mizar pair. (Steve Barnett)

Alcor and Mizar

The Alcor and Mizar system provides a good template into the study of double stars as there is a sequence of unravelling before the whole picture of the system is unveiled.

Mizar is the primary component in the double star system with Alcor, but when we step back to encompass other elements involved, a quadruple star system is revealed. All of which is situated at a not overly distant 82.9 light-years.

Mizar is the fourth brightest star in Ursa Major, shining at an apparent magnitude of 2.04 and of the pairing with Alcor, the more noticeable of the two. With a name derived from the Arabic meaning 'apron', 'wrapper', 'girdle' or 'waistband', Mizar is the 72nd brightest star in the sky with a mass 2.2 times that of the Sun, and a radius of 2.4 times that of our parent star.

With a luminosity 12 times that of the Sun, Alcor is a binary star in its own right, comprising a white main sequence star and a red dwarf companion. Shining at an apparent magnitude of 3.99, Alcor – named from the Arabic for the 'faint', 'forgotten' or the 'neglected' one – is an incredibly fast spinner, (228 km/s), with a mass 1.8 times that of the Sun. Alcor is moving towards us at a rate of 9 km/s.

Sidus Ludoviciana

Before looking at the historical observation of Alcor and Mizar, it is worth mentioning a star which is not actually part of the aforementioned system, but happens to sit directly in our line of sight between the pair. The star in question is the magnitude 7.58 main-sequence blue-white giant Sidus Ludoviciana, which shines from a distance of around 300 light years.

Portrait by an unknown artist of the German astronomer Johann Georg Liebknecht (1679–1749). (Wikimedia Commons/ Wdwdbot/PD-US/Gießener Professorengalerie)

On the night of 2 December 1722, German theologian and scientist Johann Georg Liebknecht (1679–1749) observed an eighth magnitude star positioned between Alcor and Mizar. Faint as it was, albeit clearly visible, Liebknecht believed it to be a new planet, and duly gave it the name Sidus Ludoviciana, in honour of his patron the Landgrave Ludwig of Hesse-Darmstadt. However, Liebknecht was unaware that this object had already been observed by the Italian mathematician and

Benedictine monk Benedetto Castelli (1578–1643) in 1616, subsequent observations revealing that it was nothing more than a background star.

Early Observations of the Alcor-Mizar Star System

Also known as the Horse and Rider in Arabic astronomy, Mizar is the middle star in the handle of the 'Big Dipper' which along with its companion Alcor, make for a much observed double star system. Long believed to be unrelated, upon further scrutiny, there are more runners and riders than can be initially viewed.

With the advent of the telescope early observers were gifted with a huge array of choice to train any newly acquired or self-built scope upon. As astronomy ascended to a new level in both how the universe was observed and humankind's development of its understanding, the telescope would also sound out or debunk a myriad of theories that had been fashioned around naked eye observations alone, proving revolutionary on a scale never before witnessed.

Alcor and Mizar featured prominently in early observations. The first discovery of a double star system points to a collection of names spanning many years. Some records show that it was Italian astronomer Giovanni Battista Riccioli (1598–1671) who made the discovery in 1650. However, decades before this, it is noted that in 1617 Benedetto Castelli wrote to his long-time friend Galileo about the possibility of Mizar being a double star, encouraging his colleague to make his own observations in the hope of confirming Castelli's belief.

American astronomer Antonia Caetana de Paiva Pereira Maury (1866–1952) was the first person to detect and calculate the orbit of a spectroscopic binary, that of Mizar A, in 1889. Although not credited with the find (Edward Charles Pickering claiming the discovery) Maury went on to discover – and be rightly credited – with finding a second spectroscopic binary, Beta (β) Aurigae, later on in the same year. (John McCue)

Photographer John Adams Whipple (1822–1891), who had previously taken a photograph of the bright star Vega in 1850, teamed up with American astronomer and mapmaker George Phillips Bond (1825–1865), at Harvard College Observatory, to take an image of Alcor and Mizar in 1857. Using the collodion process (an early photographic technique) and a 15-inch refractor, the pair captured the first ever double star to be photographed.

American astronomer Edward Charles Pickering (1846–1919) turned his spectroscope on Mizar, discovery plates taken in 1887 revealing that it was also a close binary system, making this the first ever spectroscopic binary to be identified. The finding duly created ζ^1 Ursae Majoris (Mizar A). American astronomer Antonia Caetana de Paiva Pereira Maury was then given the task by Pickering of determining the orbital period of the binary, which turned out to be 20.5 days.

Following the unveiling of the presence of Mizar A's binary system, which in itself made Mizar the first triple system of this kind to be discovered, another landmark moment occurred in 1908, with astronomers Edwin B. Frost and Hans Ludendorff discovering that Mizar B was also a spectroscopic binary; Mizar thus became the first quadruple star system to be discovered, the Mizar B pairing having an orbital period of 175.5 days.

System Components

The ability to sight Alcor and Mizar with the naked eye is often seen as a test of eyesight. Despite the separation of the two measuring some three billion miles, the two stars appear as a tight pair in the night sky, all due to their relatively close proximity to Earth.

Alcor and Mizar make up the main components of a four-star binary system. This system is composed of two pairs of double stars that are gravitationally bound to one another. Mizar is itself a quadruple system with Alcor a binary system in its own right, the components of which are thought to orbit each other over a period of 90 years. The Alcor and Mizar pairing, including all their counterparts, make for an intriguing sextuple system. This system is only the second sextuple system within 130 light years of the Sun, the other being Castor in the constellation of Gemini. Castor has an even more intricate system of stars comprising six individual stars organised into three binary pairs.

Far from being the solitary combination of a 'Horse and Rider', the Alcor and Mizar system has turned out to be rather more complex, being composed of Mizar A, a binary system of two components, (Mizar Aa and Ab); Mizar B, a binary system also with two components (Mizar Ba and Mizar Bb); and the binary star system comprising Alcor A and Alcor B.

Alcor and Mizar offer a great insight into not just the observational side of double stars, but a series of multiple lessons that go a long way to understanding the birth and evolution of stars.

The Astronomers' Stars
Across the Spectrum

Lynne Marie Stockman

Chemistry in outer space should be as easy as X-Y-Z where X is the relative abundance of hydrogen, Y is helium, and Z represents the 'metals' (everything else). The early universe is thought to have consisted of around 75% of hydrogen and 25% of helium, with perhaps a subtle hint of lithium, the third lightest element. The earliest stars, called Population III, were comprised of this primordial hydrogen and helium. Thought to be hot and massive, these stars synthesised, through nuclear fusion, the heavier elements up to iron, before going supernova and blasting these contaminants across the universe. Elements heavier than iron, like yttrium-89 used in the manufacture of your mobile phone or cobalt-59 which can be transformed into the radioactive isotope cobalt-60 used in medical radiotherapy, were formed during the explosions via various exotic nuclear processes. The next generation, the Population II stars, was formed out of the gas and dust containing traces of these heavier elements. The larger members of this class repeated the supernova cycle, further enriching the interstellar medium with metals. Our Sun is a member of the relatively metal-rich Population I generation of stars which formed afterwards; but even with all of these added ingredients our star is still over 70% hydrogen.

A spectroscope is a device which splits the light of a celestial body into a multi-coloured spectrum which is crisscrossed with lines and bands. Each line corresponds to an isotope of a certain element or molecule in a particular state of ionisation, thus allowing astrophysicists to determine the chemical composition of a star's atmosphere. Further details, such a magnetism, temperature, rotation rate, and radial velocity, can also be derived. Spectroscopy allows scientists to unlock the secrets of the stars.

William Wallace Campbell (1862–1938) was born on his family's farm in Ohio, USA. He graduated in 1886 with a bachelor's degree in engineering from the University of Michigan. Appointed a professor of mathematics at the University of Colorado, he remained there for two years before returning to Michigan to become an instructor in astronomy. He joined Lick Observatory as an astronomer

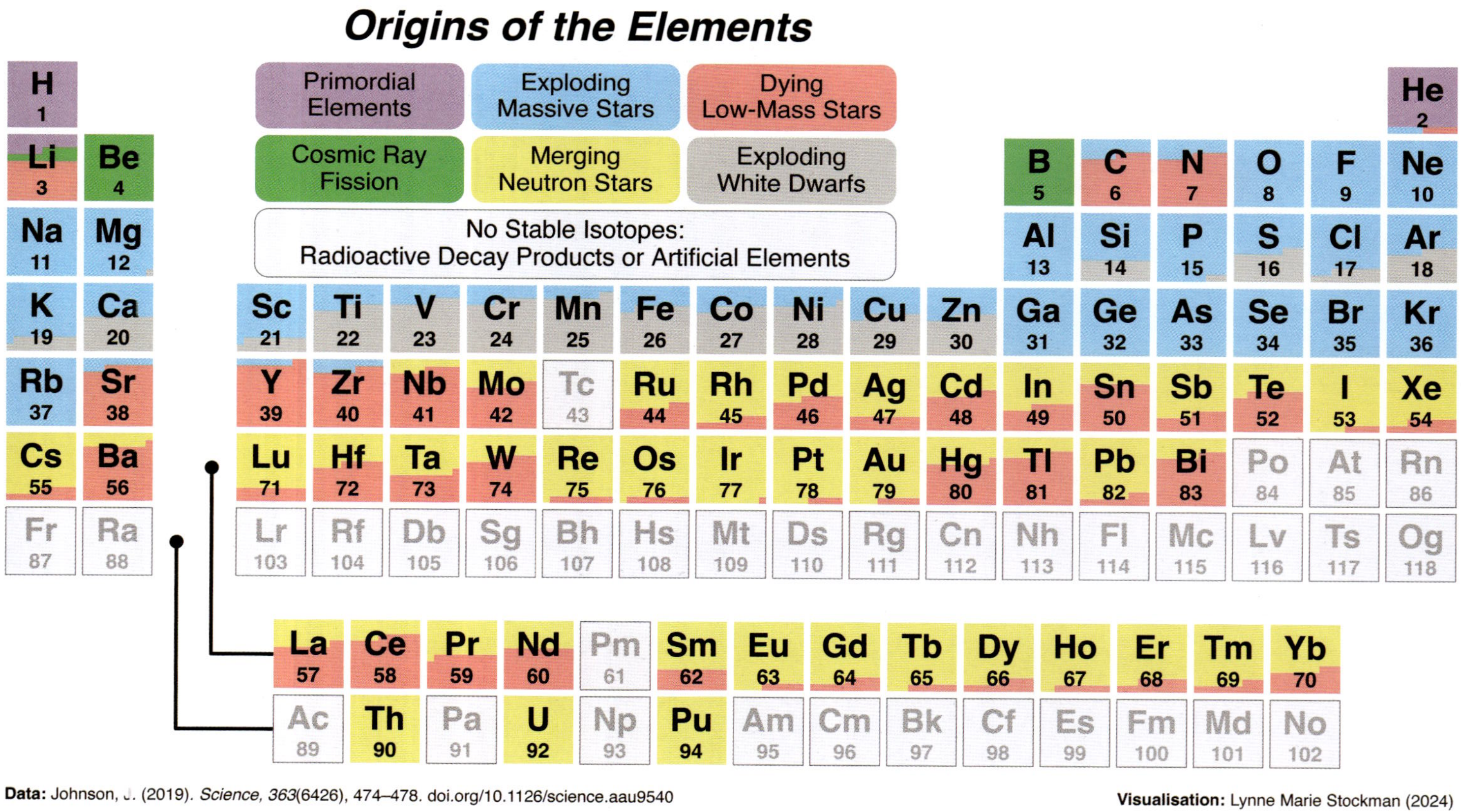

The elements that make up the universe originate from a variety of sources, from the Big Bang to modern physics laboratories on Earth. (Lynne Marie Stockman)

in 1891, specialising in spectroscopy. Campbell spent the rest of his career there and was appointed director of the observatory in 1901. His skill as an administrator was such that he was persuaded in 1923 to take up the post of president of the University of California as well, retiring from both jobs in 1930. Campbell also presided over the International Astronomical Union (IAU) for four years, as well as the National Academy of Sciences. He participated in seven solar eclipse expeditions all over the world, confirming Albert Einstein's theory of general relativity at the end of his last trip. A foreign member of the Royal Society, Campbell was the recipient of many awards, including the Prix Lalande from the Académie des sciences and the Gold Medal of the Royal Astronomical Society. Suffering from ill health, specifically blindness and aphasia, he feared being a burden to his family and tragically took his own life in 1938 (Aitken 1938) (P. 1938).

In the early 1890s, Campbell took spectrograms of an unusual faint star in the constellation of Cygnus (Campbell 1893):

> The 9.3 magnitude star BD. +30°.3639 is surrounded by an extensive hydrogen envelope. This star is of Wolf-Rayet type, and its spectrum is very rich in bright lines, about thirty having been observed by me between the wave lengths of 656 and 426...This existence of this hydrogen envelope can hardly fail to have an important bearing upon the theory of bright line stars.

Known ever since as ***Campbell's Hydrogen Star***, it and its associated nebula have been extensively observed in wavelengths from radio to X-ray for over a century. Even the *Hubble Space Telescope* has visited this well-studied object from time to time. Recent *Gaia* parallax values place it around 1700 pc away although other methods for estimating distance sometimes give rather different answers. Wolf-Rayet stars were first identified in the mid-nineteenth century. They are evolved objects and can be broken down into two main groups. Classic WR stars are hot, massive Population I stars that have lost their outer hydrogen envelope and are fusing helium or even heavier elements in their cores. The second type are post asymptotic-giant-branch stars that used to resemble the Sun but have since shed their outer hydrogen atmospheres and now sit at the centre of planetary nebulae. Unlike the classic WR stars, stars of the second type have ceased fusion and are slowly cooling down. Campbell's Hydrogen Star is of this second type. Observation of and research into this intriguing object are helping scientists answer important questions about the evolution of stars past the asymptotic giant branch, into the planetary nebula stage, and beyond.

In September 2013, the *Hubble Space Telescope* captured this image of Campbell's Hydrogen Star embedded in its fiery planetary nebula. (ESA/Hubble & NASA, image processing by Jean-Christophe Lambry).

Astrophysicist **Daniel Magnus Popper** (1913–1999) was born in California, USA. He attended the University of California, Berkley, earning his doctorate in 1938. He moved around a lot in his early professional life, holding a fellowship at Eastman Kodak in Rochester, New York, moving to the McDonald Observatory where he was a founding member of staff, spending a year at Yerkes Observatory, moving back Berkley to work in their radiation laboratory during the Second World War, then returning to Yerkes as an instructor and later an assistant professor. In 1947 he joined the new astronomy department at UCLA where he stayed for the rest of his career. He went on to become chair of the department and was made professor emeritus upon retirement in 1972. His PhD thesis dealt with spectroscopy of Nova Lacertae 1936 and his first professional paper investigated the spectrum of another nova, Nova Herculis 1934. With Carl Seyfert, Popper investigated selective absorption of the interstellar medium so that colours of stars could be used to estimate distances to stellar objects. Ultimately Popper wanted to combine the measured radial velocities of B-type stars with their distances in order to measure the rotation of the galaxy. He was also expert in determining the orbital characteristics of spectroscopic binaries. Popper was a member of several learned societies and served on the IAU commissions on stellar spectra and close binary stars. At the time of his death in 1999 of cancer, he held the record for the longest observing career at Lick Observatory (Trimble 1999).

Whilst taking spectrograms of B-type stars in the early 1940s, Popper made an unexpected find (Popper 1942):

> The star HD 124448 is listed in the *Henry Draper Catalogue* as having magnitude 9.8, spectrum B2. Spectrograms extending from Hα to λ 3530, obtained at the McDonald Observatory, show no hydrogen lines in absorption or emission, although the helium lines are sharp and strong...The abundance of hydrogen appears to be very low in the atmosphere of this star.

Popper offered three possible explanations: the hydrogen deficiency was limited to the outer layers of the star, the star was formed in the absence of hydrogen, or it had reached its current depleted state through some as yet unknown evolutionary process.

Popper's Extreme Helium Star was the first of a very select group of stars (only a few dozen have been identified thus far) which are almost entirely devoid of hydrogen. Extreme helium stars are low-mass supergiant stars with peculiar A- and B-type spectra where the Balmer hydrogen lines are missing or extremely weak and neutral helium is very strong. Carbon, nitrogen and occasionally oxygen lines are also much stronger than usual. Most are short-period pulsating variable stars and they are concentrated toward the centre of the galaxy. Perhaps most tellingly, none of them are binaries. These strange stars have proven to be a challenge to astrophysicists attempting to model their formation and evolution, but current theories suggest that an extreme helium star may be the product of a stellar merger of two white dwarf stars, one rich in helium and the other in carbon and oxygen (Jeffery 2008).

Described by his peers as 'one of the last masters of visual spectral identification', American astronomer **William Pendry Bidelman** (1918–2011) possessed an encyclopaedic knowledge of the peculiarities of stellar spectra. He was born in Los Angeles, California, USA, but moved to North Dakota as a young child where he was raised by his maternal grandparents. In 1940 he graduated alongside future American president John F. Kennedy from Harvard College. His postgraduate education took place at Yerkes Observatory, University of Chicago, where he trained in stellar spectroscopy under the guidance of such greats as Philips Childs Keenan and William Wilson Morgan (who formulated the M-K stellar luminosity classification system) and Otto Struve. During the Second World War, Bidelman worked at the U.S. Army's Ballistic Research Laboratory in Maryland, returning to Yerkes at the cessation of hostilities. He later taught and conducted research at Lick Observatory (University of California), the University of Michigan, the University of Texas, and finally Case Western Research University where he remained until retirement in 1985. Bidelman's knowledge of the vast literature on spectroscopy was unparalleled. Postgraduate astronomy students at his universities were humorously advised that if, during their examinations, they were asked how to determine the spectral type of a star, the correct answer was always 'write down its HD number and go ask Bidelman.' His last publication was in 2009, nearly 70 years after his first. During his long and productive career, Bidelman discovered hundreds of stars of unusual chemical composition. To this day, theoreticians struggle to explain the origins and evolution of many of these objects (Bond 2017).

In 1950, Bidelman described the unusual spectrum of star HD 30353:

> The spectrum of HD 30353 is shown to possess characteristics similar to those of the well-known peculiar star υ Sagittarii, but at a somewhat lower temperature…The spectrum is a most complex one, containing very strong lines of the ionized metals, as well as many additional lines which do not appear strongly in normal stellar spectra. The Hα line appears in emission.

Bidelman's Peculiar Star (also known by the variable star designation KS Persei) is a hydrogen-deficient binary; less than half a dozen of these objects are known to exist with υ Sagittarii being the brightest and best known. They are thought to be comprised of a hot hydrogen-poor primary star transferring material to its companion, and it is possible that they are the progenitors of type Ib supernovae.

Bidelman announced the discovery of another unusual object in 1965:

Principal Fraunhofer Lines in the Solar Spectrum

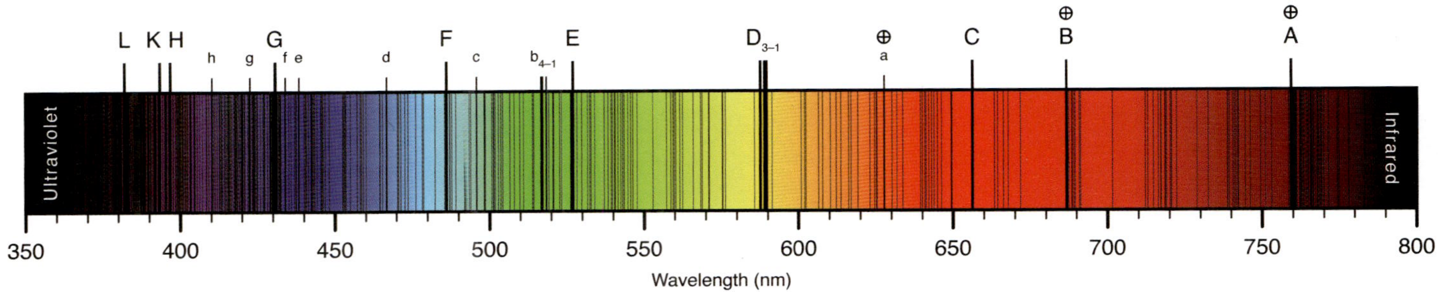

L	Fe I	382.044	G	Fe I	430.791	b_4	Mg I	516.733	D_2	Na I	588.997
K	Ca II	393.368	f	Hγ	434.048	b_3	Fe I	516.772	D_1	Na I	589.594
H	Ca II	396.849	e	Fe I	438.356	b_2	Mg I	517.270	a	O2	627.663 (atmospheric)
h	Hδ	410.175	d	Fe I	466.815	b_1	Mg I	518.362	C	Hα	656.281
g	Ca I	422.674	F	Hβ	486.134	E	Fe I	527.038	B	O2	686.719 (atmospheric)
G	Ca I	430.775	c	Fe I	495.761	D_3	He I	587.562	A	O2	759.370 (atmospheric)

Data: Mitton, J. (1975). The stronger absorption lines in the solar spectrum — an identification list. *Journal of the British Astronomical Association, 85*, 238–241.
Moore, C.E., Minnaert, M.G.M., & Houtgast, J. (1966). *The Solar Spectrum 2935Å to 8770Å.* National Bureau of Standards Monograph 61, United States Department of Commerce.
Sansonetti, J.E., & Martin, W.C. (2005). *Handbook of Basic Atomic Spectroscopic Data.* NIST Standard Reference Database 108, National Institute of Standards and Technology.
Visualisation: Lynne Marie Stockman (2024)

This diagram shows some of the stronger absorption lines present in the solar spectrum. The labelled lines are named for German physicist Joseph Ritter von Fraunhofer (1787–1826) who investigated these strange features early in the nineteenth century. (Lynne Marie Stockman)

> Several years ago HR 5378=HD 125823, a member of the Scorpius-Centaurus cluster, was found to exhibit conspicuous variation in helium-line strength on Lick Chile spectrograms...This object...does not appear to have any other spectroscopic peculiarities, but has been found to be slightly variable in light....

Also known as a Centauri and V761 Centauri, ***Bidelman's Helium Variable Star*** is proving difficult to explain. Theories abound for why the helium lines (but not the other lines) vary over a period of around eight days. One suggestion is that there is a single helium-rich spot on the surface of the star that rotates in and out of view. Another theory, the oblique rotator model, holds that the helium is concentrated at poles of an axis which is inclined both to our line of sight and to the axis of rotation. Magnetic fields are responsible for holding the helium superabundances in discrete zones. Yet another hypothesis is that the star is severely distorted into the shape of a prolate spheroid (Underhill et al 1975). However, like many of the bizarre objects found by Bidelman, this star remains unexplained to this day.

Antoni Przybylski (1913–1985) was born in Rogoźno, Poland, and took his first degree at the University of Poznań where he stayed on as a research assistant. He served as an artillery officer during the defence of Poland during the Second World War, was captured by the Germans, and spent time in a POW camp. He escaped in 1941 and made his way back to Poland, but it was too dangerous for him to remain there so he travelled overland to Switzerland. During his internment in Switzerland, he attended Zurich Polytechnic, earning a doctorate in 1949. Unwilling to return to a now-communist Poland, he emigrated to Australia in 1950. After time spent doing manual labour, he came to the attention of the director of Mount Stromlo Observatory who offered him a position, a scholarship and eventually supervision of a doctoral thesis on stellar atmospheres, the first doctorate granted by the Australian National University. After Bart Bok took over the directorship of the observatory in 1957, he encouraged Przybylski (called 'Bill' by his friends who had difficulty pronouncing his name) to take up observing. He worked in a variety of fields, including satellite orbits, comets, variable stars and stellar abundances. Upon retirement he obtained a second bachelor's degree, this time in the natural sciences. He died in 1985 (Gascoigne 1985).

Variable star V816 Centauri is the prototype rapidly oscillating Ap (roAp) star. Its peculiar spectrum was first noticed in 1960 (Przybylski 1961):

> On April 26, 1960, in the course of a spectroscopic survey of stars with large proper motions, a spectrum of the eighth magnitude star *HD* 101065 was observed with the Zeiss spectrograph at the Newtonian focus of the 74-in. Telescope at Mount Stromlo Observatory. This star, classified as a *B5* star in the Henry Draper Catalogue, proved to be a much cooler star with a peculiar spectrum of unknown type...The spectrum shows about 500 lines between λλ3922 and 4528 Å., the majority of which have not yet been identified. Preliminary identifications seem to suggest that many lines should be attributed to rare earths...The high abundance of barium and strontium—the elements with the magic number of neutrons in their nuclei—as well as the high abundance of rare earths indicate that the star *HD* 101065 is in an advanced stage of stellar evolution. The star apparently passed through a stage of evolution in which heavy elements were built up by slow neutrons.

Further investigation revealed even more surprises about ***Przybylski's Star***. The star's spectrum was covered in lines but Przybylski and his colleague P. Morris Kennedy were forced to conclude that amongst the first 53 elements in the periodic table, only hydrogen and calcium were undeniably present. No iron was found. Barium was the first heavy element detected, followed by a number of lanthanides (metallic chemical elements with atomic numbers from 57 to 71), particularly holmium. This was the first detection of holmium in any stellar spectrum, including that of the Sun (Przybylski and Kennedy 1963). The star also contained a number of short-lived actinides (Gopka et al 2008). Przybylski's Star continues to puzzle researchers today.

Another star with a decidedly peculiar spectrum is ***Sneden's Star***. It is known as BPS CS 22892-052 from its inclusion in the 1985 and 1992 surveys of extremely metal-poor stars by Timothy C. Beers, George W. Preston and Stephen A. Schectman. The K-type Population II bright giant was singled out for attention by American astronomer **Christopher Sneden** and his colleagues (Sneden et al 1994):

> The star CS 22892-052 possesses the largest overabundance of very heavy elements of any metal-poor ($[Fe/H] < -2.5$) halo star found to date; no other star approaches this level of enrichment in neutron-capture elements.

Subsequent investigations revealed a number of elements never before detected in ultra-low-metallicity stars, including hafnium, holmium, osmium, terbium and thulium. These are all much more massive than iron, the heaviest element that can be formed by nuclear fusion in a star's core, and are created through a process called neutron capture. It is likely that these unusual elements are contaminants from earlier supernovae and may prove useful in determining the characteristics of the generation of stars that preceded Population II objects (Sneden et al 1996).

Like Sneden's Star, ***Cayrel's Star*** is a Population II ultra-metal-poor star catalogued by Beers, et al. The spectrum of BPS CS 31082-0001 was analysed by French astronomer **Roger Cayrel** and an international team at the start of the millennium (Cayrel et al 2001):

> During an ESO-VLT large programme devoted to high resolution spectroscopy of extremely metal-poor stars selected from the HK survey of Beers and colleagues, the [Fe/H] = −2.9 giant CS 31082-001 was found to be as enriched in neutron-capture r-process elements as CS 22892-052 but with a much-reduced masking of molecular lines. This allowed the detection and measurement of the uranium line at 3859 Å for the first time in a stellar spectrum.

Nuclear or radioactive decay can be used to date objects. Just as radiocarbon dating uses the radioactive isotope carbon-14 to date organic samples less than 50,000 years old, other elements with longer half-lives can be used to date older objects. The detection of uranium-238 (which has a half-life of 4.47 billion years) in the spectrum of Cayrel's Star, along with thorium-232 (which has a much longer half-life of 14 billion years), allowed Cayrel and his colleagues to estimate an age of 12.5 billion years for BPS CS 31082-0001, making it one of the oldest stars yet discovered. It also sets a lower bound for the age of the universe.

SDSS J102915.14+172927.9 is an ancient ultra-metal-poor halo dwarf star and is comprised almost entirely of hydrogen and helium. It was discovered by astronomer **Elisabetta Caffau**, then at Zentrum für Astronomie der Universität Heidelberg (Caffau et al 2011):

> ...it has been suggested that low-mass stars cannot form from the primitive interstellar medium until it has been enriched above a critical value of Z, estimated to lie in the range 1.5×10^{-8} to 1.5×10^{-6}...Here we report the chemical composition of a star in the Galactic halo with a very low Z ($\leq 6.9\times10^{-7}$, which is 4.5×10^{-5} times that of the Sun) and a chemical pattern typical of extremely metal-poor stars — that is, without the enrichment of carbon, nitrogen and oxygen. This shows that low-mass stars can be formed at very low metallicity, that is, below the critical value of Z.

Caffau's Star was discovered when the method Caffau devised to obtain high precision abundances from 3D hydrodynamical computations was applied to data from the *Sloan Digital Sky Survey*. The existence of this star upsets conventional stellar formation and evolution models. Not only is it metal-deficient, but its lithium values are far below those expected from the Big Bang. Such old stars should have abundances close to those of the young universe but Caffau's Star clearly doesn't. This discovery led to Caffau winning the 2013 'Best early career researcher in astrophysics observation' prize from the European Astronomical Society (l'Observatoire de Paris 2013).

Acknowledgements

This research has made use of NASA's *Astrophysics Data System Bibliographic Services*, operated at the Harvard-Smithsonian Center for Astrophysics, Cambridge, Massachusetts, USA, and the SIMBAD astronomical database, operated at CDS, University of Strasbourg, France. The author would like to thank Dr David Harper for his enthusiastic encouragement and helpful comments.

References

Aitken, R. G. (1938). William Wallace Campbell, 1862–1938. *Publications of the Astronomical Society of the Pacific*, **50**(296), 204–209. **doi.org/10.1086/124927**

Bidelman, W. P. (1950). The peculiar star HD 30353. *The Astrophysical Journal*, **111**, 333–350. **doi.org/10.1086/145268**

Bidelman, W. P. (1965). On the spectra of three bright southern stars. *The Astronomical Journal*, **70**, 667. **doi.org/10.1086/109789**

Bond, H. E. (2017). William Pendry Bidelman (1918–2011). *Publications of the Astronomical Society of the Pacific*, **129**, 1–9. **doi.org/10.1088/1538-3873/129/971/010201**

Caffau, E., Bonifacio, P., François, P., Sbordone, L., Monaco, L., Spite, M., Spite, F., Ludwig, H.-G., Cayrel, R., Zaggia, S., Hammer, F., Randich, S., Molaro, P., & Hill, V. (2011). An extremely primitive star in the Galactic halo. *Nature*, **477**, 67–69. **doi.org/10.1038/nature10377**

Campbell, W. W. (1893). Hydrogen envelope of the star BD. +30°.3639. *Astronomische Nachrichten*, **134**, 133. **doi.org/10.1002/asna.18941340806**

Cayrel, R., Spite, M., Spite, F., Hill, V., Primas, F., François, P., Beers, T. C., Plez, B., Barbuy, B., Andersen, J., Nordström, B., Molaro, P., & Bonifacio, P. (2001). First measurement of the uranium/thorium ratio in a very old star: Implications for the age of the Galaxy. In T. von Hippel, C. Simpson, N. Manset (Eds.), *Astrophysical Ages and Time Scales, ASP Conference Series*, **245**, (pp. 244–251). Astronomical Society of the Pacific.

Gascoigne, S. C. B. (1985). Obituary—Dr Antoni Przybylski. *Proceedings of the Astronomical Society of Australia*, **6**(2), 275.

Gopka, V. F., Yushchenko, A. V., Yushchenko, V. A., Panov, I. V., & Kim, Ch. (2008). Identification of absorption lines of short half-life actinides in the spectrum of Przybylski's Star (HD 101065). *Kinematics and Physics of Celestial Bodies*, **24**(2), 89–98. **doi.org/10.3103/S0884591308020049**

Jeffery, C. S. (2008). Extreme helium stars: A decade of progress. In K. Werner, T. Rauch (Eds.), *Hydrogen-Deficient Stars, ASP Conference Series*, **391**, (pp. 53–63). Astronomical Society of the Pacific.

l'Observatoire de Paris. (2013, April 8). European prizes: awarded to two female researchers from Observatoire de Paris. **observatoiredeparis.psl.eu/european-prizes-to-two-female-resarchers-from.html**

P., H. C. (1938). Dr. W.W. Campbell. For. Mem. R.S. *Nature*, **142**, 102–103. **doi.org/10.1038/142102a0**

Popper, D. M. (1942). A peculiar B-type spectrum. *Publications of the Astronomical Society of the Pacific*, **54**(319), 160–161. **doi.org/10.1086/125431**

Przybylski, A. (1961). *HD* 101065—a G0 star with high metal content. *Nature*, **189**, 739. **doi.org/10.1038/189739a0**

Przybylski, A., & Kennedy, P. M. (1963). The spectrum of HD 101065. *Publications of the Astronomical Society of the Pacific*, **75**(445), 349–353. **doi.org/10.1086/127965**

Sneden, C., Preston, G. W., McWilliam, A., & Searle, W. (1994). Ultra–metal-poor halo stars: The remarkable spectrum of CS 22892-052. *The Astrophysical Journal*, **431**, L27–L30. **doi.org/10.1086/187464**

Sneden, C., McWilliam, A., Preston, G. W., Cown, J. J., Burris, D. L. & Armosky, B. J. (1996). The ultra–metal-poor, neutron-capture–rich giant star CS 22892-052. *The Astrophysical Journal*, **467**, 819–840. **doi.org/10.1086/177656**

Trimble, V. (1999). Obituary: Daniel M. Popper, 1913–1999. *Bulletin of the American Astronomical Society*, **31**(5), 1608–1610.

Underhill, A. B., Fahey, R. P. & Klinglesmith, D. A. (1975). The spectrum of HD 125823 (a Centauri). *The Astrophysical Journal*, **199**, 120–126. **doi.org/10.1086/153670**

Mars Attacks!
How the Red Planet Affects Earth's Climate

David Harper

Milankovitch Cycles

Long-term changes in the Earth's climate are caused by many factors. Human activity, in the form of greenhouse gases, has raised global temperatures measurably over the past century. A prolonged minimum in sunspot activity is implicated in the 'Little Ice Age', which brought unusually cold winters to northern Europe between 1300 and 1850. On a much longer timescale – tens or hundreds of thousands of years – slow periodic variations in the shape of the Earth's orbit and its axial tilt (obliquity) are known to be correlated with regular ice ages. This discovery is attributed to the Serbian astronomer Milutin Milankovitch (1879–1958), although the idea was first proposed by a Scottish scientist named James Croll in the 1860s.

A statue of Milutin Milankovitch in Belgrade, Serbia. (Wikimedia Commons/ Andrija12345678/CC BY-SA 4.0)

These orbital variations are mainly due to the gravitational influence of Jupiter and Saturn. Tiny Mars, 3,000 times less massive than Jupiter, also changes the Earth's orbit, pulling our planet backwards and forwards by a few hundred kilometres every synodic period. This is far smaller than the 4.8 million kilometre annual variation in the Earth-Sun distance due to the eccentricity of the Earth's orbit. No-one would have believed that the red planet could have a significant effect on Earth's long-term climate, but deep-sea sediments reveal a strange truth: Mars does indeed play a role on a timescale of millions of years.

Probing the Ocean Floor

The Earth's oceans are home to uncounted trillions of tiny single-celled creatures known collectively as foraminiferans. Each is a fraction of a millimetre in size. There are as many as 10,000 living species, and 40,000 more in the fossil record. They build protective shells of calcium carbonate, which sink to the ocean floor when the occupant dies. These shells survive remarkably well over millions of years, allowing individual species to be identified in sedimentary rocks such as the limestone of the Great Pyramid of Giza and chalk used in mosaics in Roman Britain.

They are also found in sediment cores that have been drilled at dozens of locations across the Earth's oceans. Chemical analysis of the ratio of oxygen-18 to oxygen-16 in the calcium carbonate of the shells tells us whether the Earth was in an ice age or a warm period when their owners lived and died. Taken together with dating inferred from the fossilised skeletons of other groups of single-celled organisms, geologists have been able to correlate the cycles of global ice ages and warm periods recorded in the sediment cores with the cycles of the Earth's eccentricity and obliquity calculated by Milankovitch and later refined in the early twenty-first century by the French astronomer Jacques Laskar and colleagues using computer models.

Gaps in the Sediment Record

However, the ocean floor sediment record contains gaps, known as hiatuses. Each hiatus represents a period when deposition stopped, or when the upper layers of the sea floor were scoured by strong deep-ocean currents. Adriana Dutkiewicz and Dietmar Müller, working at the University of Sydney, published a study in 2022 in which they reported that these hiatuses appeared to occur globally every 2.4 million years. This did not match any known Milankovitch cycle; these are typically a few tens or hundreds of thousands of years long.

The drilling ship *Joides Resolution* used by the Ocean Drilling Program (ODP). Ocean sediment samples from the ODP were among those analysed by Adriana Dutkiewicz and Dietmar Müller. (ODP/Ocean Drilling Program science operator at Texas A&M University)

They suspected that the cause of the hiatuses was an unidentified 2.4-million-year astronomical cycle which affected the eccentricity of the Earth's orbit. In a 2024 follow-up study with colleague Slah Boulila, they proposed an answer to the mystery.

Secular Perturbations

Every planet in the Solar System exerts a gravitational pull on every other planet. This causes each planet's orbit to deviate slightly from the simple Keplerian ellipse that it would follow if it were the only planet orbiting the Sun. Astronomers refer to these deviations as perturbations. They can be calculated very accurately using three of Isaac Newton's discoveries: the law of universal gravitation, the laws of motion, and calculus.

Astronomers recognise three classes of perturbation, each requiring its own mathematical treatment. Short-period perturbations act on the timescale of the synodic period of each pair of planets. It was this type of perturbation, observed in the orbit of Uranus in the decades following its discovery by William Herschel in 1781, that enabled John Couch Adams and Urbain Jean Joseph Le Verrier to infer the existence of Neptune and to calculate its orbit, allowing Johann Gottfried Galle to discover the new planet in September 1846.

Resonant perturbations occur when the orbital periods of two bodies are very close to an exact integer ratio, such as 1:2 or 3:4. The orbits become 'locked' together. This is seen in the Galilean moons of Jupiter and in the larger moons of Saturn, and also in some exoplanet systems, but not in the planets of our Solar System. Jupiter and Saturn have orbital periods which are almost in the ratio 2:5, and this close commensurability does amplify the effect that they have on each other's orbit, but it is not sufficiently close to lock the orbits, and thus does not qualify as a true resonance.

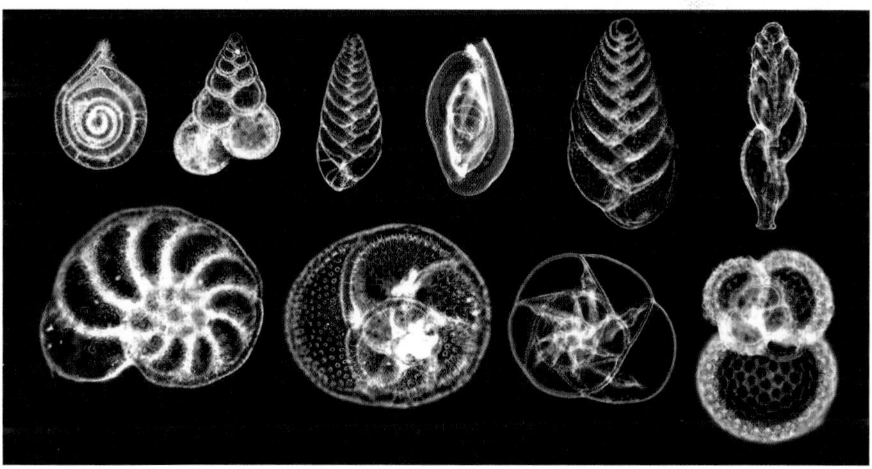

A photomicrograph showing the calcium carbonate shells of ten species of marine foraminiferans. (Randolph Femmer, U.S. Geological Survey)

Finally, there is the slow, steady drift in the direction of the perihelion and node (the point where the orbit crosses the ecliptic) of each planet, induced by the combined gravitational pull of the other planets. Astronomers refer to these drifts as secular perturbations.

According to a long-term numerical integration of the orbits of the planets carried out by Laskar and colleagues, published in 2004, the perihelion of the Earth's orbit drifts eastwards at 17.368 arc-seconds per year, making one complete revolution in 74,620 years. Its node on the fixed ecliptic of J2000 drifts in the opposite direction at 18.850 arc-seconds per year, in a cycle of 68,750 years duration.

The Grand Cycle of Earth and Mars

Mars's perihelion drift, 17.916 arc-seconds per year, differs from the Earth's by just 0.548 arc-seconds per year. The relative positions of the two perihelia thus make one cycle every 2.4 million years.

The secular drift of the node of Mars's orbit, 17.755 arc-seconds per year, is also very close to that of the Earth, so that the relative positions of the nodes make one cycle every 1.2 million years.

To Dutkiewicz and her colleagues, these numbers seemed too close to the 2.4 million year period of the hiatuses in the ocean floor sediment cores to be a coincidence. They compared the hiatuses to long-term variations in the Earth's eccentricity calculated by Laskar and colleagues in an updated 2011 study, and found a strong correlation spanning tens of millions of years of ocean sediment data.

Tiny Mars, which pulls the Earth to and fro by just a few hundred kilometres every year, had been unmasked as the culprit behind changes to the Earth's climate that have left a permanent record in ocean sediment cores. By modulating long-term variations in the eccentricity of the Earth's orbit on a timescale of 2.4 million years, Mars has an effect on the Earth's climate far out of proportion to its size.

Chaos Constrained

There is a further twist to this already surprising story. Astronomy, in the form of Milankovitch's theory, has explained why there are cycles of ice ages and warm periods in the geological record. But geology can also answers questions posed by astronomy.

One of the most significant discoveries in Solar System dynamics, made by Laskar and his colleagues, is that numerical integrations of the orbits of the planets of the Solar System break down when the time span of the integration is pushed further than about 50 million years into the past. Different integrations using almost identical starting conditions agree quite closely from T=0 to T=−50 million years, but then begin to diverge in unpredictable ways.

This is not because of errors in the mathematical formulae or in the computer program, nor is it due to the accumulation of rounding errors over billions of individual additions, multiplications, subtractions and divisions within the computer's CPU. It appears to be an inherent property of the Solar System.

Mathematicians refer to this as chaos, but it doesn't imply that the Solar System suddenly flies apart. Rather, it means that the planets continue to orbit the Sun at their present distances, but we can no longer be exactly sure where each planet was in its orbit at such distant times, nor of the precise value of the eccentricity of the orbit.

Laskar and colleagues computed multiple long-term integrations of the orbits of the planets for their 2011 study. They concluded:

> We have thus reached the time where geological data are now required to discriminate between planetary orbital solutions beyond 50 Myr.

Dutkiewicz and colleagues, in their 2024 study, noted that hiatuses in ocean sediment cores in the period from 60 million years to 50 million years before present

Earth and Mars at the same scale, showing their relative sizes. (Earth: NASA/Apollo 17 crew; Mars: ESA/MPS/UPD/LAM/IAA/RSSD/INTA/UPM/DASP/IDA)

offer evidence of chaotic changes in the orbits of Earth and Mars. In particular, the interval between hiatuses undergoes a subtle but measurable change about 56 million years before present.

That moment in the Earth's geological history also marks a major but short-lived climatic event called the Palaeocene-Eocene Thermal Maximum (PETM). For about 200,000 years – a brief moment in geological time – there was a 5–8°C rise in global average temperature, and a significant increase in carbon in the atmosphere and oceans. There is also evidence for abrupt changes to global ocean circulation patterns, and of oxygen depletion and acidification of ocean basins such as the Tethys Ocean.

Dutkiewicz and colleagues found a corresponding discontinuity in the characteristic frequencies of variations in the Earth's eccentricity at 56 million years before present in all of the integrations published by Laskar and colleagues. They suggested the intriguing possibility that the PETM event may be linked to a chaotic transition period in the orbit of the Earth, caused by Mars.

Conclusion

Mars has long fascinated astronomers and the general public alike. Writers such as H.G. Wells and Ray Bradbury have vividly imagined how life on Earth might be affected by the (fictional) inhabitants of the Red Planet. Today, thanks to astronomy and geology, we know that Mars does indeed affect our world and its climate.

References

Dutkiewicz, A. & Müller, R.D. (2022). Deep-sea hiatuses track the vigor of Cenozoic ocean bottom currents, *Geology*, **50**(6), 710–715. **doi.org/10.1130/G49810.1**

Dutkiewicz, A., Boulila, S., & Müller, R.D. (2024). Deep-sea hiatus record reveals orbital pacing by 2.4 Myr eccentricity grand cycles. *Nature Communications*, **15**:1998. **doi.org/10.1038/s41467-024-46171-5**

Laskar, J., Robutel, P., Joutel, F. & 3 others. (2004). A long-term numerical solution for the insolation quantities of the Earth. *Astronomy & Astrophysics*, **428**, 261–285. **doi.org/10.1051/0004-6361:20041335**

Laskar, J., Fienga, A., Gastineau, M. & Manche, H. (2011). La2010: a new orbital solution for the long-term motion of the Earth. *Astronomy & Astrophysics*, **532**, A89. **doi.org/10.1051/0004-6361/201116836**

What's the Matter?
It's Dark Down Boulby Mine

John McCue

Vera Rubin was born in Philadelphia in 1928, well ahead of her time…

Vera Rubin (1928–2016), American astronomer born in Pennsylvania and a pioneer in dark matter discovery. (John McCue)

The planets we have observed for millennia in our solar system are tiny in number compared to the 5,000 and more that we have found orbiting other stars. The stars we see with the unaided eye in our night skies are but a negligible fraction of the 100 billion or so altogether in our Milky Way Galaxy. Our home galaxy is only one among uncountable galaxies in the wide universe, possibly as many as 200 billion. Incomprehensible as these numbers are, astronomers estimate that these objects we know, seen by us in visible light, account for only 5% of the mass-energy content of the universe!

Twenty percent of the universe has been named dark matter; it cannot be detected in any part of the electromagnetic spectrum, let alone in visible light. Its nature is unknown, and we are only aware of it by its gravitational effect on the stars and galaxies we *can* see. Vera Rubin led the way towards discovery of this astonishing fact. After gaining her Ph.D. under George Gamow at Georgetown University, she undertook research into galactic studies at the Carnegie Institute, Washington, where a state-of-the-art spectrometer had been built by astronomer Kent Ford.[1]

1. American Museum of Natural History: **amnh.org/learn-teach/curriculum-collections/cosmic-horizons-book/vera-rubin-dark-matter**

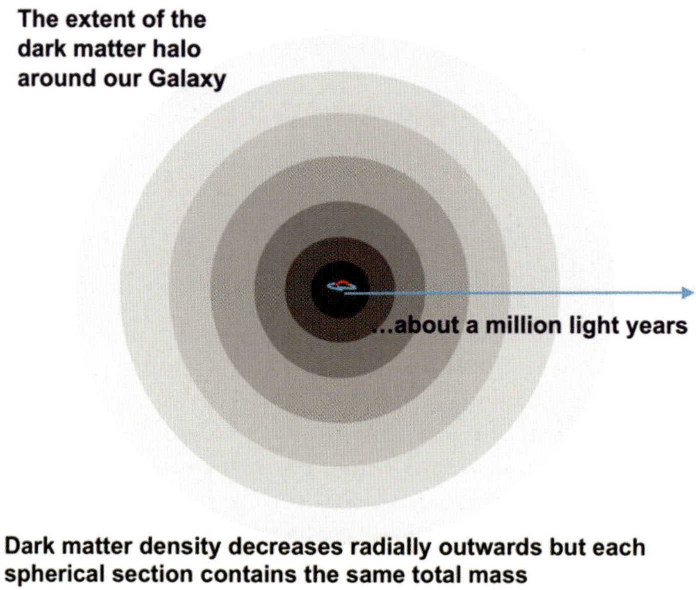

The estimated extent of dark matter around our Milky Way Galaxy, with equal masses in each of the equal surrounding shells. (John McCue)

Like everything else in the universe, galaxies rotate. Unless they are face-on to our line of view, stars on one side of the galaxy will be seen moving away from us and so their spectral lines will be red-shifted. On the opposite side, stars will show a blue-shifted spectrum. The faster the stars are whirling around the galactic centre, the greater will be the observed spectral shift. Vera used Ford's cutting-edge spectrometer to measure these rotational velocities at varying distances from the central bulge and arrived at a notable result.

She expected stars further from the bulge to travel more slowly, just as do the further planets from our sun, but no! These outer stars maintain their velocities. Vera concluded that there must be more mass in a galaxy than we can see – particularly beyond the central bulge – to account for the unexpected greater speed of the stars in the spiral arms. This led to the idea of a large halo of dark matter of an unknown nature around the bulge. If our Milky Way Galaxy is typical, this halo could extend out to about a million light years from our galactic centre. The author's impression of this halo is shown in the accompanying diagram. Calculations based on Newtonian gravitational theory suggest that the constant velocity of stars beyond the bulge leads to the conclusion that the density of the dark matter tails off inversely proportional to the outward radius from the galactic bulge. However, the surface area of a sphere of dark matter is directly proportional to that radius. The two effects hence cancel out, and so, going outwards equal shells of dark matter enclose the same mass, regardless of the falling density – remarkable.[2]

Fritz Zwicky (1898–1974), Swiss astronomer who worked at the Carnegie Institute of Technology, California. (John McCue)

The Swiss-born American astronomer, Fritz Zwicky was well known for his grumpy approach to his fellow astronomers, but was innovative and ingenious in his work, which included the compilation of a six-volume catalogue of galactic clusters using the Palomar Sky Survey of the 1950s. Vera recalled Zwicky's 1933 work when he studied the Coma Galaxy Cluster and saw that the velocities of the galaxies in the cluster were higher than expected given the visible tally of the mass of galaxies in the cluster. He had likewise surmised that unseen matter was there. His colleagues were sceptical at the time, but Vera could see that he had been right.

The major challenge is somehow to detect dark matter, a task made no easier by the fact that astronomers and particle physicists struggle to understand even what type of particle they must be. They can't be baryons because the galaxies we see now would not have had time to form.[3] (From the Greek *barys*, meaning 'heavy', a baryon is made up of three quarks, the proton and neutron being well known examples.) The reader will also know of the standard model, constructed over many modern years, which describes the structure of particle types explaining the properties of matter we see, but dark matter may not even be of a type described within that model.[4] Neutrinos have been eliminated from the search as they travel

2. Zeilik, M., Gregory, S. A., & Smith, E.V.P. (1992), *Introductory Astronomy & Astrophysics*, Saunders College Publishing.
3. Coughlan, G. D., Dodd, J.E., & Gripaios, B.M. (2006), *The Ideas of Particle Physics: An Introduction for Scientists*, Cambridge University Press.
4. Rovelli, C. & Lane, A. (2016), *Reality Is Not What It Seems: The Journey to Quantum Gravity*, Penguin.

Boulby Polyhalite Mine, located near the fishing village of Staithes, North Riding of Yorkshire. (John McCue)

at speeds close to that of light and so would likewise not have helped the early accumulating formation of the galaxies.

Notwithstanding all this, we are left with several hypothetical particles, none of which have been detected so far. The most promising is the axion, an extremely light elementary particle, and it should have formed in huge numbers during the hyperinflation that followed the Big Bang. It would also be slow moving which strengthens its claim for dark matter. As a bonus, it might even explain the imbalance between matter and antimatter in the early universe, which must have occurred otherwise there would be no matter left after they annihilated each other!

Over a kilometre down in the depths of the earth sits the Dark Matter Laboratory in the Boulby polyhalite mine on the northeast coast. This is the deepest mine in the UK and one of only four mines in the world searching for dark matter, the others being in Italy, Canada and the United States. Depth under the earth's surface ensures that cosmic rays from the wider Milky Way background are absorbed to almost zero and so will not give false detections of dark matter at that depth.

The author first visited the sterile-pure DM lab over two decades ago when work first started and was privileged to go underground again recently, but success in

Underground at the Boulby Mine Dark Matter Laboratory. (Jonathan Gutteridge / Boulby Mine)

finding these dark matter particles is still elusive. The most sensitive detector so far built is the LUX-ZEPLIN (LZ) experiment at Sanford Underground Research Facility (SURF) in South Dakota. The outer cylindrical container, as big as a house, holds a smaller cylinder containing ten tons of liquid xenon, only in this state at a temperature of −100° C. Chosen as the target for dark matter particles because it is well known for scintillation – emitting photons which constitute luminescence – xenon is the key to this detection process. An incoming particle will collide with a xenon atom, which will recoil. Xenon's outer electrons absorb this energy, and then re-emit it as the scintillation photons. One of the experiment's 488 photomultiplier tubes (PMTs) will detect these photons by converting them to electrons within the tube, amplifying them and reading them out as an electrical current. When this occurs, it will make international headlines!

Dark energy makes up the remaining 75% of the mass-energy of the universe. Astronomers know of its existence by, as with dark matter, indirect observations of external galaxies. They show that the expansion of the universe is, to everyone's astonishment, speeding up, so dark energy is, in essence, anti-gravity. Likewise, the nature of dark energy is unknown.

In June 2025 came first light on the newest telescope to look skyward, the Simonyi Survey Telescope (after the private donors Charles and Lisa Simonyi). Sitting on a mountain top in Chile, it has been fittingly renamed the Vera Rubin Observatory. With a diameter of 8.4 metres, the mirror is one of the largest in the world, and is served by a 3.2 gigapixel sensor, the biggest digital camera ever built. This ground-breaking combination will survey the entire southern hemisphere sky every three to four nights, and will keep on doing so, probing and tracking the universe in space and time.

The Unluckiest Astronomer Who Ever Lived

Neil Haggath

We have all had the disappointment of having astronomical events ruined by the weather. I travelled to both Hawai'i and China to *not* see total eclipses! But if you think you have been hard done by, think again!

I'm going to tell you the sad story of the man who was surely the unluckiest astronomer who ever lived. But first, we need some background.

Portrait of Johannes Kepler by an unknown artist, from a lost original of 1610 in the Benedictine monastery in Kremsmünster (Wikimedia Commons)

In 1618, Johannes Kepler (1571–1630) – often described as the first astrophysicist – published his famous three laws of planetary motion, which were a forerunner of Newton's work on gravity. His Third Law proved vital in measuring the scale of the Solar System.

Kepler's Third Law states that the square of a planet's orbital period is proportional to the cube of its orbital radius. If the period is measured in years, and the orbital radius as a multiple of that of the Earth, then the constant of proportionality is one, and we can simply state the law as ...

$$p^2 = a^3$$

... where a is the orbital radius and p the period.

It's a simple matter to measure the orbital periods of the planets by observation, so given that relationship, it is only necessary to measure the distance of one planet from the Sun – say that of the Earth – and then the distances of all the others can be deduced. By this time, the Earth-Sun distance – known as the astronomical unit – had been measured by parallax methods to an accuracy of plus or minus about 3%. But in 1677, Edmond Halley (1656–1742) realised how it could be refined.

As we all know, both Mercury and Venus occasionally undergo transits, passing directly in front of the Sun as seen from Earth. Transits of Mercury are fairly common; on average, they occur about 13 or 14 times per century. But those of Venus are extremely rare, occurring only four times in every 243 years. Due to a quirk of orbital dynamics, they always occur in pairs eight years apart, with the intervals between the pairs alternating between 105.5 and 121.5 years – so the majority of people never get to see one in their lifetimes. Many of us have been lucky enough to have a pair in ours, in 2004 and 2012.

Edmond Halley. Portrait by Thomas Murray, 1690. (Wikimedia Commons/ Thomas Murray)

The first person to realise that Mercury and Venus could transit the Sun was Kepler himself. In 1627, he predicted that both inner planets would transit in 1631 – but sadly, he didn't live to actually see them himself.

The first transit ever to be observed was that of Mercury on 7 November 1631, the event being recorded by the French astronomer and mathematician Pierre Gassendi (1592–1655) in Paris. Nobody saw the Venus transit later that month, as it wasn't visible from Europe – and there were few, if any, telescopes outside Europe at that time.

Kepler calculated that transits of Venus would be extremely rare, though he did not realise that they would occur in pairs. In 1639, the English astronomer Jeremiah Horrocks (1619–1641) realised what Kepler had missed – that another would occur that year on 24 November, and that it *would* be visible from Europe. Horrocks made this discovery only a few weeks before the transit was due, so he did not have much time to publicise it. The event was observed by all of two people – Horrocks himself and his correspondent and fellow astronomer William Crabtree (1610–1644) – but Horrocks wrote a paper on his observation.

In 1677, during his voyage to St. Helena, Halley observed a transit of Mercury. It then occurred to him that observations of transits could provide a method of refining the measurement of the astronomical unit. If a transit of Mercury or Venus was observed from two or more widely separated places on Earth, it would be seen in slightly different positions against the Sun, due to parallax. If the event was timed very accurately, then triangulation could be used to calculate the Earth-Sun distance with unprecedented accuracy.

The transit of Venus of 5/6 June 2012, photographed by Gary Gawthrope in Hawai'i. (Gary Gawthrope)

He soon realised that, though transits of Mercury are much more common, those of Venus would be far more suitable for the purpose. Mercury in transit, seen through a small telescope, appears as only a tiny black dot, and its ingress and egress are too quick to allow accurate timings. But from reading Horrocks' paper, Halley knew that Venus would appear as a distinct disc, and its motion would be much slower, so that there would be four distinct 'contacts' which could be accurately timed – the ingress and egress of its leading and trailing edges.

Halley's genius was matched by his far-sightedness. He knew that the next pair of Venus transits would not happen until 6 June 1761 and 4 June 1769, but even though he wouldn't live to see them himself, he wrote a paper, which he presented to the Royal Society, to alert the astronomers of a future generation to their vital importance. A further paper, written in Latin, was distributed across Europe.

The accompanying simplified diagram (see below) shows the principle of Halley's method. Observers at different latitudes on Earth would see Venus follow different apparent paths across the Sun, of different lengths; the path lengths could be measured by accurately timing the contacts. Then, knowing the distance between the observers' locations, a bit of trigonometry would give the distance of Venus from Earth, from which the astronomical unit could be determined by Kepler's Third Law.

When the time came, astronomers in Britain and France took up the challenge with a passion. In 1761, both the British Royal Society and the French *Académie des Sciences* organised a number of expeditions to diverse parts of the world, so as to obtain measurements of the transit from as many different locations as possible, and resolved to cooperate afterwards in analysing the data. This, of course, was in the days when intercontinental travel involved sea voyages lasting months, so the project was an enormous undertaking.

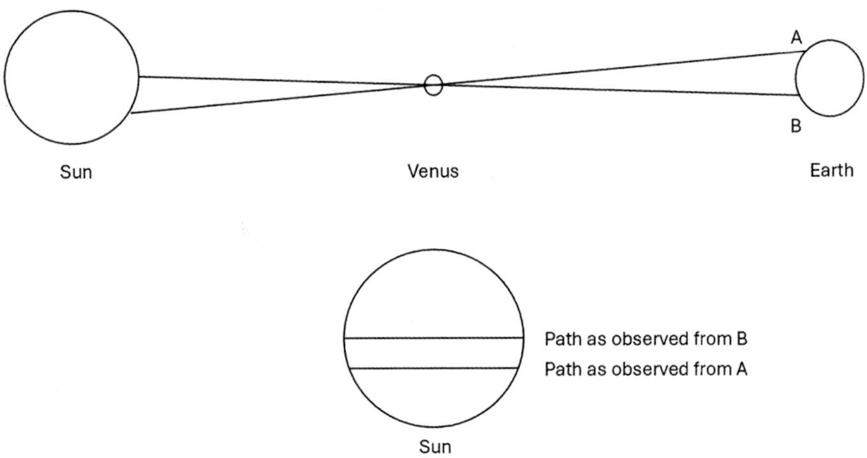

Halley's Method. (Peter Rea/Neil Haggath)

But there was a major problem; Britain and France, and their empires, were at war at the time. This was the 'Seven Years War', which was being fought across four continents, and could justifiably have been called the 'First World War'. The two governments, recognising the importance of the project, signed a treaty which supposedly guaranteed their transit expeditions safe passage through each other's territories – but this proved unworkable in practice; several expeditions were caught up in fighting, and some of their members killed.

Nevertheless, observations were obtained not only from across Europe, but also from North America, South Africa and India. One heroic French team made a gruelling overland trek to northern Siberia, where they had to fight off packs of wolves, among other hazards! While the last transit of Venus had been observed by two people, this one was observed by some 120 astronomers, from 62 locations on four continents.

Unfortunately, after all that immense effort, the data proved to be of little use! It was near-impossible to time the contacts with the required accuracy, due to an unforeseen problem which came to be known as the 'black drop'.

Despite the huge disappointment, the British and French astronomers resolved to try again in 1769; now that they knew about the black drop, they could work out ways of correcting for it to some degree, so as to get more accurate timings. This time, they had to travel even further afield, as the transit was visible from the other side of the world from Europe – but at least the war had ended by then, so the expeditions no longer had to worry about being attacked on the high seas!

One English astronomer, Charles Green (1735–1771), sailed with Captain James Cook aboard the *Endeavour*; the primary purpose of Cook's round the world voyage, which led to the discovery of Australia, was to observe the transit from Tahiti. Sadly, Green died during the return voyage.

This time, the data proved much more useful, and eventually led to a determination of the astronomical unit, which was pretty close to today's accepted value.

So now we come to the Unluckiest Astronomer. Next time you're upset about an astronomical event being clouded out, spare a thought for this fellow…

One of the men despatched by the French *Académie* in 1761 was an aristocrat, rejoicing in the kind of extravagant name which could *only* belong to a French aristocrat! – Guillaume Joseph Hyacinthe Jean-Baptiste Le Gentil de la Galaisière (1725–1792). We'll just call him Le Gentil for short! He was an accomplished astronomer, who had discovered three of the nebulae listed in Charles Messier's *Catalogue des Nébuleuses et des Amas d'Étoiles* (Catalogue of Nebulae and Star Clusters). Le Gentil was sent to observe the transit from Pondichéry in India, which was then a French territory. He sailed from France on 26 March 1760, so should have had more than ample time to get there.

However, the war intervened. He was delayed for several months in Île de France, which is now Mauritius, and by the time he got to India, Pondichéry had

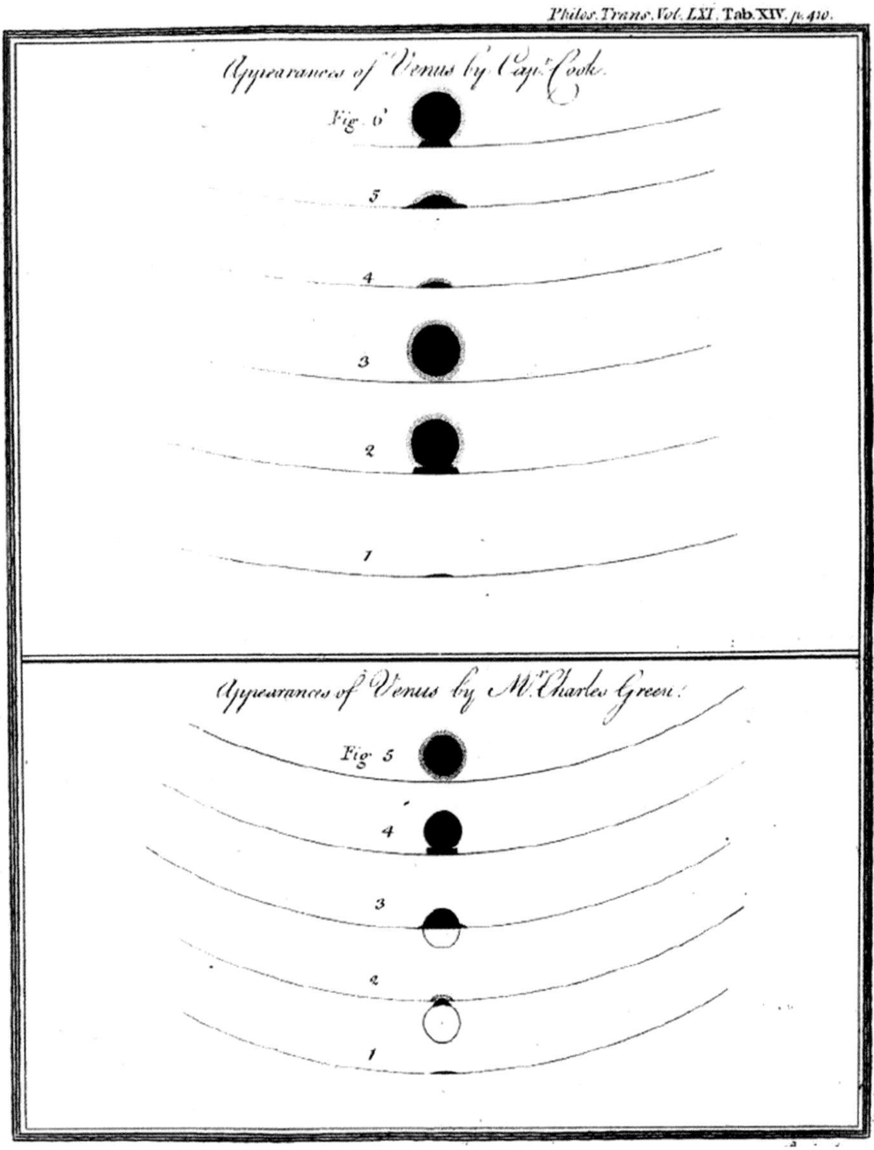

Sketches of the 1769 transit of Venus by Captain James Cook and Charles Green, clearly showing the black drop. (Wikimedia Commons/Captain James Cook and Charles Green/*Philosophical Transactions of the Royal Society*)

fallen to the British. He was unable to take advantage of the safe passage treaty, as he was sailing aboard a warship, which was obliged to return to Île de France. He was still at sea on the day of the transit; though the sky was cloudless, it was impossible to make any observations from the heaving deck of a sailing ship.

The unfortunate Guillaume Joseph Hyacinthe Jean-Baptiste Le Gentil de la Galaisière. (John McCue)

Le Gentil was nothing if not determined. He knew that the 1769 transit would also be visible from India, so he decided to wait for it! Instead of returning to France, he based himself in Île de France, and spent the next eight years travelling around the Indian subcontinent, the Indian Ocean and the Far East, where he carried out a remarkable variety of studies; far from being the 'upper class twit' that his name suggests, he was in fact a highly respected scientist, and quite a polymath. He mapped the coast of Madagascar, and carried out research in geography, natural history and anthropology.

Come 1769, he decided that Manila in the Philippines would be the best place to observe the transit, and sailed there in plenty of time. But he was met with hostility and suspicion from the Spanish Governor, so he sailed once again to Pondichéry; he could now go there safely, as the war was over. Le Gentil was made welcome by the British Governor, who not only offered to act as observing assistant, but also managed to obtain the loan of a telescope which was better than Le Gentil's own.

The day of the transit began bright and clear . . . but guess what? Shortly before Venus made its first contact with the Sun, it clouded over – and then stayed that way. Poor Le Gentil didn't see a thing!

It almost goes without saying that the sky was clear in Manila!

But our hero's troubles were only just beginning; during his years away from home, due to the war, any letters he had sent to his family and to the *Académie* had never arrived. His voyage back to France was delayed by a serious illness, storms and a whole sorry saga of lesser mishaps; he eventually obtained passage to Spain, and then travelled overland. When he finally returned home in 1771, after more than 11 years away, he discovered that he had been legally declared 'missing, presumed dead'; his wife had remarried, and his relatives had divided up his estate.

When he suddenly turned up again, they conveniently failed to recognise him, and denounced him as an impostor! His place in the *Académie* had also been given to someone else.

He then spent the next several years, and most of his remaining wealth, engaging in lawsuits against his family to get his property back – and it took the intervention of the King to get him reinstated in the *Académie*.

In 1789 came the Revolution, when France became a distinctly unhealthy place for the aristocracy. Le Gentil escaped the gruesome fate of many of his peers, only because he died of natural causes in 1792, just before the 'Reign of Terror' began.

I can empathise with poor Le Gentil; in 2012, I may have set an unenviable record among British astronomers, for travelling the greatest distance to *not* see that year's transit of Venus! I went 6,000 miles to Thailand, where on the day, I saw precisely nothing, and got very wet! But while his journey took months, mine took a mere 18 hours!

A History of Observatory Designs
The Modern Era

Katrin Raynor

I feel somewhat sad that I am writing my third, and last, instalment of 'A History of Observatory Designs'. I have thoroughly enjoyed researching the history, art and science behind the construction and use of observatories from ancient times to the present day. Astronomical observatories have transformed drastically over the centuries – from rudimentary watchtowers to precision-engineered facilities that pierce the depths of the universe. While ancient structures like Stonehenge hinted at an early awareness of celestial cycles, the birth of the modern observatory has ushered in a fusion of science, architecture, and technology designed to serve both research and imagination. From terrestrial domes that cradle massive mirrors to orbiting telescopes sailing through space, the twentieth and early twenty-first centuries witnessed extraordinary leaps in observatory and telescope design. These advancements didn't just improve resolution – they redefined our very perception of the observable universe and beyond.

We are now living in one of the most exciting astronomical technological eras. Who would have thought that in the 55 years since we successfully landed humans on the Moon, we would have created one of the most powerful telescopes to have ever been constructed? The *James Webb Space Telescope* equipped with a 21-foot gold-coated beryllium primary mirror can detect objects much earlier in the history of the universe than the previous *Hubble Space Telescope*.

For the last article on this subject, I have chosen some of the most iconic observatories including the designs and capabilities of modern-day observatories from the dawn of the twentieth century to the present day.

Against All Odds (1918)

The Dominion Astrophysical Observatory, situated on Observatory Hill in Saanich, British Columbia, was completed in 1918 by the Canadian Government. The observatory is home to the 72-inch (1.83 metre) Plaskett telescope, a bold idea which was proposed, and subsequently designed, by Canadian astronomer John Stanley Plaskett (1865–1941). Construction of the observatory was undertaken during the First World War by McAlpine-Robertson and, despite the conflict, the building work was not hindered and construction continued at a rapid pace. Luckily the somewhat heavy 5,000-pound (2,268 kilogramme) glass mirror was delivered to the site a week before the war broke out, having been shipped from the Saint-Gobain company in Antwerp, Belgium and ground in Pittsburgh, Pennsylvania by the renowned John A. Brashear Company.

Completed in 1918 amid the First World War, the Dominion Astrophysical Observatory is located on Observatory Hill in Saanich, British Columbia and houses the 72-inch Plaskett telescope. Designed by Canadian astronomer John Stanley Plaskett, materials to build the observatory were hauled up the hill by horse and cart. (Wikimedia Commons/Colin Stepney/CC BY-SA 4.0)

Using horses and wagons, materials for the observatory were transported to the top of the hill and by the end of 1916 the building and telescope mounting were ready. However, problems experienced during the grinding of the mirror together with delays in the delivery of the telescope optics saw the completion of the observatory pushed back by two years. The vision had been for the Plaskett telescope to be the largest telescope in the world at that time, although these delays led to it achieving 'first light' on 6 May 1918, a full six months later than the 100-inch Hooker telescope at Mount Wilson Observatory on 2 November 1917.

The circular observatory building is made entirely of steel and measures 20.2 metres in diameter and 9.75 metres high. External and internal coverings of galvanized iron allow free circulation of air from the base to the top of the building allowing perfect temperature conditions to exist. The building is capped by a circular plate which is supported by girders on which is a curved, circular railroad iron rail enabling the 24 wheels mounted on roller bearings to rotate the hemispherical dome. The telescope can be used to observe any part of the sky thanks to the double-motor operated shutters.

Palomar Observatory (1928)

A stunning example of an observatory created by George Ellery Hale (1868–1938) is the Art Deco inspired Palomar Observatory located in the Palomar Mountains of San Diego County, California. A visually pleasing design, American engineer, architect (and amateur astronomer) Russell W. Porter was recruited as the chief architect. Porter chose the Art Deco look due to its popularity at the time, designing the building to be as tall as it is wide to create a visually pleasing observatory. Construction spanned the Great Depression and World War II, with the telescope finally achieving 'first light' on 26 January 1949, a decade after Hale's death. Now owned and operated by the California Institute of Technology (Caltech), the observatory was designed to house the 200-inch (5.1 metre) Hale Telescope which at the time was the largest optical telescope in the world. The huge mirror is made of Pyrex, a material chosen for its low thermal expansion which maintained optical precision. The mirror was by Corning Glass Works, and took years to grind and polish. The reflector telescope uses an equatorial mounting, allowing it to track celestial objects smoothly as the Earth rotates. Hale was a driven individual who, with funding, worked to found several significant astronomical telescopes including those for Mount Wilson Observatory where he supervised the building of the 60-inch (1.5 metre) telescope in 1908 and the 100-inch (2.5 metre) Hooker telescope in 1917.

This gorgeous and iconic blue three-cent stamp depicts Palomar Observatory. Issued in 1948 by the U.S. Post Office Department to mark the observatory's dedication, it is classed as one of the first space-themed stamps issued in the U.S. The vignette (the central design or picture) features an exterior view of the observatory, and was engraved by Canadian-born stamp designer George Arthur Gundersen (1910–1975). (U.S. Postal Service/National Postal Museum/Bureau of Engraving and Printing)

An Iconic Radio Telescope (1957)

Designated as a UNESCO World Heritage Site in 2019, Jodrell Bank Observatory is located at Lower Withington, Cheshire, England, and plays host to the giant 250-foot (76-metre) Lovell radio telescope, widely regarded as one of the most iconic landmarks in the United Kingdom. As you approach the observatory, it is hard not to let out a 'wow' as this iconic radio telescope comes into view.

The English physicist and radio astronomer Bernard Lovell was born in Bristol in 1913, and is best known for founding the Jodrell Bank Observatory in 1945. In 1957, he completed the iconic 250-foot Lovell Telescope which successfully tracked Sputnik 1 as it orbited the Earth. (Unknown Author/ Jodrell Bank, University of Manchester)

Jodrell Bank Observatory is not the typical astronomical observatory that springs to mind when we think about what an observatory should look like. Founded in 1945 by physicist Bernard Lovell (1913–2012), the observatory has been instrumental in discovering quasars, pulsars and gravitational lenses. After the outbreak of the Second World War, Lovell worked for the Telecommunications Research Establishment (TRE) where he was instrumental in developing radar technology for aircraft. At the end of the war, Lovell returned to the University of Manchester, keen to utilise his knowledge of radar technology and its use in studying cosmic rays.

He began by using an ex-military radar detector unit based in the centre of Manchester. However, background interference from the electric tram system was problematic. Lovell decided to relocate his equipment to a remote area near

From humble beginnings to a radio giant; the Jodrell Bank Hut, owned by the Botany Department at Manchester University, was the first location outside of Manchester that Lovell used to set up his radio equipment in the hope of studying cosmic rays. He instead detected meteors entering the Earth's atmosphere, which inspired Lovell to go on to successfully create a research team and build the Jodrell Bank Experimental Station. (Unknown Author/Jodrell Bank, University of Manchester)

Is there any telescope more recognisable than the Lovell Telescope? At a whopping diameter of 250 feet, the telescope was fully steerable and built using equipment from defunct battleships. Jodrell Bank is now a UNESCO world heritage site, and you can peacefully eat your lunch just feet away from the ginormous telescope! (Mike Peel, Jodrell Bank Centre for Astrophysics, University of Manchester/CC-BY-SA 2.0-UK, 2.5, 4.0)

Goostrcy in Cheshire, used by the University's botany department, and it was here that the Jodrell Bank Experimental Station was established. Over the following years, scientists and technicians at the station built their own instruments using excess army equipment, and in 1947, the 218-foot (66-metre) Transit Telescope – which at the time was the world's largest radio telescope – was constructed. Although this pioneering instrument successfully detected radio waves from the Andromeda Galaxy, Lovell wanted something even bigger.

Despite being told that building an even larger radio telescope was impossible, Lovell persevered and with the help of Charles Husband (1908–1983), a consulting engineer from Sheffield, construction of what became known as the Lovell Telescope began in 1952. The telescope was built using riveted and bolted steel and recycled material, including salvaged rack and pinions from two defunct battleships. The 250-foot, 1000-ton steerable telescope was completed in 1957 and was the largest radio telescope in the world. Within days of its completion, the telescope had successfully tracked the rocket that blasted *Sputnik 1* into orbit on 12 October 1957.

One of the Great Observatories (1993)

The *Hubble Space Telescope* is one of NASA's four Great Observatories – alongside the *Compton Gamma Ray Observatory*, *Chandra X-ray Observatory*, and *Spitzer Space Telescope* – and is renowned for its breathtaking images and space-based vantage point. Built in the 1970s by NASA and the European Space Agency (ESA), Hubble faced numerous setbacks before its launch on 24 April 1990 aboard STS-31 into low Earth orbit, approximately 515 kilometres above the Earth's surface.

Originally set for launch in 1983, delays due to technical issues, budget constraints, and the fatal *Challenger* disaster in 1986 pushed the launch back. A flaw in the mirror's construction further postponed operations until astronauts corrected it during a 1993 servicing mission. The idea for the *Hubble Space Telescope* dates back as early as the 1920s, gaining traction in 1946 when American theoretical physicist Lyman Spitzer (1914–1997) outlined the benefits of a space-based telescope. He realised two advantages: seeing conditions would be much more favourable due to the lack of turbulence in space (unlike on Earth where our atmosphere interferes) and the telescope could carry out observations at infrared and ultraviolet wavelengths. Spitzer later led the committee that defined its scientific goals, leading to project approval in 1969.

Hubble's design and construction involved numerous companies and institutions. The Marshall Space Flight Center in Alabama was instructed to design, develop and construct the telescope; Perkin-Elmer Corporation (now

The *Hubble Space Telescope* as seen from the Space Shuttle *Discovery* as it is returned to orbit during the second servicing mission in February 1997. (NASA Hubble Space Telescope)

Hughes Danbury Optical Systems) developed the optics and sensors; and Lockheed Missiles and Space Company (now Lockheed Martin) were tasked with designing and building the telescope housing. Of course, a telescope in space would need to be controlled, the ground-based control being handed to the Goddard Space Flight Centre in Maryland. Building the 2.4 metre primary mirror with a honeycomb interior significantly reduced its weight. The primary and secondary mirrors are both made from ultra-low expansion glass and coated with aluminium and magnesium fluoride. If you hadn't already guessed, the telescope was named after the pioneering American astronomer Edwin Hubble (1889–1953) to honour his important contributions to astronomy and cosmology.

A Very Large Telescope in Chile (1998)

With the ability to observe at both visible and infrared wavelengths, the Very Large Telescope (VLT) is a flagship astronomical facility operated by the European Southern Observatory (ESO) at an altitude of 2,635 metres atop the Cerro Paranal

This image captures the Very Large Telescope (VLT) at Paranal Observatory in Chile's Atacama Desert. The four massive telescope units stand in formation, surrounded by the four 1.8-metre Auxiliary Telescopes and the VLT Survey Telescope (VST), all of which can be seen in this picture. Set against a stark, mountainous backdrop, the facility's remote location ensures minimal light pollution and crystal-clear skies – ideal for deep-space observations and groundbreaking discoveries in astrophysics and cosmology. (ESO / Gerhard Hüdepohl (**atacamaphoto.com**))

mountain in the Atacama Desert of northern Chile. The facility, which became operational in 1998, represents a significant leap forward in ground-based astronomy, combining the power of multiple telescopes to create a single, massive instrument. It is a sobering thought that, when all four of the main telescopes are combined, the VLT facility can achieve an angular resolution of around 0.002 arcseconds. Given this fact, it is perhaps no wonder that some of the pioneering observations made using the VLT include the first direct image of an exoplanet.

Beginning in 1991, the VLT's construction was a massive undertaking. The project's unique design consists of four individual 'Unit Telescopes' (UTs), each with a primary mirror measuring 8.2 meters in diameter. The four main telescopes were each given a name from the local Mapuche language: *Antu* (Sun), *Kueyen*

Unit Telescope 1 (Antu) is one of the four main telescopes at the Paranal Observatory. As sunset approaches, technicians retract its windshield and complete final preparations for the night's observations. During daylight hours, the enclosure remains sealed to safeguard the sensitive equipment and maintain thermal equilibrium with the outside atmosphere. Visible to the left of the main mirror housing is the Infrared Spectrometer And Array Camera (ISAAC), formerly mounted on Antu, but now relocated to Unit Telescope 3 (Melipal). (ESO/Gerhard Hüdepohl (**atacamaphoto.com**))

(Moon), *Melipal* (Southern Cross) and *Yepun* (Venus). The construction involved the careful engineering of each telescope to function independently, while also making them capable of working together as a single unit. This was achieved through a process called interferometry, where the light collected by each telescope is combined in a special laboratory. This creates a single image with a resolution equivalent to that of a telescope with a combined light-gathering capability equal to that of the entire array.

Hubble's Replacement (2021)

Another visually iconic space observatory is undoubtedly the *James Webb Space Telescope (JWST)* designed and built to study such areas as the formation and evolution of the first galaxies and the births of stars and planets. The *JWST* is the most complex and powerful telescope to be launched into space.

I don't think any of us will forget seeing its 18-segment array of gold hexagonal mirrors on news channels or magazines – it really is an unforgettable design. Launched on 25 December 2021 aboard an *Ariane 5* rocket from French Guiana, the *JWST* is an engineering marvel designed to be the successor to the *Hubble Space Telescope*. However, rather than orbiting the Earth, the JWST orbits the Sun 1.5 million kilometres away from our planet, close to the Sun–Earth L_2 Lagrange point.

This image showcases the mirror assembly for the *James Webb Space Telescope*, taking place inside the pristine specialized environment of a cleanroom. The striking gold-coated, hexagonal mirror segments form a honeycomb pattern, optimized for infrared observation. The technicians in cleanroom suits emphasize the iconic mirrors huge scale. (NASA)

Putting the *JWST* together was a feat of ingenuity; because the telescope was too large to fit into any existing rocket in its operational state, it had to be built to fold like a

The first full-colour operational image captured and delivered by the *JWST* – known as 'Webb's First Deep Field' – was of SMACS 0723, a galaxy cluster situated around four billion light years from Earth. The image revealed an amazing amount of detail and augured well for the operational future of this truly iconic telescope. (NASA/ESA/CSA/STScI)

huge piece of origami so that it could be 'collapsed' for launch. Construction of the telescope started in 2004 and required a large international collaboration, similar to the building of the *Hubble Space Telescope*. Its construction was a complex and painstaking process, driven by the need to create an observatory that could operate in the frigid vacuum of space, far from any possibility of repair. The primary mirror was completed by 2011 and all four of the telescope's scientific instruments were delivered to the Goddard Space Flight Centre by the end of 2015. The mirror segments, as well as the massive, five-layer sunshield – which is the size of a tennis court – had to be built with a series of hinges and motors that would automatically and precisely unfurl once in space.

Vera Rubin Observatory

The Vera Rubin Observatory, located in Chile, is a revolutionary facility designed to capture the dynamic universe in unprecedented detail. Its 8.4-meter Simonyi Survey Telescope, paired with a 3.2-gigapixel camera, will conduct the Legacy Survey of Space and Time (LSST), imaging the entire southern sky every few nights. This decade-long survey will track billions of stars, galaxies, and transient phenomena, helping scientists to study dark matter, dark energy, and near-Earth asteroids. Named after astronomer Vera Rubin (1928–2016), the observatory champions open data access, inviting global collaboration in exploring the ever-changing cosmos. For further information on the observatory, see the article 'The Vera Rubin Observatory' elsewhere in this edition.

The Flight Log of Exoplanet Explorer 9
The First Landing on an Exoplanet

Peter Rea

Flight Log: 21 July 2169

Altitude 25km: Optical scanners and laser altimeter have located landing site.
Altitude 15km: We are precisely following the descent curve, 5 minutes to touchdown.
Altitude 10km: Optical scanners have located a large boulder at the landing site and have requested the guidance system overshoots the touchdown point by 150 meters.
Altitude 5km: Guidance has redesignated our touchdown point. 3 minutes from touchdown.
Altitude 2km: *Exoplanet Explorer 9* is now descending vertically, landing site clear of hazards, 2 minutes to touchdown.
Altitude 800 meters: Terminal guidance active, landing legs deployed.
Altitude 400 meters: 30 seconds to touchdown.
Altitude 100 meters: 15 seconds to touchdown.
Altitude 0 meters: Touchdown. Just sent message, "Exobase[1] here, the Explorer has landed."

Introduction

There has never been a landing on an exoplanet (planets beyond our solar system) and this article will tell you all about it! The following is a mixture of current (2023) science fact, with fictional details from the imagination of the author. The details of that imaginary first landing are given to the reader in the form of a diary or flight log, using the artificial intelligence of the uncrewed *Exoplanet Explorer 9*.

---

1. Prior to the launch of Exoplanet Explorer 9, the word *exobase* had only one definition, this being the lower boundary of the exosphere, the highest region of the Earth's atmosphere. An additional meaning was allotted to the word during the twenty-second century with the successful landing of *Exoplanet Explorer 9*. The additional definition appeared in dictionaries across the solar system and reads, 'Exobase entered the interplanetary lexicon as the first word sent from the surface of Gliese 581c, the first exoplanet to be successfully landed on. Its origin pays homage to the first words spoken from the surface of the Moon in 1969 when Neil Armstrong used the expression Tranquillity Base.'

The Explorer has landed. After a journey of 40 years the *Exoplanet Explorer 9* spacecraft successfully lands on the planet Gliese 581c, known by the mission controllers as Primus. (Peter Rea / Roger Laishley)

Flight Log: 15 October 2125

Today I was activated. My name is AIMEE (Artificial Intelligence Module for Exoplanet Explorer). My controllers have tasked me with keeping a full account of my journey to Gliese 581c, known as *Primus*.[2] I will be at the centre of the *Exoplanet Explorer 9* mission. My primary mission is to watch over all systems of the Mother Ship and control the landing module of *Exoplanet Explorer 9*. My controllers are running diagnostics daily to verify that all systems are communicating with me correctly. It is only 5 months and 8 days until we leave Earth.

The First Exoplanet Discoveries

The concept of the Plurality of Worlds, often referred to as Cosmic Pluralism is an idea almost as old as astronomy itself. It is a concept that Earth with all its

2. The name *Primus* is not the official name of Gliese 581c but is a made up name by the author for this article. It is Latin for 'first'.

This artist's view shows the 'Hot Jupiter' type exoplanet 51 Pegasi b which orbits a star about 50 light-years from Earth in the northern constellation of Pegasus. This was the first exoplanet around a main sequence star to be discovered, in 1995. Twenty years later this object was also the first exoplanet to be directly detected spectroscopically in visible light. (ESO/M. Kornmesser/ Nick Risinger (skysurvey.org))

diverse flora and fauna is not alone in the Cosmos. This can be dated back to at least Greek times and continues to the present day with questions like 'are we alone in the Universe?' At the time of writing there is no evidence for life elsewhere in the galaxy. To this writer it seems somewhat implausible that in our own Milky Way galaxy, with its estimated 100 billion to 400 billion stars, life has not arisen on planets around some of those stars. We have been looking, however. The Search for Extraterrestrial Intelligence (SETI) has been pointing radio telescopes at the sky listening for signals from sentient beings since about the mid-1980s, but with no verifiable signals as of this writing. The search for exoplanets, proving the Plurality of Worlds idea has been far more successful. In 1995 an exoplanet was discovered around a main sequence star 51 Pegasi, located in the constellation of Pegasus. This was the first of many. The ground based telescopes searching for exoplanets were supplemented by Earth orbiting telescopes like *Kepler* and *Transiting Exoplanet Survey Satellite (TESS)*. With the introduction of these telescopes the exoplanet count rose sharply. More recently the *James Webb Space Telescope* has discovered its first exoplanet which I doubt will be its last.

As I write these words the total number of confirmed exoplanets stands at nearly 6,000. That figure is changing almost weekly and interested readers should go to the NASA Exoplanet Website shown at the end of this article for the current count.

Flight Log: 23 March 2126

I sense forward motion. I am leaving Earth orbit where I was assembled and starting my long journey toward Gliese 581. Acceleration will be slow at first. I will climb to 1,000km above the Earth with two of the first stage engines before the rest of the first stage engines ignite to start the inner solar system acceleration phase. My target star lies within the constellation Libra which straddles the ecliptic, that band in which our solar systems planets reside. It means also that *Exoplanet Explorer 9* will have to pass through the asteroid belt so rapid acceleration using the Hyperdrive cannot begin until I am on the other side. The chances of a collision are very small, but if I accelerate rapidly before passing through the belt it gives me little time to make a course correction should an asteroid appear along my predicted path.

Gliese 581

The star Gliese 581 lies 20 light years (6.3 parsecs) away in the constellation of Libra. The magnitude 10.5 star lies 2 degrees north of Beta Librae as seen from Earth. The name comes from the *Gliese Catalogue of Nearby Stars* published in 1957 by the German astronomer Wilhelm Gliese. Gliese 581 has a spectral classification of M3V and appears as a main sequence red dwarf star at the bottom right hand

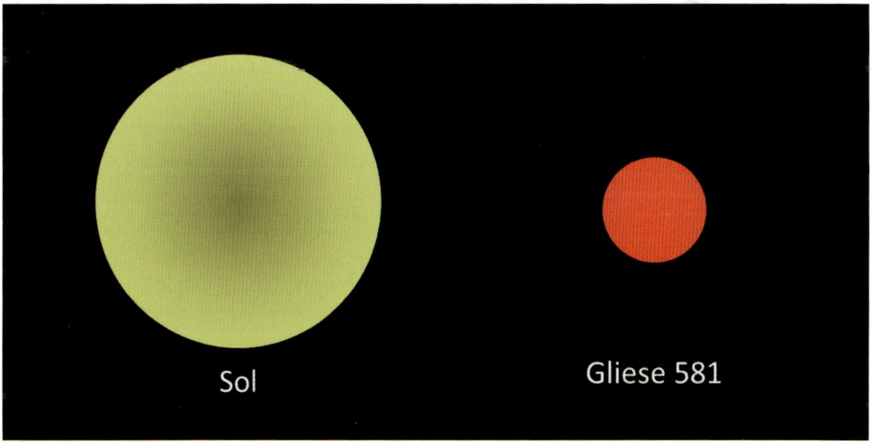

The star Gliese 581 is a red dwarf star, smaller and cooler than Earth's Sun. This diagram shows the relative sizes of Sol on the left and Gliese 581 on the right. (Peter Rea)

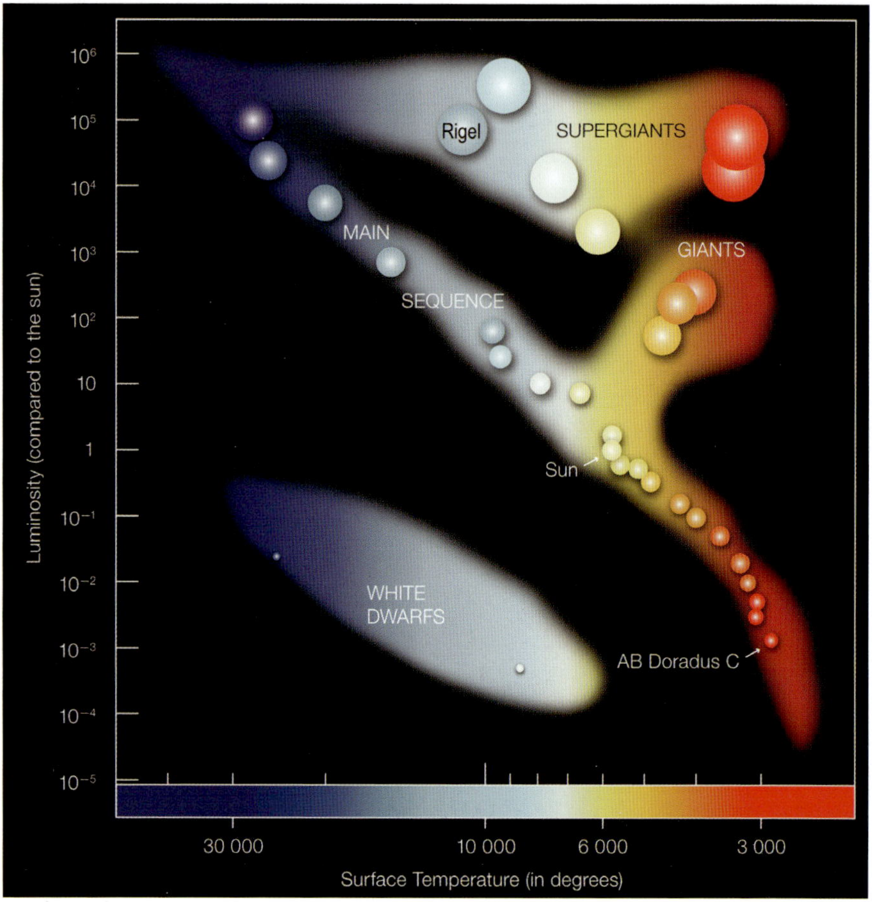

In the Hertzsprung-Russell diagram the temperatures of stars are plotted against their luminosities. The position of a star in the diagram provides information about its present stage and its mass. Red dwarf stars like Gliese 581 sit at the bottom right of the main sequence next to the star AB Doradus C. Our own Sun sits further up the main sequence. (Wikimedia Commons/Casliber)

corner of the Hertzsprung Russell diagram. As of this writing there are three confirmed exoplanets orbiting the main star. In order from the star they are Gliese 581e discovered in 2009, Gliese 581b discovered in 2005 and Gliese 581c discovered in 2007 by the European Southern Observatory facility at La Silla in the Atacama Desert. It completes a full orbit of its parent star in just 13 days, and is 14 times closer to its star than the Earth is to our Sun. As Gliese 581 is smaller and colder and less luminous than the Sun, the planet nevertheless lies in the 'habitable zone',

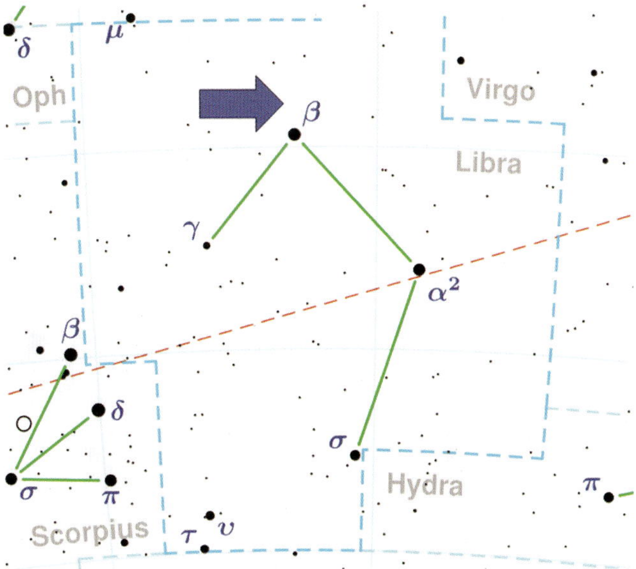

The star Gliese 581 lies 20 light years away in the constellation of Libra. From Earth it appears as a 10.5 magnitude star two degrees north from Beta (β) Librae. (Wikimedia Commons/ Torsten Bronger)

the region around a star where liquid water could exist. The use of lower case letters are used to denote exoplanets around a star. They are lettered in the order of discovery not distance from the planet.

Flight Log: 15 December 2126

I am now well clear of the asteroid belt and heading for the orbit of Jupiter, though that gas giant planet will not be close to my outbound trajectory. The Hyperdrive will start today to begin my rapid acceleration toward the star Gliese 581. It will take around 6 months at 1g (Earth gravity) acceleration to achieve 0.5c,[3] or 50 percent of the speed of light, my cruising velocity during most of our journey.

The Habitable Zone

The habitable zone of a star is a region where liquid water can exist as long as the atmospheric pressure is suitable. Liquid water is a key ingredient for life as we know it, so if we are looking for exoplanets where life could have started then this is where we should be looking for suitable planets. In our solar system, revolving

3. The lower case letter c is used to denote the speed of light, which is 299,792,458 metres per second, so 0.5c is half of that. It is perhaps most famously used in the Einstein equation $E=mc^2$ (energy = mass × speed of light squared)

The habitable zones of our Sun and Gliese 581 compared. Being a cooler red dwarf star the habitable zone of Gliese 581 is closer to the star than with our own solar system. (European Southern Observatory)

around a G-type yellowish star, the habitable zone extends from mid-way between Venus and Earth out to almost the orbit of Mars. For Gliese 581, being a much smaller and cooler M-type star, that habitable zone is much closer to the star. Gliese 581c is about 11 million kilometres from its parent star. Compare this to Earth which is 150 million kilometres from our much hotter Sun.

Flight Log: 14 June 2127

Today I reached a major milestone on my long journey to Primus. My velocity just reached 0.5c. I will stay at this velocity for 39.5 years, as measured on Earth, but only 35 years as measured by myself. My own onboard clock is now running about 15% slower than the clocks used by my controllers. It is a consequence of moving at 50% of the speed of light. The Interstellar Medium through which I am travelling is comprised of gas (mostly hydrogen) dust, and energy. The density is extremely low and far less than the strongest vacuum ever created on my home planet. Yet I must be ready to resist any possible collision. Special shields protect *Exoplanet Explorer 9* against any possible collisions. At this velocity even a speck of dust would carry dangerous amounts of kinetic energy.

Time Dilation

Time dilation is the slowing of time itself as perceived by one observer compared with another, depending on their relative motion or positions in a gravity field. It is a direct consequence of Albert Einstein's 1905 paper on Special Relativity. The closer an object travels to the speed of light the slower will be the onboard clock relative to a stationary observer. Interested readers could visit the Time Dilation website listed at the end of this article. The reader could also search 'Special Relativity.'

Flight Log: 3 January 2167

It has been a while since my last log entry. As I race toward Gliese 581 at 0.5c there has not been much to report. Although activities onboard *Exoplanet Explorer 9* have been at very low levels, all that is about to change. The deceleration to my target solar system started today. It will take 6 months to reduce my velocity sufficiently to allow *Exoplanet Explorer 9* to reach the solar system of Gliese 581. I will then make a roughly one year long survey of the planet looking for a suitable landing site. It should be noted that the use of 'year' in my logs refers to an Earth year.

Flight Log: 15 June 2167

Today I have entered the solar system of Gliese 581. *Exoplanet Explorer 9* is beginning to feel the gravitational pull of the red dwarf star. I will spend around a year slowly

Artist impression of the Gliese 581 solar system. Gliese 581c is the bluish looking planet just in front of the crescent planet at bottom left. (ESO/L. Calçada)

traversing this solar system before approaching primus. My spectrometers are recording a different part of the electromagnetic spectrum to what I recorded when I departed Sol, Earth's parent star. Red dwarf stars have a surface temperature in the 3,200 to 3,400 Kelvin range compared to Earth's G-type star whose surface temperature is in the 5,500 to 6,000 Kelvin range.

Flight Log: 3 September 2168

Today *Exoplanet Explorer 9* entered orbit around Primus. It has been a long way. I will survey this planet learning as much as I can. My principal task is to search for a suitable landing site. A quick look shows that water exists in abundance in the liquid state. The atmosphere is rich in nitrogen, carbon dioxide, and sulphur. There are three large land masses and many islands. No discernible pole caps can be seen.

Flight Log: 20 July 2169

A prime landing site has been found, with two backup sites. I am preparing the *Exoplanet Explorer 9* landing module for departure from the mother ship. This is where I have been since I was installed. I have controlled all functions of the mothership and lander from here. I carry a suite of instruments to study the surface and three autonomous wheeled rovers to extend the range of exploration. After all the years of preparation and the 8 *Exoplanet Explorers* that preceded me, it all comes down to what happens in the next few hours.[4]

Flight Log: 21 July 2169

Touchdown. Just sent message, "Exobase here, the Explorer has landed." It is the first landing on a planet beyond the solar system I left forty years ago. The date is significant being the 200th anniversary of the first human footprints on the moon revolving around Earth.

Flight Log: 22 July 2169

My cameras have started a panoramic survey of the landing site. These images and other spectroscopic data will be sent back to Earth when this hemisphere is facing that region of the sky where Sol and the Earth can be found. When I am facing away from Earth, I will store the data for later transmission. The cameras are more

4. Throughout its long mission *Exoplanet Explorer 9* continued to use the Earth based clock system of hours, days, and years, although they became out of sync with Earth whilst travelling at 0.5c. Whilst at Primus they were back in sync, but all data and messages took 20 years to get back to Earth.

Gases emanating from the area surrounding the landing site; mostly sulphurous but carbon dioxide too. Temperature is 27 degrees C at this location. The lake where Rover 3 is headed lies beyond the small mountain shown here. (Peter Rea/Roger Laishley)

sensitive to the redder end of the spectrum. The whole scene around the landing site looks more like a sunset on Earth yet the planet's red dwarf star is high in the sky. The region is volcanic in origin. Sulphur deposits can be seen close to the landing site but not as easy to see in this reddish light. There are vents where gases are emanating and what appears to be a caldera can be seen in the distance. This will need further investigation when I launch some aerial vehicles in a few days.

Flight Log: 24 July 2169

Today I deployed the three rovers. They will now explore this region independently. Large lakes of water have been seen from orbit 4 km distant. Rover 3 will depart the landing site tomorrow and head toward these lakes. It will report back to me every hour. After the lakes it will continue further out. It will collect samples from suitable sites along the way. These will eventually be brought back to *Exoplanet Explorer 9* for analysis. All the rovers are fitted with a small instrument that can analyse the collected samples, but *Exoplanet Explorer 9* is fitted with a much bigger suite of instruments and is better suited for a detailed investigation. The mother ship has been positioned in orbit to be directly above the landing site. Its altitude above Primus means that it rotates around Primus at the same rate as the planet rotates on its axis allowing it to function as a relay station for the three rovers that will travel up to 100 km from the landing site and will therefore be beyond the horizon.

Flight Log: 4 September 2169

Rover 1 made a monumental discovery today. In a large pool close to a geothermal area, single and multicellular life was found. People on Earth have long speculated about life in the universe. Are we alone? Here was the answer. Earth is about 4.5 billion years old. Scientists think that by 4.3 billion years ago, Earth had developed conditions suitable to support life. The oldest known fossils are only 3.7 billion years old. It was a long and difficult journey to modern humans. The Earth went through many natural catastrophes that killed off most of life on Earth for it to start again the long road to sentience. This newly found pool of life may well have many natural catastrophes to endure. The saying 'life finds a way' was true for humans. It may well be true for the simple life on Primus. One day it may stand proud and send its sons and daughters into the Galaxy.

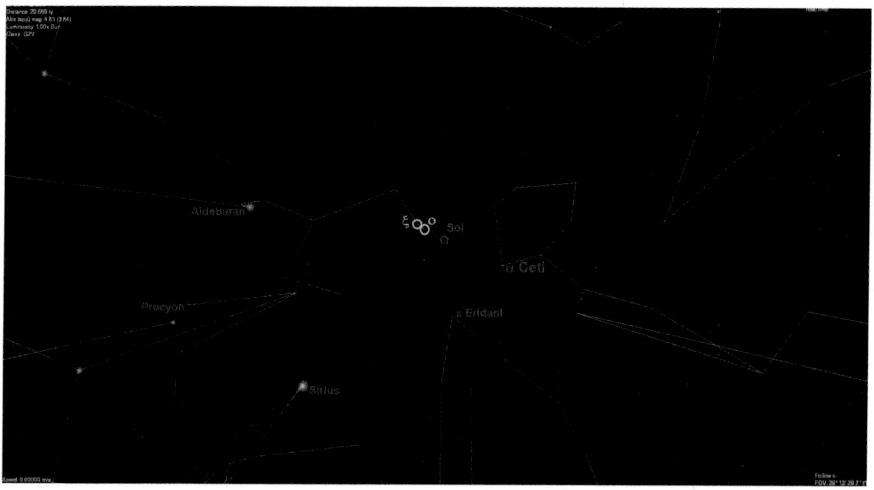

Looking into the night sky of Primus, Earth's sun can be seen as a magnitude 3.7 star in the centre of this image. Labelled as Sol. (Paul Money)

Flight Log: 8 September 2169

Today I turned a camera towards the sky and pointed it back toward Earth's Sun. Sol could be found as a 3.7 magnitude star close to the border with Cetus and Taurus. *Exoplanet Explorer 9* having answered one of the oldest questions took a look back to a solar system that once evolved sentient life, and my point of origin. If I were human, I would be in awe of the discovery of early life and the image that has just been taken. Yet the only thing I feel is satisfaction at completing my given duties. In 20 years my controllers could also be in awe. What will I find tomorrow?

Artwork Information

The original artwork purportedly showing the surface of Primus from *Exoplanet Explorer 9* was taken by the author on White Island or *Whakaari* in the Māori language. This active volcanic island lies 48 kilometres off the east coast of the North Island of New Zealand.

Acknowledgements

Thanks go to David Todd for his constructive critique. I would like to thank Roger Laishley for colour modifying the White Island images to look like the planet Primus orbiting a red dwarf star. I would further like to thank Neil Haggath for his mathematical skills on time dilation and for working out a realistic time to reach Gliese 581. My thanks also go to Paul Money for creating the chart that shows where our Sun (Sol) would be in the skies of Primus. Finally I would like to thank Brian Jones the *Yearbook of Astronomy* editor for throwing the idea of this article my way and letting me run with it. It is amazing what thoughts come into your head on a warm summer evening whilst holding a glass of lightly chilled Pinot Grigio.

Further Information

NASA Exoplanet Website: **exoplanets.nasa.gov**
Exoplanet Archive: **exoplanetarchive.ipac.caltech.edu**
List of Exoplanets: **en.wikipedia.org/wiki/Lists_of_exoplanets**
Time Dilation: **en.wikipedia.org/wiki/Time_dilation**

Minor Planet Genesis
The Celestial Police to John Russell Hind

Neil Norman

The great German astronomer and mathematician, Johannes Kepler (1571–1630), created a theory that a planet should exist between the orbits of Mars and Jupiter, an idea based on his ideas about planetary orbital harmony. He even named the hypothetical planet 'Phaeton' in his book *Mysterium Cosmographicum* (The Cosmographic Mystery) in 1596. However, despite extensive searches nothing was discovered to exist in the region predicted by Kepler.

Kepler's belief that a planet existed in this region came about because he had noted a pattern in the layout of the Solar System, with the gap between Mars and Jupiter being too large to fit his mode; something was missing yet nobody could find it. It was to be another 170 years before the renowned German astronomer and mathematician, Johann Daniel Titius (1729–1796) formulated a mathematical law in 1766 (known today as the Titius-Bode law). This law not only predicted a planet between Mars and Jupiter, but also predicted the possibility of further planets residing in the outer regions of the Solar System beyond Saturn, the farthest known planet at that time.

Johann Daniel Titius is perhaps best known for his law – drawn up in 1766 – which expressed the distances between the planets and which predicted the existence of a celestial object travelling around the Sun between the orbits of Mars and Jupiter. (Wikimedia Commons/PD-US)

Again, nothing was found; all the evidence appeared clear and mathematically sound yet searches remained fruitless. In 1772, another German astronomer, Johann Elert Bode (1747–1826), investigated and expanded the laws predicted by Kepler and Titius, and less than a decade later, on 13 March 1781, the German-born astronomer William Herschel discovered Uranus. Yet, although this was exactly where the law had predicted a planet should be, still nothing was found in the region between Mars and Jupiter despite further searches.

In 1798, Hungarian astronomer Franz Xaver von Zach (1754–1832) organised and hosted the first European congress of astronomers at his observatory in Gotha. The second congress was held on 20 September 1800 at Lilienthal, Germany – albeit with a smaller attendance – at the observatory of Johann Hieronymus Schröter (1745–1816). Top of the agenda was the founding of the group mainly known as the 'Celestial Police', their official title being the *Vereinigte Astronomische Gesellschaft* (United Astronomical Society); a total of six astronomers were present to form the group, the founding members being …

Johann Hieronymus Schröter (Lilienthal, Germany) President
Franz Xaver von Zach (Gotha, Germany) Director / Secretary
Adolf von Ende (Celle, Germany)
Johann Gildemeister (Bremen, Germany)
Heinrich Wilhelm Matthias Olbers (Bremen, Germany)
Karl Ludwig Harding (Lilienthal, Germany)

The main goal of *Vereinigte Astronomische Gesellschaft* was to create more advanced star catalogues as well as generating a greater understanding of spherical astronomy and coordinate systems, which was used for two main reasons:

- To identify and locate fainter objects.
- The precise determination of the orbits of recently discovered celestial objects.

A portrait / engraving by an unknown artist of the astronomer Franz Xaver von Zach who organised and hosted the first European congress of astronomers. (Wikimedia Commons / Magnus Manske / PD-US)

Their other tasks included tackling the question of stellar parallax and the question of just how far away the stars really are. They also paid close attention to novae and variable stars. However, the undertaking that the group is probably best known for was that of seeking out the planet that was thought to orbit the Sun between the orbits of Mars and Jupiter.

The sky is a large area and in order to share out the workload a system was devised whereby an area of 15° width centred on the ecliptic was catalogued. The ecliptic would be divided into 24 zones, each extending 15° in longitude and 7° or 8° degrees in latitude on either side.

With this task of charting the skies in 24 zones in mind, the next requirement was to find 24 members who could be allocated a zone each. The astronomers who were invited to become members of the group are as follows:

Johann Elert Bode (Berlin, Germany)
Johann Tobias Bürg (Vienna, Austria)
Thomas Bugge (Copenhagen, Denmark)
Johann Karl Burckhardt (Paris, France)
Adolf von Ende (Celle, Germany)
Johann Gildemeister (Bremen, Germany)
Karl Ludwig Harding (Lilienthal, Germany)
William Herschel (Slough, England)
Johann Sigismund Gottfried Huth (Frankfurt (Oder), Germany)
Georg Simon Klügel (Halle, Germany)
Julius August Koch (Gdańsk, Poland)
Nevil Maskelyne (Greenwich, England)
Daniel Melanderhjelm (Stockholm, Sweden)
Pierre François André Méchain (Paris, France)
Charles Messier (Paris, France)
Heinrich Wilhelm Matthias Olbers (Bremen, Germany)
Barnaba Oriani (Milan, Italy)
Giuseppe Piazzi (Palermo, Italy)
Johann Hieronymus Schröter (Lilienthal, Germany)
Friedrich Theodor von Schubert (Saint Petersburg, Russia)
Jan Śniadecki (Kraków, Poland) (later replaced by Jöns Svanberg (Uppsala, Sweden)
Jacques Joseph Thulis (Marseille, France)
Johann Friedrich Wurm (Blaubeuren, Germany)
Franz Xaver von Zach (Gotha, Germany)

The main task of the Celestial Police was to seek out the planet that was thought to orbit the Sun between Mars and Jupiter, yet positive results came purely by chance – within just three and a half months – with the first discovery …

On 1 January 1801, during an unrelated search from Palermo, Sicily, Giuseppe Piazzi came across an object that had appeared to move from one night to the next. He initially believed it to be a comet and even reported it as such but upon further examination he could not detect the usual telltale signs; there was no tail and no fuzzy coma and the object appeared star-like. He began to speculate that this could be the missing planet, but it would be September 1801 before he could publish his complete observations.

Oil on canvas portrait by Italian painter Costanzo Angelini, dating from around 1825, of Italian astronomer Giuseppe Piazzi. (Wikimedia Commons / Costanzo Angelini / Osservatorio Astronomico di Capodimonte / PD-US)

The German astronomer and mathematician Carl Friedrich Gauss (1777–1855) had by then developed his method of calculating orbits from astrometric observations. His conclusion was that this object was indeed a planet and not a comet. His calculations predicted its future movements, which allowed von Zach and Olbers to 'recover' the object once it had passed behind the Sun and re-emerged in the morning sky.

However, although calculations made by them placed the object at the right distance from the Sun as predicted by the Titius-Bode law and they believed they had found the missing planet, there was a problem – it was incredibly faint for a planet at that distance. However, they concluded that it was a planet and so Piazzi was allowed to name it. He chose *Ceres Ferdinandea*; *Ceres* after the Roman goddess of agriculture and *Ferdinandea* after Piazzi's patron, King Ferdinand III of Sicily. The somewhat unwieldy latter part of the proposed name was not accepted and the object became known as Ceres.

In March 1802, Olbers was working on the compilation of a star catalogue, this in preparation for the entry of Ceres into his designated area of sky, which would allow him

This image of Ceres, acquired by NASA's Dawn spacecraft on 4 May 2015, from a distance of 13,641 kilometres, is a far cry from the very first glimpse of this object by Giuseppe Piazzi on the opening day of the nineteenth century. The prominent, bright crater at right is Haulani. (NASA / JPL-Caltech / UCLA / MPS / DLR / IDA / Justin Cowart)

to refine the orbit. At this point he noted another object moving within his field of view. He began calculating its orbit and soon recognised it as being another object moving within the realm of the so-called 'missing planet'. As its discoverer, Olbers was given the task of naming this new world and would settle upon *Pallas*, in honour of Pallas Athena, the Greek goddess of wisdom and warfare.

William Herschel proposed that these new objects were clearly not planets, neither were they comets, but instead a new class of object altogether. He wanted to call them 'asteroids' instead of planets, the word asteroid originating from the Greek word *asteroeidēs*, which means 'star-like' or 'star-shaped'. This idea was initially rejected and the proposed term would not be used until many decades later.

The discovery of Ceres and Pallas led Olbers to hypothesise that the planet they were looking for had been destroyed in a collision with a comet. This explanation would restore the credibility of the Titius-Bode law, thereby offering hope of finding more of these smaller bodies – in particular at the crossing points of the orbits of Ceres and Pallas. Huth and von Zach, on the other hand, both believed that these 'minor planets' were exactly that – small planets which had failed to become one large single body.

Carrying on with Olbers' idea, Harding persevered with the search, and on 1 September 1804 he discovered an object that would later be named *Juno*, after the Roman goddess Juno, queen of the Roman gods and wife of Jupiter. Two-and-a-half-years later, on 29 March 1807, Olbers was to discover yet another, and his second, object, this one being named *Vesta* by Carl Friedrich Gauss (the first person to calculate its orbit) in honour of the Roman goddess of hearth, home and family.

A lithograph portrait of Heinrich Wilhelm Matthias Olbers by German portrait painter and lithographer Rudolph Friedrich Carl Suhrlandt. (Wikimedia Commons / Peter H. Richter (Hg.) / *Sterne, Mond, Kometen* / Hauschild-Verlag, Bremen 1995)

Oil on canvas portrait of Carl Friedrich Gauss painted in 1840 by Danish portrait painter Christian Albrecht Jensen. (Wikimedia Commons / Christian Albrecht Jensen / PD-US)

Following these discoveries, it became apparent that the four objects in question were all in the vicinity of the hypothesised planet, and interest from the collective participants began to dwindle. The outbreak of the Napoleonic Wars (1803–1815) did not help matters either, as they disrupted the work of many of the members. This was especially so when it came to the town of Lilienthal in Germany where Schröter's observatory was housed and which served as the home for many of the German observers working on the project. Schröter died in 1816, and other members of the Celestial Police had moved on in the direction of their work. With this, the group decided to disband, with a total of four discoveries to their name.

Although the search would seem to have ended with the disbanding of the Celestial Police, the discovery of minor planets was only just beginning. The next find would be looked on as being something of a fluke, yet it opened the door to a host of discoveries by an emerging batch of astronomers …

Karl Ludwig Hencke (1793–1866)

Born in Driesen (now Drezdenko, Poland), Karl Ludwig Hencke was by trade a post official, and eventually postmaster, until poor health forced him to take early retirement. He was also a keen amateur astronomer and had a private observatory in Driesen. It was during a search for Vesta – on 8 December 1845, some 38 years after the discovery of Vesta – that Hencke stumbled upon a hitherto undocumented object. Upon calculations of its orbit it was determined that this was a new minor planet and the fifth to be discovered. Hencke decided to name it *Astraea* after the Greek goddess of justice. On 1 July 1847 a sixth object was found, again discovered by Hencke, who invited Carl Friedrich Gauss to propose a name for the latest discovery. Gauss suggested naming it *Hebe*, after the Greek goddess of youth and cupbearer to the gods.

John Russell Hind (1823–1895)

Born in Nottingham, England, John Russell Hind would remain there until he was 17 when he went to London to serve a civil engineering apprenticeship. He would leave soon after to take up a position at the Royal Observatory, Greenwich, working under the Astronomer Royal George Biddell Airy. Hind was an exceptional observer; his discoveries included the variability of R Leporis (also known as Hind's Crimson Star) in 1845 and T Tauri (also called Hind's Variable Nebula) in 1852. He also discovered Nova Ophiuchi (V841 Ophiuchi) in 1848 the first object of its type discovered since 1670.

On 13 August 1847, while observing from London, Hind discovered the seventh minor planet *Iris*, named after the rainbow goddess in Greek mythology. Hind would discover another minor planet in the same year, on 18 October 1847, taking the number of known asteroids to eight; the latest object was named *Flora* after the Greek goddess of flowers. Hind would eventually make a total of ten minor planet discoveries, his additional finds being 12 Victoria (1850), 14 Irene (1851), 18 Melpomene (1852), 19 Fortuna (1852), 22 Kalliope (1852), 23 Thalia (1852), 27 Euterpe (1853) and 30 Urania (1854).

An image of John Russell Hind captured by the English photographer Henry Joseph Whitlock in around 1860. (Wikimedia Commons/ Henry Joseph Whitlock)

Andrew Graham (1815–1908)

Graham was an Irish astronomer and would also determine orbits of new found objects. In 1842 he was employed to work at Markree Observatory, County Sligo, and it was from here, on 25 April 1848, that he discovered the ninth minor planet while observing with a 3-inch aperture wide-field telescope that had been designed for comet hunting. The object was named *Metis*, in honour of Metis, the advisor and first wife of Zeus in Greek mythology. Graham also determined the orbit and future motion of his find, calculations for which took into account the gravitational effects of the major planets on its orbit path around the Sun.

An interesting note here is that Metis remained the only minor planet discovery to be made from Ireland until 7 October 2008 when Dave McDonald discovered (281507) 2008 TMg, a gap of 160 years.

Annibale de Gasparis (1819–1892)

De Gasparis was an Italian astronomer who would become the director of the Astronomical Observatory of Capodimonte in Naples in 1864 where he would stay until ill-health forced him to leave in 1889. He was a prolific author of scientific papers numbering 200 in total that covered subjects like mathematics, celestial mechanics, astronomy and meteorology. He was also a good visual astronomer and would discover a total of nine minor planets between 1849 and 1865. It is the first discovery from that tally that will complete our list of the first ten.

On 12 April 1849, while using the equatorial telescope of Reichenbach and Utzschneider, he discovered an unknown object. Upon hearing of the discovery, director of the observatory Ernesto Capocci chose the name *Igea Borbonica*

('Bourbon Hygeia') for the object, this in honour of Hygiea, the Greek goddess of health, cleanliness and hygiene, as well as for the ruling family at the time of the region in which Naples, and the Astronomical Observatory of Capodimonte, were located. In 1852, English astronomer John Russell Hind wrote, "It is universally termed Hygiea, the unnecessary appendage 'Borbonica' being dropped."

Annibale de Gasparis whose observational achievements included the discoveries of nine minor planets between 1849 and 1865. (John McCue)

The Celestial Police had taken a hypothesised law and subsequently discovered just four objects, although this was proof enough that objects were in the domain of the law. In less than 50 years the number of known objects would reach ten, while at the time of writing (March 2025) we have 25,000 named minor planets and a further 720,000 with orbits that have been determined and thereby qualified to receive a number.

The forthcoming (2028) edition of the *Yearbook of Astronomy* will contain the next article in this series – 'Minor Planets Golden Era: Palisa and Wolf to Tombaugh' – when we will examine the important role that photography played in the continuing discovery of minor planets.

Miscellaneous

Some Interesting Variable Stars

Tracie Heywood

You may have considered taking up variable star observing but how should you choose which stars to observe? There are so many variable stars in the night sky and you don't want to waste your time attempting to follow the "boring" ones. Your choice of stars will, of course, depend on the equipment that you have available, but also needs to be influenced by how much time you can set aside for observing.

This article splits some of the more interesting variable stars into three groups. The group that is most suited to you will depend on how often you can observe each month and for how long you can observe on a clear night. The light curves included have been constructed from observations stored in the Photometry Database of the British Astronomical Association Variable Star Section. Comparison charts for most of these stars can be found on the BAA Variable Star Section website at **britastro.org/vss**

One-Nighters

These are stars that can go through most of their brightness variations in the course of a single (reasonably long) night and would suit *people who can only observe occasionally, but can then observe well into the night.*

STAR	TYPE	RA		DEC		MAX / MIN	PERIOD
		H	M	°	′		
VZ Cancri	HADS	08	41	+09	49	7.3 / 8.0	0.17836376 days (~4.3 hours)
TW Cassiopeiae	EA	02	46	+65	44	8.3 / 9.0	1.428335 days (~34 hours)
Delta (δ) Librae	EA	15	01	−08	31	4.9 / 5.9	2.32733 days (~56 hours)
Beta (β) Persei (Algol)	EA	03	08	+40	57	2.1 / 3.4	2.86734 days (~69 hours)

VZ Cancri is a 'High Amplitude Delta Scuti' variable. Delta Scuti variables usually show very small levels of brightness changes. HADS variables are a small subset that show ranges large enough to be detected visually.

TW Cassiopeiae is an Algol-type eclipsing variable. Predictions for upcoming eclipses can be found at **www.as.up.krakow.pl/minicalc/CASTW.HTM**

Delta (δ) Librae is an Algol- type eclipsing variable. Predictions for upcoming eclipses can be found at **www.as.up.krakow.pl/minicalc/LIBDELTA.HTM**

Beta (β) Persei (Algol) is an eclipsing variable that shows deep primary eclipses but only very shallow secondary eclipses. Although circumpolar from the UK, it is too low in the sky for observation from April to June.

Most Clear Nights

These are stars that vary a bit more slowly, but which can display significant changes over a week or two. They would suit *people who can observe for a short while on (nearly) every clear night.*

STAR	TYPE	RA		DEC		MAX / MIN	PERIOD
		H	M	°	′		
Z Camelopardalis	UGZ	08	25	+73	07	10.0 / 14.5	6.96 hours (orbital)
R Coronae Borealis	R CrB	15	49	+28	09	5.8 / 15.0	None
SS Cygni	UGSS	21	43	+43	35	8.0 / 12.5	6.60 hours (orbital)
GK Persei	Nova+UG	03	31	+43	54	9.5 / 13.8	1.997 days (orbital)

Z Camelopardalis is a dwarf nova. Like many other dwarf novae, it shows frequent outbursts in brightness, but it also belongs to a small subset that also feature 'standstills' from time to time during which there are no outbursts.

R Coronae Borealis is the most famous of a small group of stars that spend most of their time near maximum but, at unpredictable times, can show dramatic fades which can be as deep as nine magnitudes.

SS Cygni is a dwarf nova. It spends most of its time near its 12th magnitude minimum brightness but, around eight times each year, it produces outbursts, which see it rapidly brighten by about four magnitudes and then fade back down over the next 1-2 weeks.

GK Persei produced a naked-eye nova outburst in 1901, but in recent decades has shown occasional dwarf nova outbursts. Recent outbursts have been seen in 2010, 2015, 2018 and 2023.

Several Times per Month

Slower variables whose brightness will change significantly over several months or a year. These variables would suit *people who can observe several times per month, but not necessarily on every clear night.*

STAR	TYPE	RA		DEC		MAX / MIN	AVERAGE PERIOD
		H	M	°	′		
V Canum Venaticorum	Semi Reg	13	19	+45	32	6.6 / 8.8	192 days
Mu (μ) Cephei	SemiReg	21	44	+58	47	3.4 / 5.1	850 days, 12 years
Omicron (o) Ceti	Mira	02	19	−02	59	1.9 / 10.1	331 days
X Ophiuchi	Mira	18	38	+08	50	6.8 / 8.8	332 days
R Trianguli	Mira	02	37	+34	16	5.4 / 12.6	267 days
SS Virginis	SemiReg	12	25	+00	46	6.5 / 9.6	320 days

V Canum Venaticorum and **SS Virginis** are semi-regular variables. 'Semi-regular' indicates that the brightness variations repeat only roughly from one cycle to the next. The observed brightness range differs between cycles, sometimes covering the whole of the listed range (occasionally exceeding it), but more often being somewhat smaller.

Mu (μ) Cephei – Herschel's Garnet Star – is a semi-regular variable. The observed brightness range differs between cycles, occasionally covering the whole of the listed range, but usually being somewhat smaller.

Omicron (o) Ceti (Mira) was the first Mira type variable to be discovered. Maxima of Mira are predicted to occur in February 2027 and in January 2028.

X Ophiuchi is a Mira type variable. Its amplitude is small for a Mira type variable due to the presence of a companion star that outshines it during minima. The 2027 maximum is predicted to occur during April.

R Trianguli is a Mira type variable. Maxima are predicted to occur in late February and November 2027.

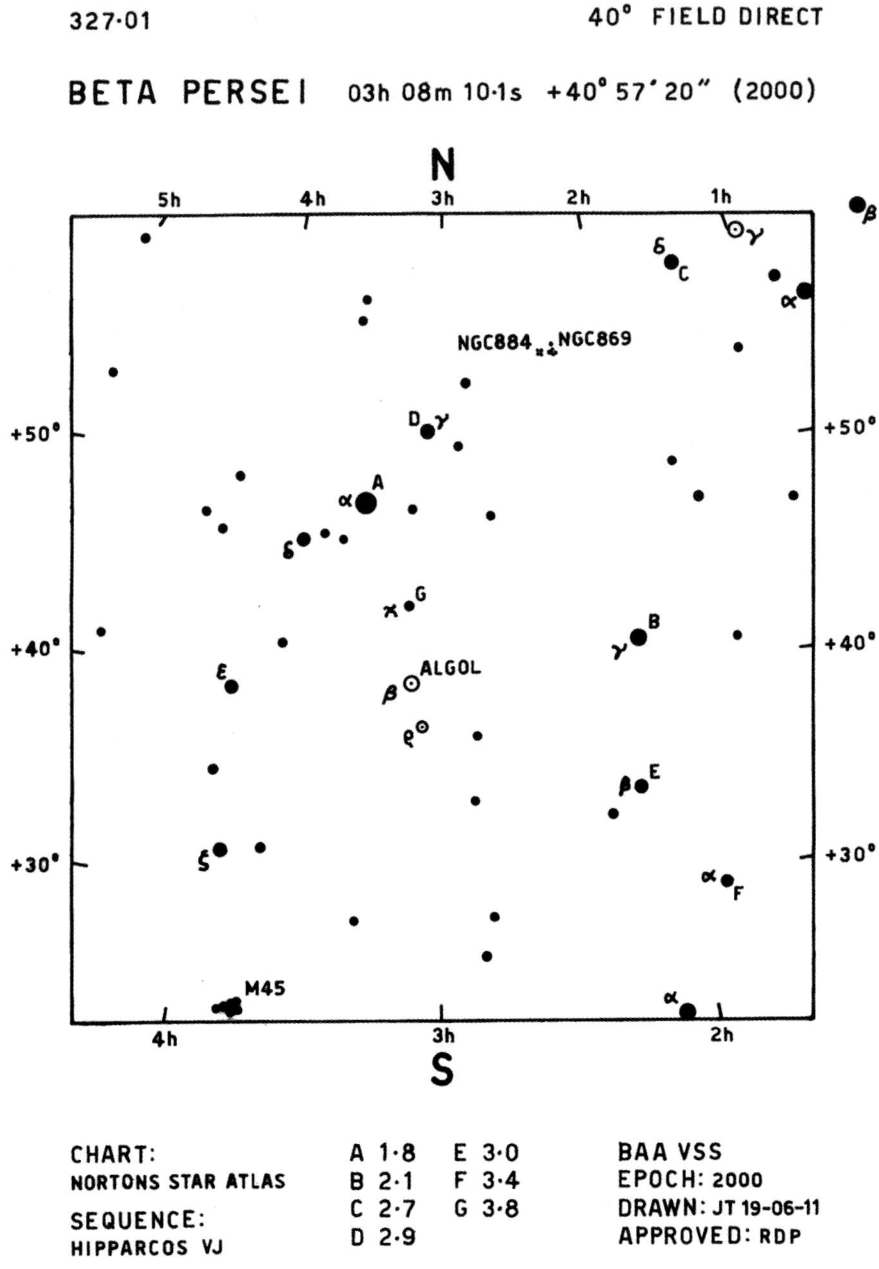

The BAA VSS finder chart for Beta (β) Persei (Algol). (BAA Variable Star Section)

Minima of Algol in 2027

Beta (β) Persei (Algol): Magnitude 2.1 to 3.4 / Duration 9.6 hours

		h				*h*				*h*				*h*	
Jan	3	11.7		Feb	1	3.8		Mar	1	20.0	*	Apr	2	9.0	
	6	8.5			4	0.6	*		4	16.8			5	5.8	
	9	5.3			6	21.4	*		7	13.6			8	2.6	
	12	2.1			9	18.3	*		10	10.4			10	23.4	
	14	22.9	*		12	15.1			13	7.2			13	20.2	
	17	19.7	*		15	11.9			16	4.1			16	17.0	
	20	16.5			18	8.7			19	0.9			19	13.8	
	23	13.4			21	5.5			21	21.7	*		22	10.7	
	26	10.2			24	2.3			24	18.5			25	7.5	
	29	7.0			26	23.2	*		27	15.3			28	4.3	
									30	12.1					
May	1	1.1		Jun	1	14.1		Jul	3	3.1		Aug	3	16.0	
	3	21.9			4	10.9			5	23.9			6	12.9	
	6	18.7			7	7.7			8	20.7			9	9.7	
	9	15.6			10	4.5			11	17.5			12	6.5	
	12	12.4			13	1.4			14	14.3			15	3.3	
	15	9.2			15	22.2			17	11.1			18	0.1	*
	18	6.0			18	19.0			20	8.0			20	20.9	*
	21	2.8			21	15.8			23	4.8			23	17.8	
	23	23.6			24	12.6			26	1.6	*		26	14.6	
	26	20.5			27	9.4			28	22.4	*		29	11.4	
	29	17.3			30	6.3			31	19.2					
Sep	1	8.2		Oct	2	21.2	*	Nov	3	10.2		Dec	2	2.3	
	4	5.0			5	18.0			6	7.0			4	23.1	*
	7	1.8	*		8	14.8			9	3.8			7	20.0	*
	9	22.7	*		11	11.6			12	0.6	*		10	16.8	
	12	19.5			14	8.4			14	21.4	*		13	13.6	
	15	16.3			17	5.3			17	18.2	*		16	10.4	
	18	13.1			20	2.1	*		20	15.1			19	7.2	
	21	9.9			22	22.9	*		23	11.9			22	4.0	
	24	6.7			25	19.7	*		26	8.7			25	0.9	*
	27	3.6			28	16.5			29	5.5			27	21.7	*
	30	0.4	*		31	13.3							30	18.5	*

Eclipses marked with an asterisk (*) are favourable from the British Isles, taking into account the altitude of the variable and the distance of the Sun below the horizon (based on longitude 0° and latitude 52°N).

All times given in the above table are expressed in UT/GMT.

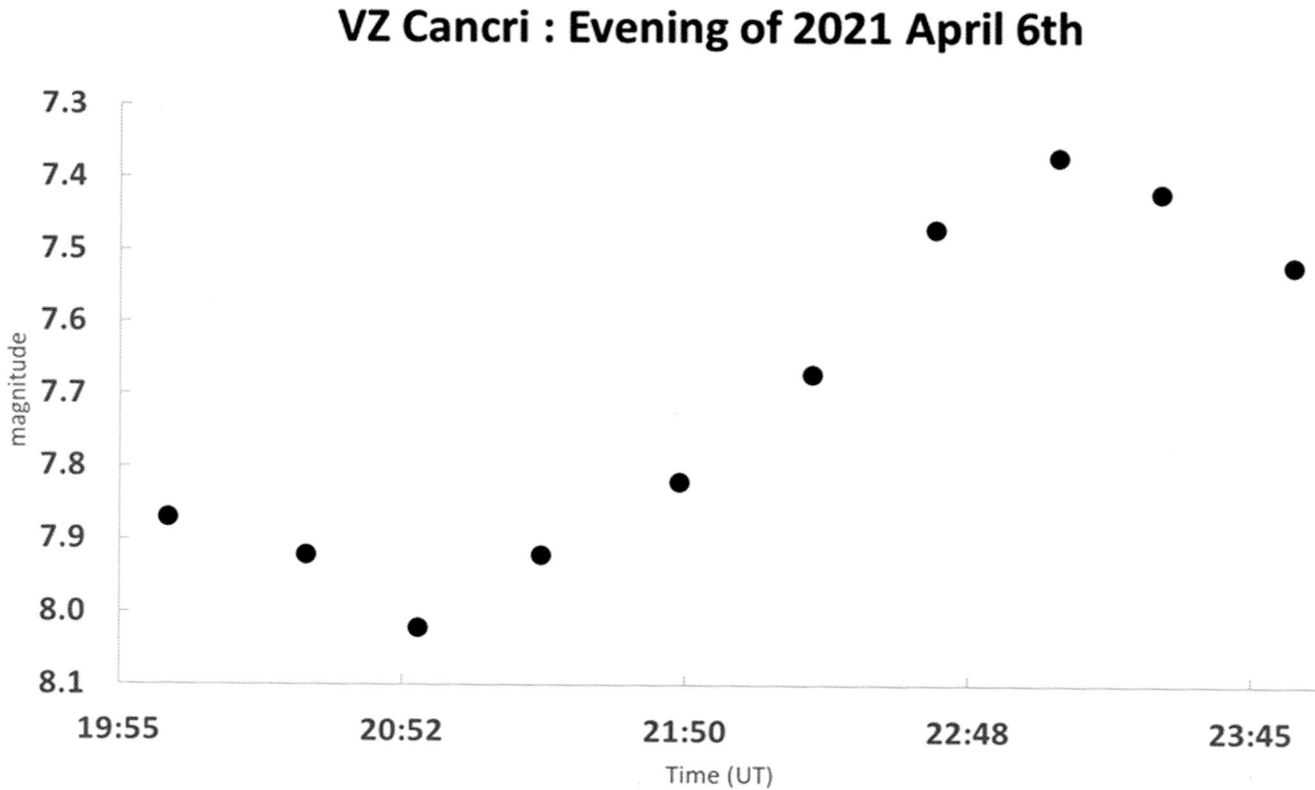

A light curve showing Tracie Heywood's observations of VZ Cancri (delta Scuti type) during the evening of 6 April 2021. (Tracie Heywood)

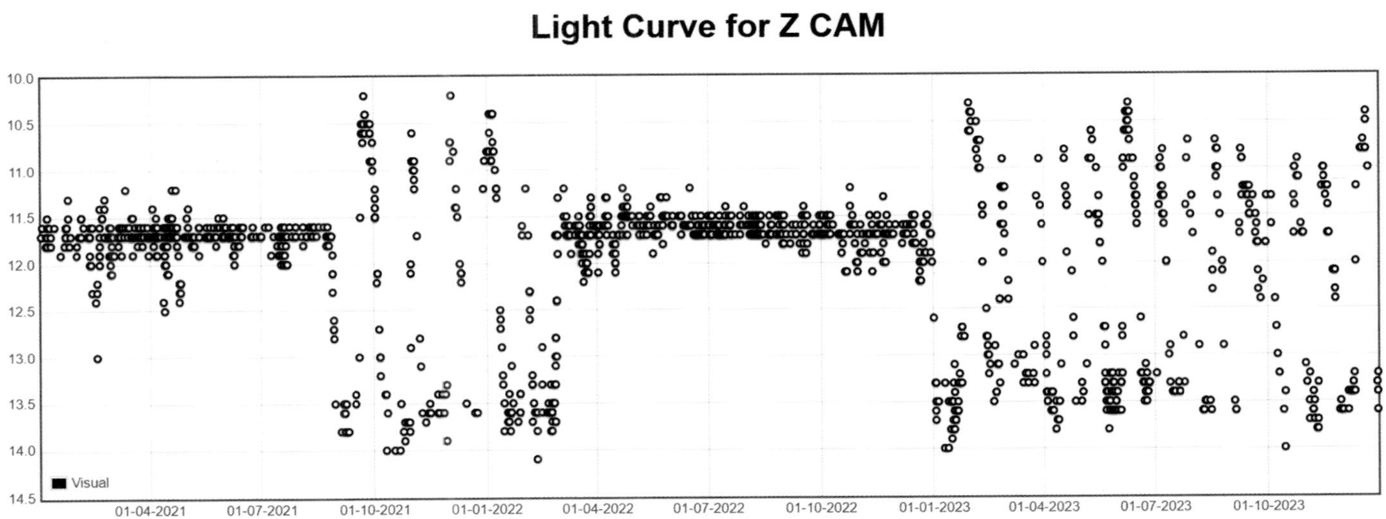

The BAA VSS light curve for dwarf nova *Z Camelopardalis* for the years 2021 to 2023. (BAA Variable Star Section)

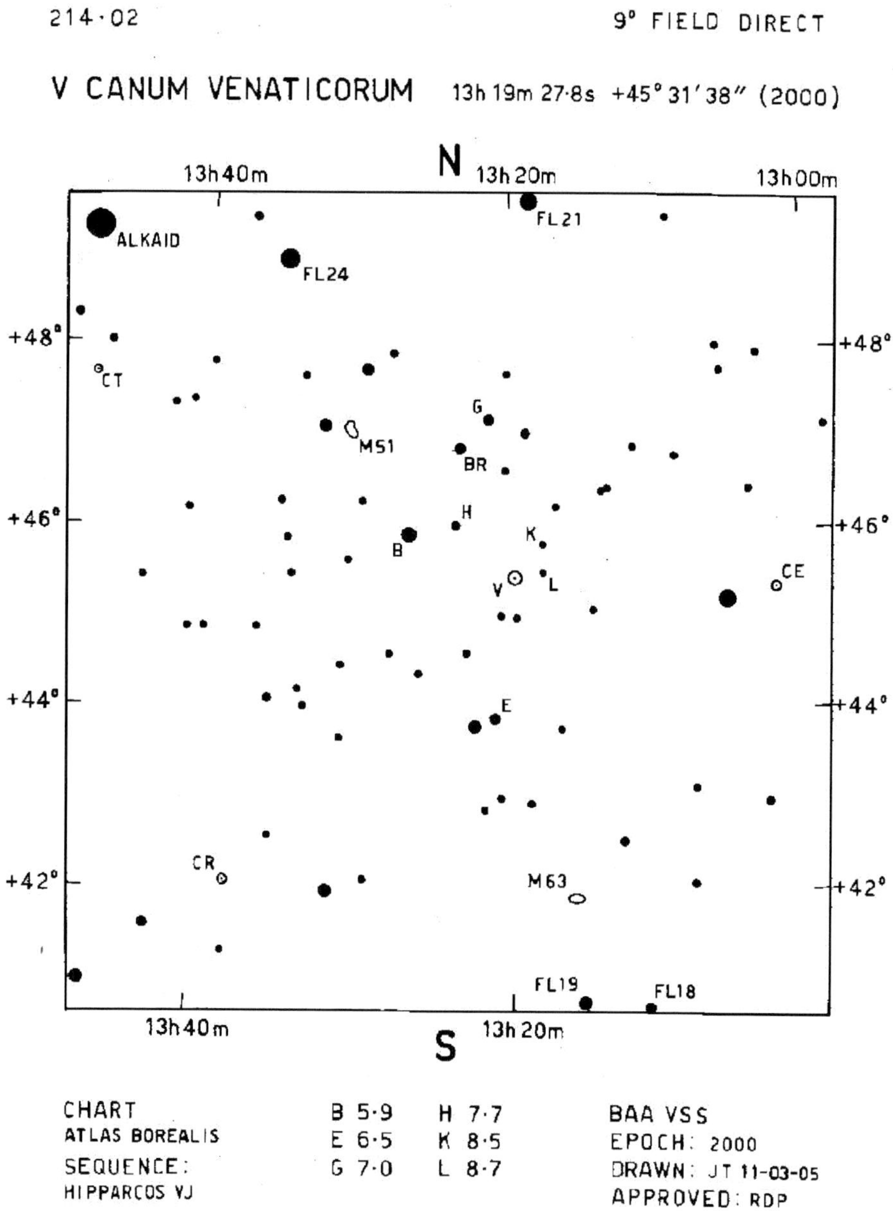

The BAA VSS finder chart for the semi-regular variable V Canum Venaticorum, showing several useful comparison stars. (BAA Variable Star Section)

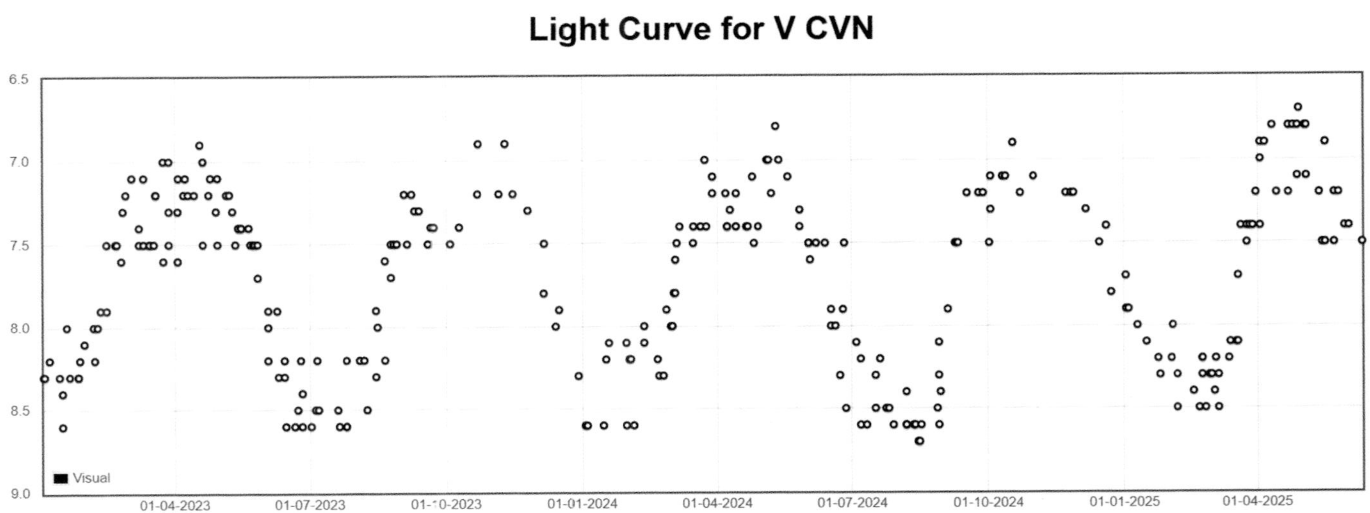

The BAA VSS light curve for the semi-regular variable V Canum Venaticorum from January 2023 to June 2025. (BAA Variable Star Section)

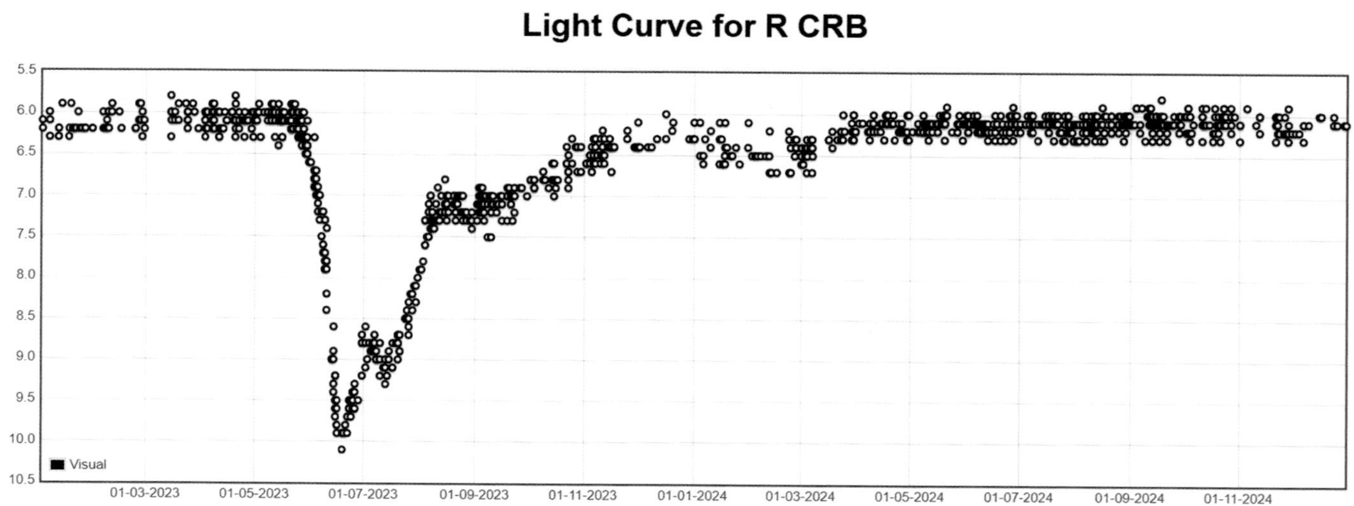

The BAA VSS light curve for R Coronae Borealis during 2023 and 2024. (BAA Variable Star Section)

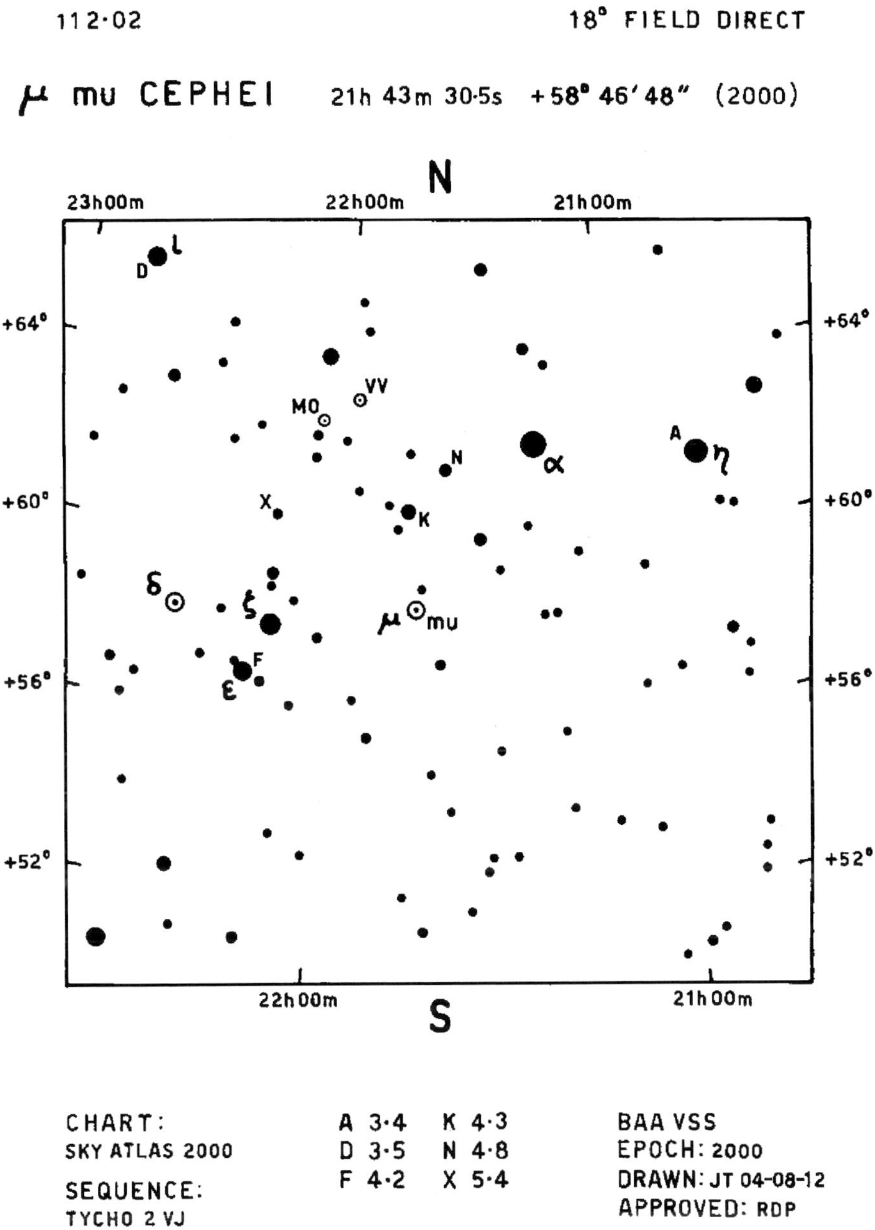

The BAA VSS finder chart for Mu (μ) Cephei, with six suitable comparison stars labelled. The same comparison stars can also be used for Delta (δ) Cephei, which is also labelled on the chart. Information relating to Delta Cephei can be found in the corresponding article in the *Yearbook of Astronomy 2023*. (BAA Variable Star Section)

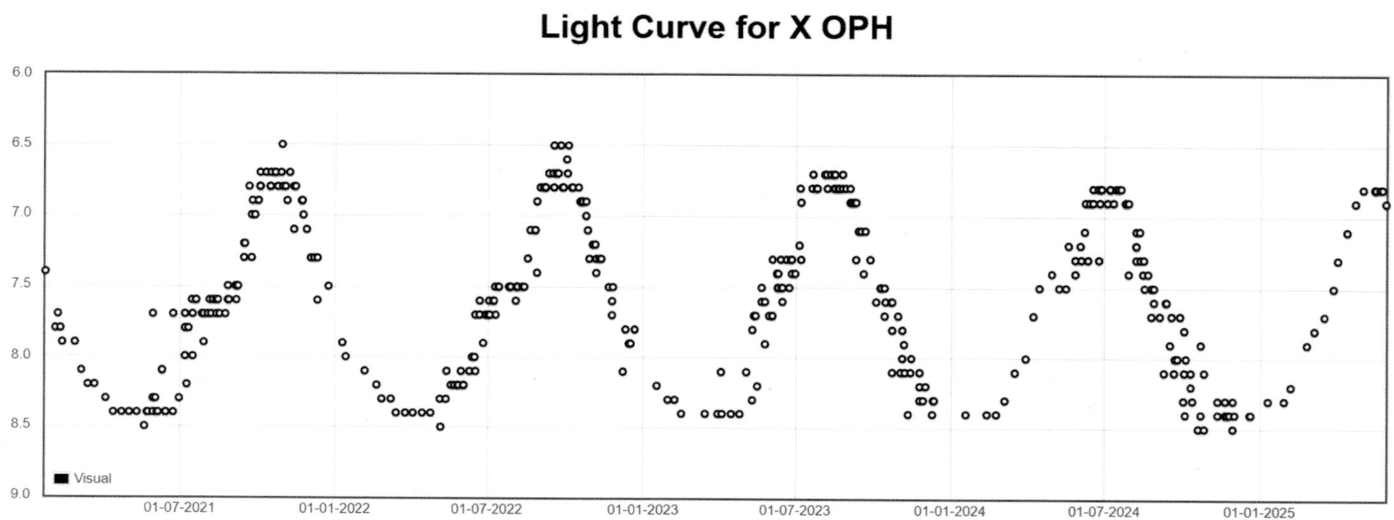

The BAA VSS light curve for the Mira-type variable X Ophiuchi between January 2021 and June 2025. (BAA Variable Star Section)

Some Interesting Telescopic Objects

John McCue and Brian Jones

Gamma (γ) Arietis (*Mesarthim*) R.A. $01^h\ 53^m\ 32^s$ Dec. $+19°\ 17'38''$ (J2000)
First spotted as a double star by the English scientist and astronomer Robert Hooke in 1664 during his observations of the bright comet of that year, Gamma Arietis is excellent target for small telescopes. Popularly known by its traditional name of *Mesarthim*, Gamma Arietis was described by Webb as consisting of, '... two small stars very near together; a like instance to which I have not else met with in all the heavens.' The system shines with a combined magnitude of 3.86 from a distance of around 165 light years, comprising components of magnitudes 4.58 and 4.64 separated by a distance of 7.61 seconds of arc.

Gamma (γ) Cassiopeiae (*Cih*) R.A. $00^h\ 56^m\ 43^s$ Dec $+60°\ 43'00''$ (J2000)
Located at a distance of around 550 light years, this eruptive variable star of spectral type B is many times bigger and brighter that our Sun. The brightness of the star usually varies between magnitudes 2.3 and 2.7 but sudden eruptions can increase this by more than a magnitude as it throws off a huge gaseous shell. Gamma Cassiopeia is believed to be nearing the supernova stage. The magnitude 2.3 Beta (β) Cassiopeiae and magnitude 2.7 Delta (δ) Cassiopeiae lie close by in Cassiopeia's famous W-shape, and are handily placed to be used to estimate the magnitude of Gamma with the naked eye. The author (John McCue), as a teenager, did this on many occasions but was not lucky enough to see an outburst!

Beehive Cluster (Messier 44) R.A. $08^h\ 40^m\ 22^s$ Dec $+19°\ 40'19''$ (J2000)
The Beehive Cluster is numbered 44 in Charles Messier's *Catalogue des Nébuleuses et des Amas d'Étoiles* (Catalogue of Nebulae and Star Clusters). Known and recorded since ancient times, it is visible to the unaided eye as a small nebulous object, being referred to in the star catalogue of the Greek astronomer Hipparchus during the second-century BCE as *Nephelion* ('Little Cloud'). Also known as *Praesepe* (a name derived from the Latin for 'cradle' or 'manger'), Messier 44 is a fine open cluster of young stars. Lying reasonably close, at a distance of around 600 light years, it is best seen in binoculars, which give an excellent view of this close group of young stars. The Beehive Cluster can be found between the two faint stars Delta (δ) Cancri and Gamma (γ) Cancri at the heart of the constellation.

Beta (β) Cygni (*Albireo*) R.A. $19^h 30^m 43^s$ Dec +27° 57′35″ (J2000)
The author never tires of viewing this beautiful double star in a telescope whenever the chance arises. All stars carry some hue, but when close to each other, any colour difference becomes more apparent, as in yellow for Albireo A and bluish for B. The pair shines from a distance of around 390 light years, though it is not known for certain whether they actually orbit each other. Albireo is the beak of the swan, but the name has suffered from mistranslations of ancient manuscripts.

Ring Nebula (Messier 57) R.A. $18^h 35^m 35^s$ Dec +33° 01′45″ (J2000)
In his book *Through My Telescope*, the English comedian and accomplished amateur astronomer Will Hay vividly described the Ring Nebula as, 'Of the planetary nebulae, perhaps the finest example is the Ring Nebula in Lyra. This nebula has the appearance of a ring of smoke such as one might puff from a cigarette.' The remnant of a dying star, the famous Ring Nebula is located south of the brilliant star Vega, and lying between Beta (β) Lyrae (*Sheliak*) and Gamma (γ) Lyrae (*Sulafat*). It was initially recorded in January 1779 by Charles Messier during one of his searches for comets, although was independently discovered a few weeks later by fellow French astronomer Antoine Darquier de Pellepoix, who was observing Comet Bode (C/1779 A1) at the time. Darquier went on to describe the nebula as being, '... as large as Jupiter and resembles a planet which is fading.' The faint, but very hot, magnitude 14.8 central star ran out of hydrogen fuel and collapsed, ejecting shells of its material as luminous outbursts into surrounding space.

Messier 65 and **Messier 66**, Spiral galaxies in Leo R.A. $11^h 18^m 56^s$ Dec +13° 05′32″ (J2000)
These two spiral galaxy twins are seen fairly close to each other (the quoted coordinates are for M65) around 20 minutes of arc apart, and so can be observed in the same telescopic field of view at low power. M65 is of magnitude 10 and M66 slightly brighter at magnitude 9. Although these objects are both spirals, M65 is more wound up – an Sa galaxy – compared with the Sb-type M66. The English astronomer William Henry Smyth described M66 as being, '... a large elongated nebula, with a bright nucleus, on the Lion's haunch.' Both galaxies were discovered by Charles Messier in 1780 and both shine from a distance of about 35 million light years.

Astronomical Organizations

American Association of Variable Star Observers

49 Bay State Road, Cambridge, Massachusetts, 02138, USA

aavso.org

The AAVSO is an international non-profit organization of variable star observers whose mission is to enable anyone, anywhere, to participate in scientific discovery through variable star astronomy. We accomplish our mission by carrying out the following activities:

- observation and analysis of variable stars
- collecting and archiving observations for worldwide access
- forging strong collaborations between amateur and professional astronomers
- promoting scientific research, education and public outreach using variable star data

American Astronomical Society

1667 K Street NW, Suite 800, Washington, DC 20006, USA

aas.org

Established in 1899, the American Astronomical Society (AAS) is the major organization of professional astronomers in North America. The mission of the AAS is to enhance and share humanity's scientific understanding of the universe, which it achieves through publishing, meeting organization, education and outreach, and training and professional development.

Association of Lunar and Planetary Observers (ALPO)

Matthew L. Will (Secretary), P.O. Box 13456, Springfield, IL 62791-3456, USA

alpo-astronomy.org

Founded in 1947 by Walter Haas, the ALPO is an international non-profit organization that studies all natural bodies in our solar system. ALPO Sections include Lunar, Solar, Mercury, Venus, Mars, Minor Planets, Jupiter, Saturn, Remote Planets, Comets, Meteors, Meteorites, Eclipses, Exoplanets, Outreach and Online, many with separate "Studies Programs" within these Sections. Minimum membership is very reasonable and includes the quarterly full colour digital *Journal of the ALPO*. Interested observers of any experience are welcome to join. Many members stand ready to improve the skills and abilities of novices.

Astronomical Society of Australia

c/o A/Prof. J.W. O'Byrne, School of Physics, The University of Sydney, NSW 2006, Australia

asa.astronomy.org.au

The Astronomical Society of Australia (ASA) was formed in 1966 as the organisation of professional astronomers in Australia. Membership of the ASA is open to anyone contributing to the advancement of Australian astronomy or a closely related field. As well as publishing a refereed journal *Publications of the Astronomical Society of Australia*, the Society runs an annual conference and several workshops and schools.

Astronomical Society of the Pacific

390 Ashton Avenue, San Francisco, CA 94112, USA

astrosociety.org

Formed in 1889, the Astronomical Society of the Pacific (ASP) is a non-profit membership organization which is international in scope. The mission of the ASP is to increase the understanding and appreciation of astronomy through the engagement of our many constituencies to advance science and science literacy. We invite you to explore our site to learn more about us; to check out our resources and education section for the researcher, the educator, and the backyard enthusiast; to get involved by becoming an ASP member; and to consider supporting our work for the benefit of a science literate world!

Astrospeakers.org

astrospeakers.org

A website designed to help astronomical societies and clubs locate astronomy and space lecturers which is also designed to help people find their local astronomical society. It is completely free to register and use and, with over 50 speakers listed, is an excellent place to find lecturers for your astronomical society meetings and events. Speakers and astronomical societies are encouraged to use the online registration to be added to the lists.

British Astronomical Association

Burlington House, Piccadilly, London, W1J 0DU, England

britastro.org

The British Astronomical Association is the UK's leading society for amateur astronomers catering for beginners to the most advanced observers who produce scientifically useful observations. Our Observing Sections provide encouragement and advice about observing. We hold meetings around the country and publish a bi-monthly *Journal* plus an annual *Handbook*. For more details, including how to join the BAA or to contact us, please visit our website.

British Interplanetary Society

Arthur C. Clarke House, 27/29 South Lambeth Road, London, SW8 1SZ, England

bis-space.com

The British Interplanetary Society is the world's longest-established space advocacy organisation, founded in 1933 by the pioneers of British astronautics. It is the first organisation in the world still in existence to design spaceships. Early members included Sir Arthur C Clarke and Sir Patrick Moore. The Society has created many original concepts, from a 1938 lunar lander and space suit designs, to geostationary orbits, space stations and the first engineering study of a starship, Project Daedalus. Today the BIS has a worldwide membership and welcomes all with an interest in Space, including enthusiasts, students, academics and professionals.

Canadian Astronomical Society

Société Canadienne D'astronomie (CASCA)

100 Viaduct Avenue West, Victoria, British Columbia, V9E 1J3, Canada

casca.ca

CASCA is the national organization of professional astronomers in Canada. It seeks to promote and advance knowledge of the universe through research and education. Founded in 1979, members include university professors, observatory scientists, postdoctoral fellows, graduate students, instrumentalists, and public outreach specialists.

Royal Astronomical Society of Canada

203-4920 Dundas St W, Etobicoke, Toronto, ON M9A 1B7, Canada

rasc.ca

Bringing together over 5,000 enthusiastic amateurs, educators and professionals RASC is a national, non-profit, charitable organization devoted to the advancement of astronomy and related sciences and is Canada's leading astronomy organization. Membership is open to everyone with an interest in astronomy. You may join through any one of our 29 RASC centres, located across Canada and all of which offer local programs. The majority of our events are free and open to the public.

Federation of Astronomical Societies

The Secretary, 147 Queen Street, Swinton, Mexborough, S64 8NG

fedastro.org.uk

The Federation of Astronomical Societies (FAS) is an umbrella group for astronomical societies in the UK. It promotes cooperation, knowledge and information sharing and encourages best practice. The FAS aims to be a body of societies united in their attempts to help each other find the best ways of working for their common cause of creating a fully successful astronomical society. In this way it endeavours to be a true federation, rather than some remote central organization disseminating information only from its own limited experience. The FAS also provides a competitive Public Liability Insurance scheme for its members.

International Dark-Sky Association

darksky.org

The International Dark-Sky Association (IDA) is the recognized authority on light pollution and the leading organization combating light pollution worldwide. The IDA works to protect the night skies for present and future generations, our public outreach efforts providing solutions, quality education and programs that inform audiences across the United States of America and throughout the world. At the local level, our mission is furthered through the work of our U.S. and international chapters representing five continents.

The goals of the IDA are:

- Advocate for the protection of the night sky
- Educate the public and policymakers about night sky conservation
- Promote environmentally responsible outdoor lighting
- Empower the public with the tools and resources to help bring back the night

The Planetary Society

60 South Los Robles Avenue, Pasadena, CA 91101, USA

planetary.org

The Planetary Society was founded by Carl Sagan, Louis Friedman and Bruce Murray in 1980 in direct response to the enormous public interest in space, and with a mission to introduce people to the wonders of the cosmos. With a global membership in excess of 50,000 from over 100 countries, it is the largest and most influential non-profit space organization in the world. The Planetary Society bridges the gap between the scientific community and the general public, inspiring and educating people from all walks of life and empowering the world's citizens to advance space science and exploration.

Royal Astronomical Society of New Zealand

PO Box 3181, Wellington, New Zealand

rasnz.org.nz

Founded in 1920, the object of The Royal Astronomical Society of New Zealand is the promotion and extension of knowledge of astronomy and related branches of science. It encourages interest in astronomy and is an association of observers and others for mutual help and advancement of science. Membership is open to all interested in astronomy. The RASNZ has about 180 individual members including both professional and amateur astronomers and many of the astronomical research and observing programmes carried out in New Zealand involve collaboration between the two. In addition the society has a number of groups or sections which cater for people who have interests in particular areas of astronomy.

Astronomical Society of Southern Africa

Astronomical Society of Southern Africa, c/o SAAO, PO Box 9, Observatory, 7935, South Africa

assa.saao.ac.za

Formed in 1922, The Astronomical Society of Southern Africa comprises both amateur and professional astronomers. Membership is open to all interested persons. Regional Centres host regular meetings and conduct public outreach events, whilst national Sections coordinate special interest groups and observing programmes. The Society administers two Scholarships, and hosts occasional Symposia where papers are presented. For more details, or to contact us, please visit our website.

Royal Astronomical Socicty

Burlington House, Piccadilly, London, W1J 0BQ, England

ras.org.uk

The Royal Astronomical Society, with around 4,000 members, is the leading UK body representing astronomy, space science and geophysics, with a membership including professional researchers, advanced amateur astronomers, historians of science, teachers, science writers, public engagement specialists and others.

Society for the History of Astronomy

Birmingham and Midland Institute, 9 Margaret Street, Birmingham, B3 3BU

shastro.org.uk

The Society for the History of Astronomy was founded in 2002 to promote the study of the history of astronomy by hosting talks by members and publishing new research into the field. One of the main objectives was to encourage research into past astronomers who have previously been neglected within the history of science. Some of its members are professional historians of science but most are amateur historians. The Society hosts several one-day conferences at venues across the United Kingdom each year. A *Bulletin* is published twice yearly containing articles and news items about astronomical history along with short reports of original research by members. The SHA also issues a quarterly electronic newsletter "e-News" which supplements the email messages from the society with updated events/meetings, and general news from council and SHA library. A library of publications of importance to the history of the science is maintained by the Society at the Birmingham and Midland Institute.

One of the Society's major activities is organising a Survey of Astronomical History in the form of lists of historical astronomers and observatories in every part of Britain and Ireland. This has been motivated by a desire to promote research into local astronomical activities that may have previously been neglected. The Society publishes annually a refereed journal called *The Antiquarian Astronomer* containing new research into the history of astronomy, particularly articles written by members. Published papers have discussed activities in major observatories, scientific research by individuals of particular note, scientific instrument makers, and the activities of prominent amateurs.

Society for Popular Astronomy

Secretary: Guy Fennimore, 36 Fairway, Keyworth, Nottingham, NG12 5DU

popastro.com

The Society for Popular Astronomy is a national society that aims to present astronomy in a less technical manner. The bi-monthly society magazine Popular Astronomy is issued free to all members.

Webb Deep-sky Society

Secretary: Steve Rayner, 11 Four Acres, Weston, Portland, Dorset, DT5 2JG

webbdeepsky.com

Founded in 1967 – and named after Thomas William Webb, author of *Celestial Objects for Common Telescopes* – the Webb Deep-Sky Society is one of the leading international deep sky organisations, and publishes a journal *The Deep-Sky Observer* together with a regular double star *Circular*. The original aim of the society was to update Webb's publications, and this was achieved through a series of eight handbooks. It still publishes material that it believes is relevant to deep sky observing. The society welcomes all levels of observers and has a number of sections dedicated to the observations of Double Stars, Nebulae and Clusters, and Galaxies.

Our Contributors

Martin Braddock is a professional scientist and project and people leader in the fields of drug discovery and development with 40 years' experience of working in academic institutes and large multi-national corporate organizations. He holds a BSc in Biochemistry and a PhD in Radiation Biology and is a former Royal Society University Research Fellow at the University of Oxford. He was elected a Fellow of the Royal Society of Biology in 2010, and in 2012 received an Alumnus Achievement Award for distinction in science from the University of Salford. Martin has published over 210 peer-reviewed scientific papers, filed nine patents, and edited two books for the Royal Society of Chemistry. He also serves as a proposal evaluator for multiple international research agencies. Martin holds further qualifications from the University of Central Lancashire and Open University. He is a member of the Mansfield and Sutton Astronomical Society and was elected a Fellow of the Royal Astronomical Society in May 2015. An ambassador for science, technology, engineering and mathematics (STEM), Martin seeks to inspire the next generation of young scientists to aim high and be the best they can be. In September 2024 he started his own consulting business helping small, medium and large business progress their new medicine pipelines. Martin can be contacted via email at **martin.braddock@genixiconsulting.com** or via his LinkedIn address at **linkedin.com/in/martin-braddock**

Matt Caplan is a theoretical nuclear astrophysicist and currently a professor of physics at Illinois State University where he studies white dwarfs, neutron stars, and microscopic black holes. Dr. Caplan received his PhD from Indiana University and his BS from the University of Virginia. Before beginning at ISU he was a Canadian Institute for Theoretical Astrophysics National Fellow at McGill University, and is an inaugural fellow of the Physicists Coalition for Nuclear Threat Reduction. Beyond academia, he is a scriptwriter for PBS Digital Studios and Kurzgesagt.

Neil Haggath has a degree in astrophysics from Leeds University and has been a Fellow of the Royal Astronomical Society since 1993. A member of Cleveland and Darlington Astronomical Society since 1981, he has served on its committee since

1989. Neil is an avid umbraphile, clocking up seven total eclipse expeditions so far to locations as far flung as Australia and Hawai'i. Five of these were successful, the most recent being in Waco, Texas on 8 April 2024. In 2012, he may have set a somewhat unenviable record among British astronomers – for the greatest distance travelled (6,000 miles to Thailand) to NOT see the transit of Venus. He saw nothing on the day … and got very wet!

David M. Harland gained his BSc in astronomy in 1977 and his PhD in computer science in 1981. He has lectured in computer science, worked in industry and managed academic research. In 1995 he 'retired' in order to write on space themes.

David Harper, FRAS has had a varied career which includes teaching mathematics, astronomy and computing at Queen Mary University of London, astronomical software development at the Royal Greenwich Observatory, bioinformatics support at the Wellcome Trust Sanger Institute, and a research interest in the dynamics of planetary satellites, which began during his PhD at Liverpool University in the 1980s and continues in an occasional collaboration with colleagues in China. He is married to fellow contributor Lynne Marie Stockman.

Tracie Heywood is an amateur astronomer from Leek in Staffordshire and is one of the UK's leading variable star observers, using binoculars to monitor the brightness changes of several hundred variable stars. Tracie currently writes a monthly column about variable stars for *Astronomy Now* magazine. She has previously been the Eclipsing Binary coordinator for the Variable Star Section of the British Astronomical Association and the Director of the Variable Star Section of the Society for Popular Astronomy.

Rod Hine was aged around ten when he was given a copy of *The Boys Book of Space* by Patrick Moore. Already interested in anything to do with science and engineering he devoured the book from cover to cover. The launch of Sputnik I shortly afterwards clinched his interest in physics and space travel. He took physics, chemistry and mathematics at A-level and then studied Natural Sciences at Churchill College, Cambridge. He later switched to Electrical Sciences and subsequently joined Marconi at Chelmsford working on satellite communications in the UK, Middle East and Africa. This led to work in meteorological communications in Nairobi, Kenya and later a teaching post at the Kenya Polytechnic. There he met and married a Yorkshire lass and moved back to the UK in 1976. Since then he has had a variety of jobs in electronics and industrial controls, and until recently was lecturing part-time at the University of Bradford. Rod got fully back into astronomy in around 1992 when his wife bought him an astronomy book, at which time he joined Bradford Astronomical Society. He is currently working part-time at Leeds University providing engineering support for a project to convert redundant satellite dishes into radio telescopes in developing countries.

Brian Jones hails from Bradford in the West Riding of Yorkshire and was a founder member of the Bradford Astronomical Society. He developed a fascination with the night sky at the age of five when he first saw the stars through a pair of binoculars, his interest in astronomy eventually taking him into the realms of writing sky guides for local newspapers, appearing on local radio and television, teaching astronomy and space in schools and, in 1985, to becoming a full time astronomy and space writer. His books have covered a range of astronomy and space-related topics for both children and adults and his journalistic work includes writing articles and book reviews for several astronomy magazines as well as for many general interest magazines, newspapers and periodicals. His passion for bringing an appreciation of the universe to his readers is reflected in his writing, his books having covered a range of astronomy and space-related topics for both children and adults. The minor planet 45689 Brianjones is named after him.

John McCue graduated in astronomy from the University of St Andrews and began teaching. He gained a PhD from Teesside University studying the unusual rotation of Venus. In 1979 he and his colleague John Nichol founded the Cleveland and Darlington Astronomical Society, which then worked in partnership with the local authority to build the Wynyard Planetarium and Observatory in Stockton-on-Tees. John is currently double star advisor for the British Astronomical Association.

Neil Norman, FRAS first became fascinated with the night sky when he was five years of age and saw Patrick Moore on the television for the first time. It was the Sky at Night programme, broadcast in March 1986 and dedicated to the Giotto probe reaching Halley's Comet, which was to ignite his passion for these icy interlopers. As the years passed, he began writing astronomy articles for local news magazines before moving into internet radio where he initially guested on the Astronomyfm show 'Under British Skies', before becoming a co-host for a short time. In 2013 he created Comet Watch, a Facebook group dedicated to comets of the past, present and future. His involvement with Astronomyfm led to the creation of the monthly radio show Comet Watch, which is now in its fourth year. Neil lives in Suffolk with his partner and three children. Perhaps rather fittingly, given Neil's interest in asteroids, he has one named in his honour, this being the main belt asteroid 314650 Neilnorman, discovered in July 2006 by English amateur astronomer Matt Dawson.

Jonathan Powell is a contributor to the *Sky at Night* magazine and has also had articles published in *Astronomy Now*. He has written three books on astronomy; *Rare Astronomical Sights and Sounds* (which was selected by *Choice* magazine as an Outstanding Academic Title for 2019); *Cosmic Debris: What It Is and What We Can Do About It;* and *From Cave Art to Hubble: A History of Astronomical Record Keeping*, all of which are available from Amazon. Jonathan worked at BBC Radio Wales as their astronomy correspondent, and was astronomy and space correspondent for *The National*, (an online newspaper for Wales). He is currently a columnist at the *South Wales Argus*, and a contributor to *CAPCOM*, an online magazine which promotes astronomy and spaceflight to the public. He has also presented on commercial radio at Sunshine FM in Worcester, Brunel FM in Swindon, and Bath FM, and has been a presenter on the astronomy and space dedicated radio station Astro Radio (UK). In addition, he is a former panellist for the YouTube channel 'Space Oddities'. Jonathan is currently at 107.9 GTFM in South Wales. He has also written a book on castles, *Fortress Wales*, and was part of the writing team for the BBC Television programme, 'The Fast Show', which won a BAFTA.

Katrin Raynor lives in Pontypridd, Glamorgan, South Wales and is a Fellow of the Royal Astronomical Society and Royal Geographical Society. She is an amateur astronomer and philatelist, and in her spare time writes articles and interviews for popular astronomy magazines including *Sky at Night* magazine and the Society for Popular Astronomy magazine, *Popular Astronomy*. Katrin is also a co-presenter on the Sky at Night's podcast, Star Diary. The minor planet 446500 is named after her.

Peter Rea has had a keen interest in lunar and planetary exploration since the early 1960s and frequently lectures on the subject. He helped found the Cleethorpes and District Astronomical Society in 1969. In April of 1972 he was at the Kennedy Space Centre in Florida to see the launch of Apollo 16 to the moon and in October 1997 was at the southern end of Cape Canaveral to see the launch of Cassini to Saturn. He would still like to see a total solar eclipse as the expedition he was on to see the 1973 eclipse in Mali had vehicle trouble and the meteorologists decided he was not going to see the 1999 eclipse from Devon. He lives in Lincolnshire with his wife Anne and has a daughter who resides in Melbourne, Australia.

Andrew Santarelli is theoretical astrophysicist and a graduate student at Illinois State University studying black holes, stellar evolution and populations, and a combination of the two in Hawking stars and quasi-stars. Andy received his BS in Physics from Illinois State University, and prior to that worked as a music recording engineer in Los Angeles.

Randall Stevenson is the author of many studies of literature, most recently *Reading the Times: Temporality and History in Twentieth-Century Fiction* (2018). He is Emeritus Professor of English Literature in the University of Edinburgh – a position allowing more scope, these days, to study and photograph the night sky, and to recall distant days when he was briefly a student of astrophysics, before turning to literature.

Lynne Marie Stockman holds degrees in mathematics from Whitman College, the University of Washington and the University of London. She has studied astronomy at both undergraduate and postgraduate levels, and is a Friend of the Astronomical Society of the Pacific and a member of both the Society for the History of Astronomy and the International Meteor Organization. A native of North Idaho, Lynne has lived in Britain since 1992. She was an early pioneer of the World Wide Web; with her husband and fellow *Yearbook* contributor David Harper, she created the web site **obliquity.com** in 1998 to share their interest in astronomy, computing, family history and cats.